OFFENSIVE CYBER OPERATIONS

DANIEL MOORE

Offensive Cyber Operations

Understanding Intangible Warfare

HURST & COMPANY, LONDON

First published in the United Kingdom in 2022 by
C. Hurst & Co. (Publishers) Ltd.,
New Wing, Somerset House, Strand, London, WC2R 1LA
© Daniel Moore, 2022
All rights reserved.

The right of Daniel Moore to be identified as the author of
this publication is asserted by him in accordance with the
Copyright, Designs and Patents Act, 1988.

A Cataloguing-in-Publication data record for this book
is available from the British Library.

This book is printed using paper from registered sustainable
and managed sources.

ISBN: 9781787385610

www.hurstpublishers.com

Printed in Great Britain by Bell and Bain Ltd, Glasgow

CONTENTS

ABSTRACT

Cyber-warfare is often discussed and rarely seen. Operations against networks mostly fall below the required threshold, and instead are varying criminal acts or peacetime intelligence gathering missions. To use cyber-warfare effectively, it is essential to distinguish where it begins and ends. The spectre of cyberwar can and should be turned into a spectrum of offensive cyber operations, or OCOs. There is so much to learn by piecing together operational history, public analysis of capabilities, and existing military thought. By exploring the idea of *intangible warfare*—conflict waged through non-physical means such as the information space and the electromagnetic spectrum—existing operational and strategic concepts can be adapted rather than reinvented.

While OCOs are often discussed as a monolithic operational space, they can usefully be divided into *presence-based* and *event-based* operations. The former are strategic capabilities that begin with lengthy network intrusions and conclude with an offensive objective. The latter are directly-activated tactical tools that can be field-deployed to immediately create localised effects. This top-level distinction is abstract enough to be usable by military planners and researchers, but specific enough to create two meaningful categories. This book seeks to show how OCOs can contribute to the overall military effort. The first four chapters explore theory, strategy, and the practicalities of different OCOs. The next four chapters are then dedicated to an in-depth examination of OCO strategy demonstrated by the United States, Russia, China, and Iran. Each of the four coun-

ABSTRACT

tries exhibits a unique approach to intangible warfare stemming from differences in culture, resources, history, and circumstance. The case studies help clarify the relative advantages and disadvantages of each country and how it stands to benefit by better employing OCOs within a military context.

ACKNOWLEDGEMENTS

This book was originally submitted as my PhD thesis at the War Studies Department at King's College London. I owe a great deal of my approach to tackling academic issues in network operations to Thomas Rid, who inspired me through his own writing to carve my own unique space where I could feel my contribution would actually matter. I took his Masters level Cyber Security class in 2012—which I would later end up teaching—where I first encountered "cyber" as an academic discipline sorely in need of thoughtful, cautious contributions. I'm also deeply thankful to Ben Buchanan, who I originally met as a fellow PhD researcher under Thomas, for all his help in turning interesting ideas into coherent analysis.

Over the years numerous people have engaged with and provided critique of my ideas. I couldn't possibly include them all, but I'd like to thank Richard Bejtlich, Rob Lee, Max Smeets, Alessio Patalano, David Omand, and Tim Stevens for fascinating conversations that helped to shape key portions of this work. My entire team at iDefense, including Valentino De Sousa, Annabel Jamieson, Christy Quinn, and Charlie Gardner, had to endure numerous rants on offensive operations, whether willingly or otherwise. They not only humoured me but contributed thoughtful ideas, advice, and their own expertise.

One of the greatest experiences I've had was teaching the Cyber Security class at King's College London for three years. Throughout those years I met some incredible students who would often go on to find their own places in information security. They were often unwitting but extremely helpful participants in an elaborate experiment to

ACKNOWLEDGEMENTS

test my ideas against versatile cohorts of critical thinkers, and my work was all the better for it.

Though perhaps not a personal acknowledgement per se, I owe thanks to all those who cover network operations across all angles. This book's bibliography is a testament to how multi-disciplinary the field has to be. The sources range broadly between journalists, policy makers, academics, strategists, threat intelligence analysts, and many others. There is simply no other way for the field to thrive, and it is worth acknowledging just how essential the different perspectives are for complete (or near-complete) analysis.

I owe the biggest thank you of all to my wife, Nitzan. The word support does not do her role justice; she is the main reason I maintained my sanity as I attempted to navigate first a full-time job alongside my PhD, and later alongside completing a book, all the while parenting our baby daughter Natalie through a global pandemic. Beyond all that, our long conversations around the key themes in this book resulted in some of the most profound changes to its structure. Her candid feedback allowed me to truly test if my concepts were reasonable to those who do not live and breathe "cyber" every day.

LIST OF ABBREVIATIONS

A2AD	Anti Access / Area Denial
C4ISR	Command, Control, Communications, Computers, Intelligence, Surveillance, Reconnaissance
CNA	Computer Network Attack
CND	Computer Network Defence
CNE	Computer Network Exploitation (espionage)
CNO	Computer Network Operation
EMS	Electromagnetic Spectrum
EW	Electronic Warfare
FSB	Federal'naya Sluzhba Bezopasnosti, Federal Security Service (Russia)
GCHQ	Government Communications Headquarters (UK)
GRU	Glavnoje Razvedyvatel'noje Upravlenije, Military Intelligence (Russia)
IRGC	Islamic Revolutionary Guard Corps
NCW	Network Centric Warfare
NSA	National Security Agency (US)
OCO	Offensive Cyber Operation
SIGINT	Signals Intelligence
SSF	Strategic Support Force (China)
SVR	Sluzhba Vneshney Razvedki, Foreign Intelligence Service (Russian)

INTRODUCTION

Computing is an indispensable facet of modern military operations, but attacks against computers have yet to deliver on the promise of revolutionising warfare. Intelligence collection, command and control, guidance, and weapons platforms themselves are all aided by networks. Even as networks have become pivotal in enabling joint operations, the spectre of cyberwar—envisioning battles waged between and against networks—has yet to come to fruition; war remains innately kinetic. The arrival of the twenty-first century heralded the explosive rise of the cyber-warfare narrative, but its actual utility in warfare is often unclear. It is imperative for strategic intent to lead the use of technology, rather than have technology create de facto strategies. In essence we ask, what limits military forces from realising the potential of cyber-warfare, and how could these limitations be mitigated?

"I have given Cyber Command really its first wartime assignment... and we're seeing how it works out",[1] former US Secretary of Defense Ashton Carter half-heartedly claimed in April 2016, referring to the use of offensive cyber operations against the Islamic State in Iraq. He added that "Even a few years ago, it would not have occurred to a secretary of defense to say, 'let's get cyber in the game', but here we have real opportunities".[2] Such broadly ambitious claims introduced confusion rather than clarity, as the contribution from "cyber" to the campaign against the Islamic State seemed initially murky. Practitioners and researchers immediately suspected hyperbole from Carter's overtures on military integration of cyber-offensives. How useful were the so-called "cyber bombs"[3] against the Islamic State?

Fast forward to eighteen months later. A now-retired Carter candidly admitted in a Belfer Center special report on the Islamic State campaign that he was "largely disappointed in Cyber Command's effectiveness against ISIS. It never really produced any effective cyber weapons or techniques."[4] Tension arose from the ownership of such capabilities by intelligence agencies, principally the National Security Agency: "When CYBERCOM did produce something useful, the intelligence community tended to delay or try to prevent its use, claiming cyber operations would hinder intelligence collection."[5] Finally, he lamented: "none of our agencies showed very well in the cyber fight."[6] The first declared US attempt at network warfighting was deemed a failure by the very individual that spearheaded it. Coverage of CYBERCOM's efforts against ISIS in the years that followed described a more complimentary landscape of capabilities and successes, with the purpose-created Joint Task Force ARES disabling ISIS networks and disrupting their command and control.[7]

Nations now openly incorporate cyber-warfare into their military planning. The People's Republic of China and the United States— among many others—have declared doctrine, formed units and invested considerable funds towards conducting operations through and against networks. The United Kingdom acknowledged that it "has conducted a major offensive cyber campaign against Daesh"[8] and later in 2020 officially trumpeted that it "has been a world-leader on offensive cyber operations, with GCHQ pioneering the use and development of these cyber techniques."[9] Chinese military doctrine highlights information as a significant new aspect of warfare, accompanying the internal rise of networked combat capabilities.[10] Russian forces have deployed offensive network capabilities against Ukrainian critical infrastructure by causing power outages concurrent to a kinetic campaign in Ukraine.[11] Nations are increasingly realising that the potential in targeting networks ranges from manipulating news organisations to crippling military hardware; the usefulness of network operations lies within a broad spectrum of possibilities. Yet cyber-warfare did not appear in a vacuum; it is rooted in military history, technological progress and the development of modern doctrine. Considering how such capabilities can best be made useful in warfare is more than just an academic exercise.

Some network attacks push against the established normative boundaries of conventional warfare. In 2017, NotPetya, a destructive strain of malware flimsily masquerading as ransomware, spread virulently around the world, wiping devices and inflicting steep financial damages to numerous organisations and corporations.[12] The sum global harm inflicted by NotPetya was unprecedented. Production and operations were affected in multiple industries as companies scrambled to reimage computers and restore lost data. The dissemination vector soon pointed to a small Ukrainian software company used locally to pay taxes,[13] though other infection methods were later discovered. It was suspected that the original intent was to wreak digital havoc within Ukraine, yet it was eminently clear that the malware had escaped its original boundaries. Whether that was intentional or not remains uncertain.

NotPetya is now publicly attributed to the Russian military by the United States,[14] United Kingdom,[15] and others. This unusual attribution was conducted publicly, and to a startling level of specificity. The message sent by Western governments was anything but subtle: this was a military operation sanctioned by the Russian government. Yet nuance is needed to identify the utility of attacks such as NotPetya, how they contribute to a strategic narrative, and whether they can even be fundamentally classified as warfare. As the models presented in Chapter 1 will suggest, based on targeting, impact, identity of the attackers, the underlying goals and existing relationships between victim and attacker, NotPetya can be viewed as an act of cyber-warfare against Ukraine, but not against the collaterally affected nations.[16] Warfare or not, it was an offensive cyber operation of accidentally massive scale.

Incidents against networks occur daily, in droves. The overwhelming majority of these attempted intrusions are no more than exploratory probes for weaknesses, easily shrugged off by automated defences. Beyond those, many successful compromises occur, leading to an unprecedented loss of sensitive data. Fewer still seek not just to extract data, but also to influence it and the systems that host it, resulting in attacks. Only a sliver of intrusions is carried out under a military mandate and seeks to achieve political-strategic goals by way of network attacks. Often conflated with intelligence operations or criminal activity, this slice of malicious network activity has distinct characteristics that are explored within this book.

The primary goal is therefore to address how offensive cyber operations (OCOs) can best contribute to battlefield success at all levels of operation. Rather than discussing the spectre of cyberwar, the goal is to piece together the spectrum of cyber-warfare. Military doctrine is built on accrued experience and historical analysis that can contribute immensely towards crafting a modern joint warfighting approach that incorporates the intangible; the introduction of cyber does not necessarily mean abandoning conventional wisdom. By examining the intersection of established military strategy, information security, and the technical characteristics of military-adopted technology, it is possible to construct practical models that help determine both what constitutes an attack and how these could reliably be integrated into planning across all three key facets of warfighting: the strategic, operational, and tactical. A combination of doctrinal, technical, and strategic analysis helps bridge the gap between established practice and the seemingly new circumstances of cyber-warfare.

Typologies already exist for offensive network activities. US doctrine divides by purpose; OCOs may disrupt, destroy, degrade, deny, or manipulate their targets.[17] This is a useful distinction when attempting to distinguish between potential impacts, but purposes are less compelling when used to create overarching categories for the operations themselves. Each of the potential five purposes may just be a different payload at the end of identical processes. Healey and Rattray suggested a dozen parameters with which to categorise offensive network operations, creating a granular framework that may best be applied to individual cases.[18] When comparing strategic implementation of OCOs between nations at scale, the framework becomes more unwieldy. A simpler solution would offer straightforward, easily identifiable categories with impactful distinctions.

This book will argue that all OCOs—military or otherwise—can be usefully divided into two primary categories; *event-based operations*, and *presence-based operations*. The former represent immediate attacks against networks and equipment, while the latter involve lengthy intrusions that culminate in an attack. This distinction helps divide between areas of responsibility in military forces and intelligence agencies, and between capabilities that can be designated as field-deployed weapons and those that would require high-echelon political approvals. The differences between the two categories manifest across the entire operational life cycle, and it is therefore an instructive way of classifying offensive activities, for both practitioners and researchers.

The categorical division presented here helps prevent both over-simplification and over-complication of offensive network operations. The tendency to lump all network intrusions as "cyber" strips away crucial distinctions, making clear analysis immensely difficult. Intelligence operations are not automatically attacks and grouping influence campaigns with destructive malware leaves much to be desired. At the same time, while offensive network capabilities introduce numerous intricate variables and technological circumstances, these considerations must be abstracted to the level where researchers, strategists, and policy makers can make sense of them with their existing toolsets.

The analysis is presented through four layers. The first layer will provide a five-step model towards assessing if a given incident should be classified as cyber-warfare. The model allows all assessment of offensive activities to be independently standardised to the same scale, excluding those that do not meet required criteria. The second layer will then offer a historical analysis demonstrating that OCOs draw heavily from accrued experience in adjacent fields, essentially showing how the marriage of signals intelligence and electronic warfare gradually evolved to *intangible warfare*. The third layer will contend that distinguishing between immediate-effect *event-based operations* and time-consuming, clandestine *presence-based operations* can help form more coherent roles and processes for each. The fourth and final layer will apply the above models to show how it all translates to combat operations at various levels and scenarios.

It is possible and desirable to disambiguate between military OCOs in wartime and operations in peacetime. Hacks that are weaponised to sway public opinion[19] may be an egregious violation of sovereignty, but do not necessarily meet the threshold of warfare. Unless carried out by sponsored offensive forces and for a commensurate objective, shaping public perception by manipulating news and social media is not an inherently in-conflict venture. To give policy makers, strategists, doctrine-crafters, and battlefield commanders a robust understanding of what they can and cannot expect from offensive network operations, we must first dispel the "grey areas" currently afflicting such capabilities. While it is tempting to leave "cyber" as a porous concept where wartime and peacetime inherently bleed into each other, doing so introduces risk and waste. The risk entails undue escalation between nations, if peacetime operations are misidentified as having conflict-like intent.[20] Waste may stem from the misapplication of OCOs where they are unlikely to achieve any meaningful objective.

Cyber-warfare has the potential to provide a set of capabilities that act as a force multiplier in armed conflict, yet these will not supplant but rather complement existing military doctrine. Uniquely, OCOs primarily allow attackers to weaponise an enemy against itself by subverting its systems, networks and weapons, thereby contributing to—but not single-handedly generating—victory. The more advanced and interconnected the adversary is, the more it may be susceptible to this form of targeting. History informs us of the many similarities between the advent of cyber-warfare and the introduction of other forms of warfare, specifically manoeuvres and tactics employed via the electromagnetic spectrum throughout the twentieth century.

Concepts

Clearly defined concepts are crucial to the scoping agenda at the heart of this book. Information security professionals often skew negatively towards "cyber" as a term of art. In its most abstract, appending cyber as a prefix simply means "involving a computer or network". A reasonable concern is that as most human functions and interactions become more reliant on some form of digital involvement, the term itself becomes largely redundant. Yet for now, "cyber" is unavoidable. The term appears in policy documents, media coverage, private sector services, Western military strategy, and official public reports.

INTRODUCTION

Irrespective of the sentiment towards it, cybersecurity is a meaningful concept because we ascribe it as such. As of now, using "cyber" as a linguistic qualifier effectively reflects the intersection of all other topics with networks and computing.

However, for the purpose of this work, "cyber" and "network" will be used nearly interchangeably. Both entail the use of interconnected computing resources, and thus effectively mean the same. As a result, network-warfare and cyber-warfare are viewed here as analogous, as are network operations and cyber operations. This substitution is not particularly new; the National Security Agency has relied on the term "computer network operations" (CNO) for several decades.

The concept most fundamental to this book is "military offensive cyber operation". For the purpose of this research, it shall be defined as any means of digitally affecting adversary systems and networks for a military goal or objective; affecting data by using data. This definition includes a wide swathe of possible offensive vectors while excluding non-offensive operations or kinetic operations against network equipment.[21] It also includes civilian intelligence agencies and directly co-opted third parties conducting operations in service of a military objective or campaign. Cyber-warfare, in essence, refers mainly to military OCOs and defensive operations occurring within conflict. Most of what is characterised as network-warfare today is in fact routine intrusion operations conducted by intelligence agencies for broader peacetime national security objectives. When espionage and corporate sabotage are intermixed with actual network attacks, it becomes increasingly difficult to distinguish what passes the threshold of warfare. If countries were to adopt the wider view of cyber-warfare that incorporates espionage, the escalatory ramifications for international diplomacy would be dire.

Targeting of corporate and government entities occurs frequently in peacetime. While tempting, it is dangerous to insinuate that the world is engulfed in constant unrelenting cyber-warfare as a result. By itself, cyberwar is an awkward term attempting to depict a conflict wholly waged through networks, detached from other forms of political contest. Cyberwar as a term of art will only be briefly acknowledged and discussed, and soon after discarded in favour of a more integrative perception of cyber-warfare and offensive operations. To accurately frame what falls within the remit of cyber-warfare, it is necessary to reach a coherent depiction of offensive cyber capabilities that exceed acts of espionage or localised sabotage. It is useful to generate proper boundaries and thresholds.

No widely accepted term currently exists to describe the evolution of forms of warfare that do not have a kinetic, physical manifestation. Jamming, electronic warfare, computer network operations, cyber-warfare, and information warfare all share several common characteristics: they rely on the unseen transmission and manipulation of data to achieve an effect. As military forces become increasingly physically distant from the violence they inflict upon people and property, data and its many uses become significantly more meaningful to the conduct of war. Data affects communication, telemetry, coordination, targeting, command, intelligence, navigation, and planning. Striking at the channels which silently enable these functionalities is an understandably important undertaking, one that has commensurately evolved as the functionalities themselves have. The term offered by this book to encompass all efforts to undermine transmission, reception, and processing of data is *intangible warfare*. While it is a descriptive instead of technical label, exploring how it practically evolved over the last eight decades will illustrate the continuity it embodies. Intangible warfare may differ in technique, approach or effect, but there is strong historical scaffolding that ties all such operations together; this will be explored throughout.

A Structure of Theory and Practice

Direct evidence of state-backed offensive operations against networks is relatively lacking, though mounting. In lieu of vast datasets, we can instead rely on different sources to create a tapestry of complementary information. This work relies on critical assessment of sources across three axes; *technical, operational*, and *strategic*. Until history provides a richer offering of case studies to examine, those seeking to understand network operations must rely on cautiously informed assessments based on limited evidence.

The technical axis is an examination of how networks and devices may be targeted for attacks. Some effects are either infeasible or unrealistic to carry out. Whereas disabling a tactical communication network is both possible and plausible, causing a nuclear submarine's reactor to undergo a critical failure is a remote possibility due to difficulty of access, numerous fail-safes, and mitigation procedures. Aligning expectations to reality is a key part of any analysis that seeks to chart what warfare may realistically look like when carried out through and against networks. Technical assessment is carried out by inspecting the specifications of military equipment and networks, and the potential vulnerabilities that these may be afflicted with. Examples of this include over-reliance on ageing and hard-to-update technology, or the increasing tendency to introduce remote command and control directly into weapons.

Sourcing on technical specification of military hardware and software includes freely available manuals on military standards, officially published reports, leaked sensitive data, and even promotional materials published by private sector contractors entrusted with designing and manufacturing military equipment. US government accountability reports shine a fascinating light on assessed vulnerabilities and limitations of newly developed combat platforms. When these sources are combined, a fairly robust mesh of military-deployed technologies emerges, supporting an analysis of how these may be targeted in conflict.

The F-35 Lightning II Joint Strike Fighter is offered as an example in Chapter 3. This ambitious project includes numerous contractors and participating nations spanning well over a decade of research and development. By design, the project is meant to integrate well into existing orders of battle, while simultaneously offering state-of-the-art sensors, onboard software, and peripheral logistics and maintenance systems. Sources on the F-35 platform include official accountability reports detailing its many software flaws,[22] public coverage of the F-35's recurring errors,[23] official specifications of the F-35's "ALIS" semi-autonomous logistics system,[24] and even leaked classified documents pertaining to BYZANTINE HADES,[25] the network operation in which crucial intelligence pertaining to the F-35 project was exfiltrated—presumably by Chinese threat actors. By and large, the F-35 exhibits deep flaws that may be exploited for effect by a determined adversary. Possible operations may include disruption of the onboard radar suite, interfering with communication protocols used by the aircraft, or even a presence-based operation to disrupt logistics and maintenance by corrupting regionally-deployed ALIS units. Such attack vectors are not merely theoretical when the plane itself exhibits numerous issues, and its own auditors express scepticism at the craft's software readiness for sustained operations in a contested airspace. This is more than conjecture; the very people who prepare the F-35 for combat are concerned about its ability to survive it.

Network attacks of varying goals are ubiquitous. There is no dearth of evidence when it comes to mapping a multitude of techniques for targeting networks and devices. The unprecedented public scrutiny of network intrusion tools resulted in an explosion of available analysis of offensive network operations and capabilities. These vary in quality and relevance, but uniquely provide a glimpse into how networks are targeted by countries for national security objectives. As the intelligence agencies behind network intelligence campaigns will often also precipitate network attacks, learning about their craft and methodology by dissecting and analysing their detected intrusions is paramount. There are lessons to be learned from analysing where they excel, where they make errors, and perhaps most importantly, how they adapt to challenges and evolve their capabilities.[26] Visibility into network operations spans numerous countries, threat actors, and underlying goals. In some cases, coverage of intelligence operations reveals complex, modular toolsets[27] that could be applied to a variety of offensive purposes should there be an inclination to do so. In other cases, reviewing how nations have successfully degraded adversary systems suggests how similar operations may materialise against equivalent military targets. The toolsets apply to both generic event-based attacks and the most targeted and expansive of presence-based attacks.

Sourcing for network intrusion analysis extends beyond the information security industry. Several batches of leaked materials pertaining to compromised intelligence agencies provide intimate access to internal documentation and assessments of operational capabilities within some of the most capable network aggressors. Disclosures include the expansive Snowden documents leaked from numerous US agencies and units including the National Security Agency (NSA) in 2013, the leak of sensitive information from the NSA's Tailored

Access Operations unit in 2016,[28] and the leaks codenamed Vault-7 that allegedly contain a vast repository of information on CIA intrusion and attack capabilities.[29] Russia[30] and Iran[31] similarly suffered leaks providing a glimpse at their practices.

The strategic axis is evaluated by relying on official publications and de facto state behaviour. Most simply, countries often publicise their relationship with offensive network operations within their core official documents, such as national military strategies. The detail level of these documents varies greatly based on the country analysed, with the United States arguably engaged in the most significant public discourse around shaping its operational capabilities. However, in order to assess doctrinal elements of offensive network operations, this book will rely on documents, reports, and speeches from several countries.

Countries vary greatly in their approaches to network warfare doctrine. Extensive US literature on the topic reveals an evolutionary approach which increasingly views "cyber" as an independent domain of warfare. If taken at face value, the new domain then requires distinct doctrine and allocation of resources. Evidence for this is most immediately reflected in Joint Publication 3–12: Cyberspace Operations,[32] which then has complementary implementations in branches such as the US Air Force[33] and Army.[34] Documents including manuals on joint operations[35] also shed light on how existing strategies could be updated to reflect the inclusion of novel capabilities, and other declassified documents similarly contribute complementary elements.

For other countries, some high-level documents allow identification of how policy makers and military strategists view the role of network operations. The evolution of official military doctrine from Russia[36] and China[37] is highly indicative of the role of network operations in altering conventional asymmetries and creating unique advantages. Views of network warfare vary based on the overall perception of information and its role in conflict. This in turn affects how nations seek to weaponise information against their adversaries.

Limitations

Research undertaken to explore military network operations immediately encounters two related difficulties; capabilities are often highly classified, and evidence of use is scarce. These difficulties are not insurmountable, but merely complicate attempts at fashioning an inclusive framework that would accurately reflect military OCOs as they may be used. The diverse sourcing employed when dissecting such capabilities is the answer; assembling a mosaic of network warfare from disparate fragments of information reveals a rather compelling result. Available technical data informs about military equipment vulnerabilities. Leaked information and public-domain analysis of state-nexus network operations instruct on the techniques and strategy used by nations. Strategic documents, official statements and academic analysis help fashion assessments on the potential use of OCOs to military campaigns.

There is still much that we do not see. The above approach does not fully alleviate the concern regarding the partiality of coverage. Considering the rapid pace of development in the field, commenting on modern capabilities may be difficult, and many of the top-tier forces have not yet tipped their hand as to their full capacity to do harm.[38] However, that should not preclude attempts at analysis. While information technology evolves, many of its underpinnings have remained unchanged for the last five decades. The modern internet has inherited an architecture first developed in the 1960s.[39] Military equipment today must often still support communication protocols first introduced in the 1970s. Military equipment is often designed to last decades, with internal software and hardware gradually evolving ponderously over time. Endeavours to introduce new networks and equipment often lead to new reported vulnerabilities. Thus, looking at the military network space reveals that much of it relies on technology incepted decades ago, now fully understood and accessible to researchers.

Using information to alter public perception at scale is not the same as achieving an effect through a network attack. This book intentionally excludes such information and influence operations from the scope of analysis. As previously said, the impetus for doing so is clear; information operations cover an immense swathe of non-violent activities that muddle analysis of capabilities employed in warfare. While combat operations may certainly incorporate information operations, and some nations such as Russia fuse them seamlessly with other facets of intangible warfare, they remain sufficiently distinct as to merit exclusion. Without excluding information operations from the analysis, attempting to scope where military responsibility lies becomes near-impossible. How can network warfare capabilities be relegated to battlefield commanders if the distinction between them and psychological operations is hazy? Rather than being a detriment, segmenting information operations into a separate frame of observation lends clarity to the ontologies offered in this book. Different frameworks exist to discuss information warfare.

1

PRINCIPLES OF CYBER-WARFARE

How can warfare be waged with software? The practice of military combat evokes imagery of opposing forces colliding with each other in increasingly sophisticated ways. The spear, the horse and the shield gave way to the bow, the rifle and the missile. Throughout the ages, war charted an evolutionary course in which the finest technology of the time was wielded for organised acts of violence. Efforts to maintain an advantage in warfare had a tremendous effect on other walks of life. The same forge used to craft the blade was also used for the work tool. The innovations in rocketry devised to fuel Cold War ballistic missiles were used to send people into space. A tight-knit symbiosis formed between military technology and its civilian counterparts; one bred the other, the former fed the latter, and vice versa. The development of combat waged over networks—what is colloquially called cyber-warfare—is just one modern iteration of that same cycle of innovation.

The threat landscape is replete with network intrusions, ranging from theft of information to tangible asset loss reaching millions of dollars, and even physical damage. Where many of the tools of warfare are distinctly operated by militaries, many of the most influential network intrusions have notably been perpetrated by civilian intelligence agencies. Increasingly, we see such operations used to pursue offensive objectives, including incapacitating networks, corrupting

15

data, and incurring damage. Should they be assessed on the spectrum of intelligence operations, or warfare?

We can improve how we identify when operations qualify as warfare, or even attacks. This chapter will seek to address these questions by offering five cumulative parameters with which network attacks can be individually assessed, those being *target, impact, attacker, goals* and *relationships*. Together, the parameters form a model that excludes most incidents which are out of scope for analysis of military-aligned offensive cyber operations. Discerning that the affected targets (1) are of significant quality or quantity is the first milestone in identifying a warfare-threshold activity; a think tank does not equal a military target. Impact (2) includes both the initial observable effects of the attack and its wider consequences, as the vast majority of known incidents have little or no physical effects on the afflicted party. Identifying the attacker (3) establishes attribution to a state or sub-state entity directing the attacks. Goals (4) relates to assessing that the agenda of the perpetrators is military-strategic, crucial in an ecosystem where most intrusions are motivated by criminality or loose ideology. Finally, relationships (5) addresses the larger geopolitical and strategic considerations in which the attack takes place, and is often the key differentiator between warfare and other adversarial situations. All five parameters must be met if an incident is to qualify as within the spectrum of cyber-warfare.

Standardised assessment based on the above five parameters can help discern between combat and intelligence campaigns, or distinguish a criminal enterprise from a cautious precursor to a military attack. In software-based attacks, these distinctions are often less trivial than they may seem. In the wake of mass exfiltration of sensitive information and nation-sponsored attacks against the banking sector, as has occurred in the US,[1] the most resonant question is often "what does this intrusion mean?" If the answer is the devolution of the political situation into armed conflict, it is far more consequential than an embarrassingly successful espionage campaign. This book will address several operations that blur the boundaries between warfare and peacetime, often intentionally. It will also exclude the broader scope of influence campaigns through a focus on offensive operations.

The Venn diagram of military OCOs and cyber-warfare is not a circle. Some offensive cyber operations are carried out in conflict by non-military actors, while militaries may pursue OCOs in peacetime. This book focuses most heavily on the intersection of these two concepts, but through this also looks at borderline cases and instances where countries purposefully blur the lines or pursue mixed strategies. These are educational as they teach us about the flexibility and limitations of attacking information systems for political gain.

The second process included in this chapter is disentangling cyber-warfare and cyberwar. This matters beyond being a bothersome semantic quirk—the two terms are not one and the same. While cyber-warfare can be established as a distinct part of warfare, cyberwar is the quintessential bogeyman used to terrify policy makers into funding inscrutable plans. Cyberwar and cyber-warfare are often used interchangeably in literature and coverage alike to denote any friction between two parties over the internet; this lack of distinction has proven harmful to the overall quality of the discussion. Cyberwar is simply not a meaningful construct and may therefore be replaced with other appropriate labels based on the underlying context and motivation. That may be crime, espionage, attacks, or actual warfare-threshold incidents.

Although it is a thoroughly Western concept, a framing of cyber-warfare must extend beyond its perception in the United States and its allies. Awkward and misappropriated analogies have only detracted from generating agreed-upon standards for what is acceptable and what constitutes a red line. The difficulties in crafting a cohesive notion of offensive cyber operations were suggested at least as early as 2001, when US Air Force veteran Gregory Rattray noted in his book, *Strategic Warfare in Cyberspace*, that "Frameworks for evaluating the capabilities of international actors to conduct conflicts based on attacking information infrastructures remain underdeveloped".[2] He continued to aptly warn that the label of information warfare was being too broadly applied.

This was exemplified in April 2016 when US Deputy Secretary of Defense Robert Work exclaimed, "We are dropping cyberbombs!"[3] as he discussed the ongoing military campaign against the Islamic State. One could almost hear the collective groans of cybersecurity experts worldwide as they witnessed a further sliver of nuance wither away from their craft. The particular phrasing drew criticism, while still serving to reflect the military perception of the value of offensive network capabilities. Time has not improved on such rhetoric, as seen in the wake of the 2020 SolarWinds campaign. Within days of revealing that numerous federal agencies and other organisations had been targeted through a supply chain compromise, Democratic Colorado congressman Jason Crow exclaimed his belief that the incident "could be our modern day, cyber equivalent of Pearl Harbor,"[4] a staggering analogy for a campaign that cost no lives and incurred no publicly known effect beyond information loss. Identifying where espionage ends and attacks begin matters a great deal in national security. Across the Atlantic, the Russian-language term for cyberwar—*kibervoyna*—mostly only exists as an acknowledgement of Western thinking rather than a centrepiece of Russian doctrine.[5] As a rising number of nations openly or tacitly acknowledge the significance of various offensive network capabilities to their strategy, cyber-warfare must appropriately expand as a concept to encompass its different varieties. Concurrently, expanding the scope of offensive operations risks diluting analysis beyond utility. Striking a balance between inclusiveness and cohesion is crucial.

This chapter will explore several examples of malicious network activities that—despite their elevated public profiles—should not be depicted as warfare. In 2014, Sony Pictures was the victim of a destructive network attack that resulted in data loss, exfiltration of sensitive information, severe disruption in daily operations and many millions of US dollars in damages.[6] In what was at the time a relatively rare act of public attribution, then FBI Director James Comey publicly pointed an accusatory finger at the attackers: "we know who hacked Sony. It was the North Koreans who hacked Sony".[7] The offensive was widely indicated to be a continuation of North Korean policy by cyber means, a strike at the heart of the American studio that dared to publish the parody movie *The Interview*. The movie presented revered North Korean leader Kim Jong Un in a comical manner with the entire film centred on his attempted assassination. Public attribution efforts of the attack pointed to Unit 121 of the North Korean Reconnaissance General Bureau, one of the nation's more notorious hacking units[8] often associated with offensive network activities.[9] By 2018, the US government had indicted an operative affiliated with the North Korean government for this attack and others.[10] In this sense, the Sony attack straddled the grey area between warfare and non-warfare activities, with its unique set of capabilities applicable to both. While the visible, high-profile attack held the possibility of escalation, it resulted in no apparent kinetic or virtual countermeasures save heated rhetoric. As the offered model will show, the attack against Sony failed to meet several criteria necessary to qualify as a warfare-threshold incident, even if it was an OCO.

The Boundaries of Cyber-Warfare

When the virtual medium itself is mostly intangible, communicating mutually agreeable limitations on the conduct of war becomes even more significant, even if this communication is tacit. The barrier of entry for conducting some forms of offensive action over the internet has decreased; it is arguably easier to generate a noticeable effect against a military network than it is to physically harm a missile battery. In this sense, the global internet has provided both opportunity and capability, reducing the barrier of entry somewhat.[11] However, as we will see, even if it is easier to attempt offensive network operations, the truly impactful operations remain challenging blends of intelligence, operations, and technical skill.

As indicated before, five criteria can help distinguish warfare-level attacks from other offensive incidents, such as financially-motivated criminality, illegal ideological incidents, or even peacetime offensive activities. The parameters are *targets, impact, attacker, goals* and *relationships*. They are not ordered by importance, as all five parameters must be met, but rather by increasing difficulty of assessment. Put differently, while identifying the exact intended victim of an attack (step 1) is often the easiest endeavour, surmising the significance of the underlying strategic and political relationship between attacker and target (step 5) is the most daunting of tasks in the process. There is a steep increase in the complexity of analysis required to meet each subsequent parameter. Importantly, the model here has no claim on establishing legality of offensive action within the international order.

OFFENSIVE CYBER OPERATIONS

Figure 1: Five-step model for assessing offensive cyber operations

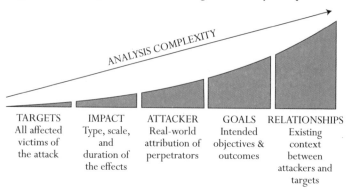

TARGETS	IMPACT	ATTACKER	GOALS	RELATIONSHIPS
All affected victims of the attack	Type, scale, and duration of the effects	Real-world attribution of perpetrators	Intended objectives & outcomes	Existing context between attackers and targets

Though we discuss existing legal frameworks, they are provided as a perspective into assessing incidents.

In information security, incident response to network compromise is an incremental process in which forensic evidence is analysed and assessed against available knowledge of offensive activities. The sequence of criteria in the model reflects the spirit of this process. Identifying the victim is the genesis of any investigation. Once the exact target, targets, or parts thereof have been identified, it becomes possible to deduce the incident's impact on it, by observing deviations from the target's normal state of affairs. Based on the victim and the evidence of the attack itself, it then becomes plausible to attempt to identify the incident's instigators by way of a meticulous attribution process. Only if attribution has been reasonably successful can the observer then gauge motivations and the underlying goals. Finally, the pre-existing relationship between the target's parent country and the aggressor's country can be coupled with the overarching context in which this relationship exists.

The first examined parameter is assessing the incident's affected *targets*. Beginning the process with a victim assessment provides an early opportunity to classify the attack's immediate scope. Intrusions against military assets, infrastructure and logistics will clearly meet the threshold, as they are immediately indicative of an adversarial relationship in which at least one party is the warfighting apparatus of a nation or a nation-like entity. Almost unerringly, safeguarding the multitude of military networks and systems is an internal military responsibility, which vast resources and efforts are allocated to.[12]

A direct attack against military assets would result in a response cycle initiated by the affected military. The affected party may in turn—based on the nature of the attack and its assessed perpetrators—choose to respond in force. This is naturally different than any offensive action—destructive as it may be—which targets a private organisation. An intrusion against Coca Cola has a markedly different significance than one against the British military's Royal Logistic Corps.

Critical national infrastructure forms the second category against which attacks will meet the required operational threshold of warfare.[13] China,[14] Russia,[15] the United Kingdom[16] and the United States[17] have all separately acknowledged that attacks against networks of critical national infrastructure (CNI), such as energy, banking and communication, shall be considered as potentially indicative of an armed attack against the nation itself. Critical infrastructure has largely been defined similarly by most nations. As defined by the UK Government's Centre for the Protection of National Infrastructure (CPNI):

> Those critical elements of infrastructure (namely assets, facilities, systems, networks or processes and the essential workers that operate and facilitate them), the loss or compromise of which could result in (a) major detrimental impact on the availability, integrity or delivery of essential services—including those services whose integrity, if compromised, could result in significant loss of life or casualties—taking into account significant economic or social impacts; and/or (b) significant impact on national security, national defence, or the functioning of the state.[18]

Two incidents are worth evaluating as contrasting examples of targets. The 2014 attacks by North Korea against Sony's networks, destructive as they may have been, would not constitute warfare due to the nature of Sony Pictures Entertainment as the victim entity; it is a wholly private corporation that serves limited critical functions in the various services it offers. It therefore becomes apparent that the high-profile nature of the attack does not grant the US military recourse over the hack. Conversely, the alleged US–Israeli campaign to inflict physical harm against the Iranian nuclear program could be viable as a warfare-threshold target, solely based on targeting of the Natanz uranium enrichment facility.[19] Noting that attacking Natanz risks escalation may explain why the Stuxnet operation was waged painstakingly covertly.[20] The campaign's underlying desire may have been to avoid the risk of retaliation for choosing such a sensitive target.

Finally, a third option towards meeting the target parameter is replacing quality with quantity. Where an attack against Sony is insufficient on its own to merit an escalation cycle, a simultaneous attack against all major movie studios—Paramount, 20th Century Fox, Sony and others—may be interpreted as a fundamental attack against American soft power. Similarly, governments would be hard-pressed to ignore simultaneous destructive attacks against thousands of targets of opportunity, even if they are relatively insignificant on their own. The actual quantity to meet the target threshold can either be a meaningful enough percentage of the targeted industry, or conversely simply hundreds or more of consecutive targets.

Assessing the *impact* of network attacks is essential to understand their scope and reach. In a reality bereft of many empirical examples with which to assess "cyber armed attacks", anchoring the discussion on existing legal guidelines for conventional attacks is reasonable. Though not the focus of this work, the existing legal framework is also useful when looking to tether cyber-warfare to the laws of armed conflict, as the NATO-sponsored Tallinn Manual originally attempted in its first iteration.[21] In its second iteration, the manual defined cyber-attacks as "a cyber operation, whether offensive or defensive, that is reasonably expected to cause injury or death to persons or damage or destruction to objects".[22] This definition usefully defines attacks through their impact, but in doing so precludes operations that only affect data or software. The desire to segment attacks as only affecting the physical world is dangerous; most offensive network capabilities are not analogous in consequence to physical attacks and will not yield comparable results. Barring the most extreme cases, offensive network capabilities will not physically destroy or impair their targets.

Impact is a key source of divergence between cyber-warfare and conventional, kinetic warfare. By shackling the comparison to the physical domain, the end result suffers by excluding whole categories of attacks. One such example, data disruption attacks, would be left out of the narrow definitions requiring observable damage. By the same metric, crippling the defensive capabilities of an adversary with a software-only attack is not contained within the spectrum of warfare.

Whether an intrusion is an attack or meant for intelligence collection may in some cases only become apparent once the intruder decides to activate a malicious payload—or instead does not. It is roughly analogous to observing a burglar enter a house and inspect the rooms, waiting for him to attack the sleeping residents—or simply leave with their jewellery and cash. A defender would still want to stop the burglar regardless of their goals, but should the criminal be successful, the violent outcome is immeasurably worse than theft. As with forensic assessment of burglaries, intruders may be sloppy and leave some indicators as to their intent. To pursue the break-in analogy further, the burglar may have left a knife as he fled. Indeed, Buchanan has indicated that this very issue stands at the core of what he called the "cybersecurity dilemma".[23] Nations may inadvertently be drawn into a cycle of escalation as a result of misinterpreted network intrusions.

A basic distinction is required for intelligence operations, namely the demarcation between extraction of information and destructive sabotage. As history instructs, espionage has been a mainstay of international relations for millennia.[24] While compromise during espionage operations is nationally embarrassing and dangerous to the operatives involved, it rarely results in escalation to the level of open hostilities. Mutual espionage between rivals and allies, whilst unfortunate, is ultimately desirable conduct, at least to some degree. Successful spying may reduce the levels of uncertainty as to an adversary's capabilities, disposition and intent, potentially allowing observers to make more informed decisions.[25]

This differentiation is crucial, as no international legal framework bars non-destructive surreptitious espionage between adversarial nations.[26] A spy extracting information—as successful as they may be—does not grant the afflicted nation any legal recourse in the international arena. If captured, the spy would undoubtedly face the ramifications of breaching domestic anti-espionage laws, but therein the complications conclude. Between the two nations, the implications of such an ordeal rest squarely on the existing nature of the relationship between them. Internationally, the treaties governing non-harmful espionage and intelligence gathering are underdeveloped and vague, even if they do recognise that spying constitutes a breach in the targeted nation's sovereignty.[27]

Decoupling espionage from attacks in network operations is essential because it is a recurring point of contention. The impact of network espionage campaigns may appear daunting simply due to their unprecedented scale. The ability of an adversary to extract immense quantities of data—sometimes from multiple targets—over the course of months or years can amount to a strategic breach of national security. Naturally, lawmakers and other government officials may be tempted to classify the incident as an attack in an attempt to generate an escalation cycle or pursue certain recourses.

On 8 December 2020, security company FireEye revealed that it had been compromised.[28] Perhaps more dramatically, it reportedly had its internal red team tools copied. Red team tools are used to accurately simulate the behaviour of operational adversaries and are valuable for preparing organisational network defences for realistic threats. They are essentially internally developed malware variants and other operational support tools. While this would seem like a success for the intruders, FireEye's detection of the operation initiated its unravelling. Five days later, FireEye announced that they had identified a supply-chain compromise of global platform IT management software SolarWinds.[29] The unknown adversary had weaponised the update life cycle of the SolarWinds Orion software, infecting numerous organisational users. In the days that followed, multiple US federal agencies that had been using the software were told to urgently patch their Orion deployments or shut them down altogether.[30]

Several US lawmakers immediately lashed out at the intrusion. Congressman Jason Crow called it the equivalent of Pearl Harbor,[31] and Senator Dick Durbin said the intrusion was "virtually a declaration of war".[32] The rhetoric was neither accurate nor particularly helpful; the information known at the time only suggested a breach of agencies rather than any impact against them. With the US likely conducting similar wide-reaching operations of its own, there was little strategic or political benefit in portraying successful espionage campaigns—far reaching as they may be—as acts of warfare. Pending any disclosure of effects attempted against the victims, such operations must be viewed through the prism of intelligence collection.

A notable caveat exists when active network intelligence collection appears to be a prelude to an armed attack; such instances may be considered to threaten the use of force.[33] As an example, spy plane sorties intent on highlighting targets for a subsequent attack meet this standard and may trigger self-defence countermeasures. A spike in network intrusion attempts on tactical military networks, if intertwined with a reasonable threat context and existing heightened tensions, may similarly reach the threshold meriting self-defence. Such acts are sufficiently escalatory as to be on a spectrum of operations that may lead up to hostile engagement and war.

That the *attackers* are organs of a state or at least state-affiliated is crucial; the identity of the perpetrators helps determine culpability and accountability. Put differently, if an individual or an organised crime unit performs a hack of sensitive military networks, it does not immediately follow that the nation hosting the unaffiliated attackers is culpable and must be penalised directly. Some form of affiliation must be established, or at the very least provable negligence in prevention efforts by the host nation's law enforcement agencies. Nations do not typically engage in war against criminal organisations, nor can they do so against individuals.[34] A state or meaningful political sub-state entity must be the responsible party to the attack before a military conflict perspective can be used to assess the incident. This requires reaching an acceptable threshold of high-level attacker attribution.

Attribution is the challenging art of identifying who is behind malicious network activity. The challenge stems from the relative ease in which we can disguise ourselves online, mask our activities, and generally blur the trail of our activities. The fuzziness in linking online activity to a real-world identity is sharply contrasted to the relatively clear signals provided by most other offensive military activities. As former US Secretary of Defense William Lynn famously said in 2010, "Whereas a missile comes with a return address, a computer virus generally does not".[35] Performing attribution of an attacker to a politically acceptable degree is a contentious topic in network operations.

The government of the People's Republic of China has long been suspected of relying on a corps of a loosely-affiliated "non-governmental forces" (民间力量), capable of being tasked with pursuing national intelligence requirements and even conducting networked attacks.[36] Comparably, Russian network attacks against Estonia in 2007 and Georgia in 2008 have both been publicly depicted as ideological and criminal participation in weaponised online protest rather than an organised Russian assault, despite some indications suggesting otherwise.[37] Iran frequently employs civilian contractors and private companies to both build and deploy malicious tools.[38] The shifting levels of deniability baked into the internet's relative anonymity makes high degrees of certainty difficult—although certainly not impossible—when performing attacker attribution. As passionately echoed by Russian doctrine, the confusion sown in the wake of a network attack is one of the prominent advantages of the medium.[39]

Attributing attackers thus becomes an issue of reaching an acceptable level of confidence in their identity. As Rid and Buchanan illustrated when presenting their model, "attribution is what states make of it".[40] Accepting that there are no absolute certainties, nations can turn to several approaches towards making assessments: an accumulation of technical forensic evidence, known operational techniques of the adversaries, external intelligence sources, and political context. The combination of the above can and has previously resulted in public attributions. Since the mid-2010s, the US government has repeatedly sought to attribute network operations to state-affiliated individuals and organisations in Iran,[41] China,[42] and Russia.[43] Reaching acceptable levels of attribution in cyberspace is therefore possible, even if it is a challenging endeavour. But once we have identified an attack's perpetrators with some certainty, a tougher challenge is assessing their underlying reasons.

Goals determine attacker motivations, pivotal to assessing an attack's significance and appropriate countermeasures. The significance of a network attack varies greatly depending on the intention and underlying objectives of the attacker. An offensive campaign against banks meant to secure illicit funds through ransomware differs from one meant to cripple an adversary's financial system, even if both are indeed attacks. Similarly, a tactical attack against a military early-warning network as a prelude to conflict is different than one against the Olympics meant to destabilise the event and signal discontent with political proceedings.

Destructive acts of espionage, often referred to as sabotage, are tricky to classify. As previously stated, all aspects of modern life are reliant upon networks and computers. Adversely subverting their normal operation in a purposeful act can inflict heavy financial losses, loss of critical functionality and in rare cases even manifest as physical damage and loss of life. Yet not all such cases are created equal. Attacks designed to restrict access to services and disturb the flow of data occur frequently, even with state sponsorship. At the time of writing, no single purported case has publicly culminated in open hostilities between nations, unless the attacks have transpired in the context of an existing conflict. Most directly, if an aggressor acts in pursuit of military objectives or to prevent the accomplishment of those by the target, this meets the cyber-warfare threshold for the "goals" parameter. This subset of goals has the clear determination of being directly carried out by armed forces or subordinate groups in order to achieve military goals—whether within active combat or without. US Department of Defense Joint Publication 3–12, titled 'Cyberspace Operations', makes clear the boundaries in which the armed forces operate in cyberspace to this effect. It states: "Commanders conduct cyberspace operations (CO) to obtain or retain freedom of maneuver in cyberspace, accomplish [joint force commander] objectives, deny freedom of action to the threat, and enable other operational activities."[44]

The viability of network capabilities as part of a military objective is not unique to Western doctrine. Although there is no single cohesive Russian doctrinal document cementing the definition of network warfare, abundant official and unofficial Russian texts separately refer to information as crucial to modern battlefield dominance. As several Russian military theorists have posited in the journal *Military Thought* in 2009:

> [The] main objectives will be to disorganize (disrupt) the functioning of key enemy military, industrial and administrative facilities and

systems, as well as to bring information–psychological pressure to bear on the adversary's military–political leadership, troops and population, something to be achieved primarily through the use of state-of-the-art information technologies and assets.[45]

The further important subset of goals relevant to cyber-warfare includes any and all operations intent on disrupting the sovereignty or integrity of the targeted nation and its affiliated organs. The Russian approach to examining what constitutes an act of cyber-warfare is purposefully broad and can be inferred from the nation's overall strategy on utilisation of such operations by the armed forces and intelligence agencies. Russian military doctrine from 2014 is fairly explicit on the perceived threat of cyber-warfare when it enumerates the top external threats to Russia:

> Use of information and communication technologies for the military-political purposes to take actions which run counter to international law, being aimed against sovereignty, political independence, territorial integrity of states and posing threat to the international peace, security, global and regional stability.[46]

This is then mirrored by an equally broad internal military threat perception: "Activities aimed at changing by force the constitutional system of the Russian Federation; destabilizing domestic political and social situation in the country; disrupting the functioning of state administration bodies, important state and military facilities, and information infrastructure of the Russian Federation."[47] Combining the internal and external threat equates to the claim that any attack of significant potency against the state, its various organisations, its citizens or any infrastructure that serves them could constitute an attack worthy of a military response. The irony of this approach is palpable, as Russia has been frequently alleged to conduct such attacks against its neighbours and global peers.

A proper example to the significance of this parameter would be the hack of the US Democratic National Convention (DNC), initially publicised in June 2016.[48] In the incident, a hacker operating under the moniker Guccifer 2.0 released a trove of internal documents and emails allegedly exposing the inner workings of the Democratic campaign. While the hacker declared himself to be a Romanian hacktivist, extensive analytical commentary by security organisations such as

ThreatConnect,[49] academics such as Thomas Rid,[50] and media outlets such as Motherboard[51] and the *New York Times*[52] quickly raised the assessment that intruder was not, in fact, Romanian, and that the breach originated from state-affiliated Russian perpetrators. Rather unusually at the time, attribution to the Russian government itself was publicly acknowledged by the US Department of Homeland Security: "The US Intelligence Community (USIC) is confident that the Russian Government directed the recent compromises of e-mails... from US political organizations".[53]

The DNC hack satisfies the target, impact and attacker parameters. However, when observing the assessed goals of compromising a national election race, while they certainly appear political they do not seem military-strategic. While there is extensive potential utility in favourably shaping the US political landscape to better accommodate Russian political grand-strategy, the incident itself does not appear to be motivated by a distinct military agenda. Conversely, assessing the 2014 attacks against the Ukrainian voting infrastructure[54] appears markedly different. As military conflict was already in progress, in part around the issue of Ukrainian government legitimacy over disputed territories, upsetting the tenets of Ukrainian democracy fuels destabilisation efforts that may then convert to reduced military resolve.

The fifth parameter, *relationships*, accounts for the differences in how nations respond depending on the political context between victim and perpetrator. Even the gravest of incidents can be overcome if they occur between friendly nations, or otherwise between two parties which have a vested interest in avoiding conflict. Therefore, in order for a network attack to qualify as a warfare-level incident we must establish the existence of appropriate context; one in which the incident is weaved into the larger political climate and relationship between the involved parties.

In 2013, online access to multiple US banks was intermittently cut off by a distributed denial of service attack (DDoS)[55]—a form of attack characterised by a large subset of computers overflowing the victim with frivolous data to prevent legitimate access. Denial attacks are inherently transient, usually incapable of causing permanent damage save financial losses incurred from the victim's temporary lack of access to the targeted services. In this case, the outrage expressed by US officials at the continuous attacks was palpable, with the desire for attribution and retribution mounting.

"There is no doubt within the US government that Iran is behind these attacks,"[56] claimed James Lewis, a former hig US government official now working for the Center for Strategic and International Studies. As compelling as the attribution against Iran may have been, it did not result in any significant military action or meaningful retaliation. As expected, the attacks were handled by the judicial system, resulting in the indictment of seven Iranian nationals.[57] Notably, this was an action taken against the involved individuals, not the sovereign entity that had seemingly set them on their disruptive course to begin with. As a limited disruptive attack with a political but ultimately non-military goal, it was contained and not pursued further in the international arena.

Examining the first four parameters, at least the first three appear to be distinctly met. It was a successful attack against US critical financial infrastructure, conducted by state-affiliated Iranian actors. Arguably, the claim could be made that the impact parameter was not met, but sustained denial of functionality to critical financial institutions may indeed pass the required threshold. The fourth parameter does not appear to be easily met, as the goals do not readily translate to a military-strategic agenda. But indeed, even if the banking attacks were military-oriented, it is the existing US–Iran dynamics, the deferential character of the Obama administration, and the pre-established behavioural patterns by the US government that all cumulatively prevented the fifth and final relationship parameter from being met.

Political analysis of the United States simply does not corroborate interpretation of the banking operation as an act of warfare. A powerful regional actor, the United States acted upon its political agenda to prevent Iranian entrenchment in current and future negotiation of its nuclear capabilities.[58] The US government was simultaneously roped into a sizeable role in regional hotspots in Syria, Afghanistan, Iraq, Yemen and Libya.[59] As a nexus of Shia influence, and source of funding, military technology and training for many of the United States' adversaries, Iran's brinkmanship proved its potency in the way the United States contained the attack. Iran had succeeded in fashioning a political climate in which conflict was to be avoided unless absolutely necessary. Similarly, the Obama administration had at that point established a prior history of non-reaction, even when faced with adversaries crossing its self-defined red lines for mobilization. Such was the case when embattled Syrian president Bashar Assad crossed US President Obama's red line by deploying chemical weapons against rebel forces and the civilian population.[60] The US, in turn, did not implement its own threat, thereby further establishing itself as a noncommittal defender of its own policies and interests. The context of a sabotaging network attack is highly meaningful.

Also involving Iran, one of the only confirmed acts of physical damage perpetrated through protracted network operations is Stuxnet, the network operation used to retard the Iranian nuclear program by damaging its centrifuges between 2009 and 2011. The case was heavily lauded as the first militarised "cyber-weapon"[61] and a herald of a new era of cyber-warfare.[62] However, deeper observation of the Stuxnet campaign reveals how little it had to do with facilitating war. Perhaps it even sought to prevent it.

As reported extensively by journalists[63] and information security companies,[64] Stuxnet was a clandestine campaign that was allegedly waged by the US and Israel together. While the operation itself was complex, the underlying goals were seemingly straightforward: stunt the accumulation of enriched uranium by the Iranian government. The goals in turn supported efforts at keeping the possibility of nuclear militarisation distant, thereby obviating the need for a military attack against Iranian nuclear facilities. The involved malware was characterised as being highly covert and self-restrictive.[65] Specifically, Stuxnet's developers made exorbitant efforts to conceal its physical disruption from technicians by falsifying maintenance signals from the equipment.[66] This allowed for rolling, incremental losses in centrifuges rather than a high-profile strike against the Natanz facility. It was an intelligence sabotage operation, managed and conducted by at least one intelligence agency,[67] within the scope of a larger political agenda keen on preventing a nuclear Iran without actual warfare taking place.

The Stuxnet attack induces more confusion than clarity. If such a brazen, damaging campaign against a declared enemy nation does not constitute warfare, what does? The incident seemingly ticks most relevant boxes; it was physically damaging, targeted critical national infrastructure, was conducted by nation-level actors and served a coercive military–political goal. It is often cited as the perennial example of cyber-warfare, and yet it was neither predicated nor followed by any other acts of war between the parties involved; the fallout from the incident was still somehow reined in.

The crucial element in which Stuxnet's Operation Olympic Games[68] fell short of warfare is relationships. The adversarial relationship between Israel, the United States and Iran did not merit the event devolving into a state of armed conflict. Nor was this attack conducted within pre-existing conditions of open hostilities. It was a clandestine sabotage campaign occurring between global adversaries. Without such indicators, even a physical network attack may not reach the threshold of warfare if it is not sufficiently damaging or high-profile. However, the attack conducted by US Cyber Command in 2019 in response to Iranian attacks against shipping containers is a whole other matter. The attack targeted elements of Iranian intelligence; caused significant impact; was attributed to the US military; was pursued as a military objective; and occurred between nations on the brink of armed conflict.[69] Across all five parameters, this was a warfare-threshold offensive cyber operation in all but name, undertaken by the US against Iran. The context of any network attack is critical to their classification as acts of war.

When can equipment sabotage be considered warfare? The 2007 Israeli operation against the Syrian nuclear reactor, allegedly including a network attack against the Syrian air defence grid, is more thoroughly discussed later in the chapter. It is—if the reported details are correct—the most easily palatable instance of cyber-warfare in the public domain. When appraising the five elements presented above it becomes apparent that the operation meets all criteria. It was an attack against the military assets of a state actor (target) with significant consequence (impact), by an adversary military force (attacker), for a military–political objective (goal) within the context of a larger kinetic military operation (relationship) designed to hamstring any attempts by Syria to attain nuclear status. Thus, tactical cyber-operations launched in conjunction with a kinetic attack are perhaps the embodiment of easily discernible cyber-warfare. But they are not the only such cases.

Borrowing from the Russian doctrinal playbook, mass-impact network attacks on civilian infrastructure may be a notable subset of warfare-level operations. If a government-affiliated actor strikes at a likewise-affiliated adversary with an attack designed to severely impact civilian life, the conduct of warfare may appropriately apply. It must, however, meet all parameters to properly qualify. As a result, such an attack must either be a prelude to or in tandem with other acts of violent aggression, and be sufficiently impactful as to cause severe disruption or harm. Finally, there must be an underlying military political goal rather than a criminal one, a goal reflecting the national agenda espoused by the attackers.

Re-examining the Russian network attack against the Ukrainian power station with the aid of the five accumulative parameters shows it may indeed qualify as warfare. The attack targeted Ukrainian critical infrastructure, resulted in meaningful outages, and was conducted by Russian security services.[70] As assessed by private sector reporting, the goal was in fact to generate disruption on a far larger scale, with their analysis indicating that "the number of control systems identified for manipulation is large and more widespread than the actual outage".[71] Finally, as there were active hostilities between Russia and Ukraine over Crimea and its adjacent territories,[72] the relationship parameter is also fulfilled.

To conclude, offensive network operations embody the continued evolution of warfare as a result of the adoption of networks into every aspect of modern living. Demonstrable acts of cyber-warfare are rare, but the discussions conducted by various nations on the role of cyber-warfare within their national doctrines and the threats emanating from cyberspace are highly educational when attempting to frame cyber capabilities as tools of war. By utilising the five provided parameters, assessing if network intrusion incidents are acts of warfare becomes a more direct endeavour. Were it applied to most of the instances widely panned as warfare, this method would show that most incidents are wholly outside its boundaries. The global state of cyber-warfare is comparably calm.

The spectrum of cyber-warfare includes attacks which are either directly supporting kinetic campaigns or are otherwise generating an impactful influence upon the adversary within a larger military-political context. When viewed this way, only a very specific—but still significant—subset of operations is included. Reaching this realisation helps illuminate the advantages and disadvantages of cyber-warfare and its larger role within conflict. War remains a violent struggle between groups vying to upset the existing power dynamics, and offensive network capabilities do not radically alter this underlying truth.

Cyberwar and Cyber-Warfare

The inescapably political nature of war presents the first challenge of assessing cyberwar as a term of art. As Kenneth Waltz once posited, "War begins in the minds and emotions of men, as all acts do".[73] Warfare has changed, but its fundamental tenets have not. While warfare may rely on increasingly complex tools, it remains at heart a contest between people. Computers, devices or networks cannot declare and wage war against one another; at least not yet. Therefore cyberwar—if it is a valuable concept at all—is simply war as waged by actors through computers and networks.

It is curious to note how prevalent the use of the term cyberwar is, as if wars can be waged and won through network attacks alone. War does not routinely restrict itself to a single domain—be it land, sea, air, or indeed cyberspace. As NATO realised in Kosovo, as the United States experienced in Iraq, and as Israel internalised in Lebanon,[74] avoiding entire domains in favour of safer, indirect forms of warfare merely delays and intensifies future cross-domain warfare. In this respect, Waltz's remarks prove prescient once more: "In reality, everything is related to everything else, and one domain cannot be separated from the others".[75]

Cyberwar is a ubiquitous term because it is often conflated with cyber-warfare. Cyberwar too often encompasses acts of aggression that do not equate to war, threading these together in a narrative of conflict limited to digital means. Much like previously overstated theories of complete battlefield dominance through overwhelming airpower, offensive network capabilities are unable to singularly achieve political goals. In the opening chapter of their oft-cited book *Cyberwar*, Richard Clarke and Robert Knake artistically depict the suspected use of offensive network capabilities in 2007, when Israeli Air Force (IAF) warplanes supposedly bypassed thick Syrian air defences to clandestinely strike at the Kaibar nuclear reactor in Deir-Azzor.[76] While this attack is arguably one of the most easily classifiable instances of publicly suspected cyber-warfare,[77] it falls short as a depiction of cyberwar. Irrespective of the role of offensive network capabilities in the strike, it was guided bombs and missiles that actually reduced the facility to ruin. The Israeli incident is a textbook depiction of a combined arms package, in which an alleged network attack enabled an immediate physical strike for lasting effect. The political objective underpinning the conflict, which was to prevent non-conventional abilities that would upset the strategic balance in the Middle East, was supposedly achieved by jointly operating cyber and air power.

Viewed independently, the Syrian–Israeli incident does not validate the existence of cyberwar; there was no reciprocal conflict waged through networks. Where war is an exercise of aggression to attain political goals, offensive network capabilities are poorly positioned to single-handedly ensure success. Comparing this to other domains of warfare, the fallacy of this perceived reality becomes readily apparent; modern wars do not neatly constrain themselves to a single domain or set of capabilities. To analogise, renowned naval strategist Alfred Thayer Mahan posited heavily upon the momentousness of the naval arena in power struggles between nations.[78] He did not, however, decree that naval engagements exist in a vacuum, devoid of additional domains of warfare.

Analysis of modern warfare can benefit from setting aside cyberwar as a concept. Instead, it is more useful to observe cyber-warfare as a subset of offensive network capabilities used to engage an adversary in a period of conflict. By emphasising cyber-warfare as part of a broader tapestry of armed conflict, we discount the notion that network intrusions are immediately indicative of an overall, war-level conflict between two or more parties. Focusing on operations rather than war then leads to a range of critical questions about the utility and role of networked capabilities. Can network attacks trigger war? What forms of network attacks fall within the remit of warfare and which are merely acts of espionage or sabotage? As the next chapter details, many military OCOs are in fact more similar to electronic warfare than they are to the other distinct battlefield domains—air, sea, land or even space.

"It is not correct to call every bad thing that happens on the internet 'war'...," James Lewis rightly exclaimed in his 2010 paper, "Thresholds for Cyberwar."[79] Around that time, many network intrusions and even rudimentary scanning attempts were problematically and incorrectly labelled as attacks.[80] Tallies were inflated to enormous, unfathomable numbers to present an alarming image of constant digital warfare occurring between global political adversaries. Thankfully, coverage of digital warfare has since vastly improved, though the problematically hyped-up narrative persists somewhat.

A well-established approach is to classify network warfare as a subset of information warfare conducted through computers and networks.[81,82] As a notable example, the Russian military has no independent conceptualisation of "cyber" except when remarking on Western doctrine.[83] Instead, warfare conducted via the internet or against enemy networks is folded into the struggle for information dominance, one pursued aggressively by Russia for decades.[84] By adopting this approach, networked attacks operate within similar parameters to electronic warfare, information warfare and psychological operations.

Lumping electronic warfare with cyber-warfare is attractive. Much like electronic warfare, offensive network operations seek to disrupt, alter, corrupt or otherwise influence the operation of the targeted system. While classic electronic warfare most often seeks to achieve this by emitting radio frequency (RF) transmissions, cyber-warfare attempts to achieve the same by interfacing directly with the targeted system's hardware and software. The attack vector may differ; the underlying logic and desired effect often remain the same.

The Russian government has visibly adopted this approach in its political and military manoeuvres, yet it is not singularly Russo-centric in nature. In 1999, John Arquilla—a notable early scholar of cyber-warfare—revisited his previous statements and claimed that "information warfare is a concept that ranges from the use of cyberspace to attack communication nodes and infrastructures to the use of information media in the service of psychological influence techniques".[85]

Arquilla's then RAND colleague Martin Libicki expressed a similar sentiment as early as 1995.[86] As the United States gradually became more transparent in its acknowledgement of offensive network capabilities, the Department of Defense's 2003 Information Operations Roadmap initially outlined the perspective in which electronic warfare and computer network operations are viewed as equally significant pillars of information operations.[87] Employing motivational phrasing, the document awkwardly declared: "We Must Fight the Net."

Russia is unusual, as its approach to information warfare audaciously blends influence with coercion, the civilian with the military, and the digital with the kinetic. It does not matter whether a general seeks to influence global perception of Russian forces or achieve command and control dominance in a battlefield—it all falls under the purview of offensive information operations. Controlling the flow and shape of information is thus a key tenet of modern Russian doctrine. An approach that clusters different facets of manipulating the flow of information is understandably useful. The digital building blocks that make up civilian communication are often highly similar when broken down. Networks of different types, sizes and purposes often use the same protocols and thus can be targeted in similar ways. Where manipulating computer networks may yield military results one day, it may similarly disrupt the flow of terrorist propaganda the next. Yet that is where the similarities end.

There are important limitations to the Russian perspective. A monolithic approach to information operations may risk muddling the important distinctions between its different sub-categories. When providing the military and supporting security organisations carte blanche to conduct full spectrum information warfare across all targets and agendas, the results may bleed into one another with reduced effectiveness. The degree of finesse required to successfully influence mass global media is incomparable to disconnecting aircraft from their regional command. Where the former relies on subterfuge, nuance, and an intimate socio-cultural familiarisation with the target, the latter requires technical acumen and an operational intelligence cycle to breach hardened networks and identify vulnerabilities.[88]

A secondary risk of blurring the boundaries between information warfare and militarised cyber-warfare is undue escalation. Simply put, if information operations are within the discussion of war, even low-yield, minimally affective operations may be treated as *jus ad bellum* (cause for war). We must then consider whether an American digital influence campaign in occupied Crimea desirably constitutes an act of war against Russia. Information operations are pervasive in peacetime as a means of manipulating the political landscape to generate favourable conditions. Attaching such operations to the spectrum of war greatly increases the risk of friction between otherwise low-contact adversarial relationships.

An alternative way to observe the development of cyber-warfare is as a counter-innovation phenomenon. In this view, rather than the pristine outlook of offensive network capabilities as independent means of securing battlefield goals, these toolsets become means to offset the capability of the adversary to achieve their own goals. It is an evolutionary concept, born as a response to the rise of the interconnected battlefield. Where digitisation and networks once enabled the precise, coordinated operations of the twenty-first century, we now see the techniques that materialise to mitigate them.

The People's Republic of China (PRC) is a staunch advocate of this doctrine, and understandably so. During the First Gulf War of 1991, Chinese military leadership were in sudden full view of a revolutionary theatre-level campaign against the Iraqi military.[89] The US-led military coalition assembled to expel Saddam Hussein's Republican Guard from Kuwait was so successful as to result in near-surgical evisceration of Iraq's conventional forces, previously considered to be well-trained, well-equipped and capable. Relatively precise joint operations were made possible by the interweaving of intelligence, surveillance and reconnaissance (ISR) assets with guided ordnance. Data from ISR assets was fed into command and control centres which allowed effective operational decision-making on an unprecedented scale.

This new approach was later coalesced into the US Network-Centric Warfare doctrine. It was defined as focusing "on the combat power that can be generated from the effective linking or networking of the warfighting enterprise."[90] Integration of digital networks into the full range of wartime decision-making would significantly enhance the quality, quantity and response rate of actions taken. Sensors from assets deployed both within and without the battlefield were to provide critical mission support and assist in dispelling the fog of war. To nations such as the PRC, which were—at the time—gradually increasing investment into the modernisation of their armed forces,[91] observing joint warfare in its full capacity was alarming.

Lessons learned from observing integrated coalition forces in the Gulf War were eventually co-opted into PRC operational doctrine.[92] The People's Liberation Army was directed towards focusing on integrated joint warfare as a key means of defeating asymmetrically preferable adversaries. As explained in the key account of the Chinese People's Liberation Army (PLA) strategy, "The Science of Military Strategy", achieving this reality was made (and remains) uniquely possible by increasingly adopting information sharing on an operational level: "[s]upported by information technology… various combat factors are woven into a unity".[93]

From network-centric warfare grew the PRC's brand of network warfare as a form of military counter-culture. The PRC internalised that their conventional forces were heavily outmatched, and as a result launched a concerted effort to rebalance this asymmetry by turning the unique characteristics of the US command and control structure into a vulnerability.[94] If all battlefield operations were now reliant upon timely, consistent and numerous data inputs from multiple sensors, interrupting this flow could potentially wreak havoc.

In 2004, China devised a new strategy to counteract the advantage of C4ISR systems, pithily labelled "local wars under the conditions of informationisation".[95] A decade later, this was shortened to "winning informatized local wars,"[96] signifying the centrality of information to modern conflict. Although cyberspace was viewed holistically in the PRC's national strategy as a "new pillar of economic and social development,"[97] it was also immediately christened an altogether new area of national security.[98] Within this doctrine, degrading or disabling the adversary's information hubs became paramount towards attaining initial operational dominance, thereby facilitating victories on the battlefield. The key development from previous strategy was the significance of information both to conducting effective joint operations and countering adversary advantages.

The PLA's views on intangible warfare hold similarities to their US counterpart's.[99] Both the US and Chinese militaries view offensive network capabilities as a combined arms package—the melding of several approaches and armaments—to be employed alongside conventional operations.[100] In this sense, attacks targeting computer software are a crucial asset for the modern battlefield, as they deny adversary capabilities, reduce threat and enable kinetic operations. This perspective reflects how the PLA evolved its doctrine to its modernised adversaries. While the true potency of the PLA's militarised offensive capabilities remains uncorroborated, the focal shift reflected in official writing and organisational changes is telling. Yet however publicly visible the adoption of offensive network operations seems to be in China, it is not the PLA that had been the initial purest embodiment of the integrative approach—but rather Israel.

As previously mentioned, in late 2007, numerous international media outlets reported a surprise airstrike against a previously publicly unknown Syrian nuclear facility in Deir Azzor.[101] The strike was widely and immediately attributed to the Israeli Air Force, known for its adoption of a proactive operational doctrine bent on preventing regional powers from attaining the capability to produce nuclear weapons. The results were staggeringly successful; complete destruction of the facility, Syria widely condemned by the UN after a follow-up probe confirmed the presence of nuclear materials,[102] and perhaps most importantly—no retaliation from an incensed Syrian regime. It was a clear tactical, operational and strategic success for the aggressors.

Breaching the Syrian air defences to accomplish a surgical strike with no friendly casualties or loss of materiel, and minimal collateral damage is especially noteworthy when matched with the perceived high quality of Syrian air defences.[103] Known as the most tightly-interwoven modernised air defence network in the region, it surprised many to learn that it never fired a shot at the transgressing warplanes. The question of how this came to pass lingered in the wake of the irradiated ruins of the facility. Five years later, in September 2012, the *New Yorker* published a lengthy investigative piece on the attack in which it was claimed that the Israelis were "using standard electronic scrambling tools" to effectively blind Syria's anti-air radars.[104]

Conventional jamming against such a sprawling network is risky. Syrian forces operate a wide range of anti-air assets, ranging from older SA-2 and SA-6 batteries that are susceptible to older forms of jamming, to more modern SA-17 Buk and SA-22 Pantsyr-M1 batteries purportedly sporting significant jamming resistance.[105] A larger, "hotter" strike could have included aircraft carrying high-speed anti-radiation missiles (HARMs) designed to physically eliminate radar-emitting anti-air threats, or the use of Israeli home-grown standoff cruise missiles such as the Popeye or Delilah. High-profile kinetic attacks would have undoubtedly raised the calculus of brinkmanship, risking cornering Syrian President Bashar Assad into an otherwise undesirable escalatory reaction. Israeli decision-makers acknowledged the significance of maintaining low attack profile, opting for the limited "Thin Shkedi" operational package—so named after then air force commander Eliezer Shkedi—rather than the wider, more comprehensive "Fat Shkedi".[106]

A network attack against the Syrian air defence grid has been suggested as a crucial enabler of the overwhelming Israeli success. When the Israeli censor finally let local media outlets report more extensively about the strike, it continued to apply heavy limitations on the operational details of how it was carried out.[107] Specifically, details around penetrating Syrian anti-air defences remained scant. So, while unconfirmed by official sources, the motivations for the use of offensive network capabilities are sound. A non-lethal network attack is advantageous as it results in a low-profile temporary disruption to adversary systems, thus carving out a window of operations for incoming aircraft while avoiding meaningful collateral damage. While Israeli capabilities in this field are largely unproven, US reporting has previously covered the suspected existence of an airborne anti-platform cyber-warfare platform dubbed Suter, allegedly developed by British security company BAE Systems.[108] If a cyber-attack was indeed employed, it is one of the only publicly discussed combined-arms military use of offensive network capabilities. The attack embodies the significance of software-based strike vectors as key enablers and supporters of physical follow up attacks.

While Israel led in testing its methods directly on the enemy, the United States evolved its military bureaucracy by discourse. The US armed forces have over the last 20 years increasingly identified the importance of information systems to modern warfare. In 1998,

Cebrowski and Garstka of the US Navy notably penned a piece in which they claimed that the US was "in the midst of a revolution in military affairs (RMA)".[109] They were referring to the importance of the "fundamental shift from... platform-centric warfare to... network-centric warfare".[110] By that time, the revolutionary perspective on cyber-warfare has been a long time in the making within the US Department of Defense. The speed and quality of wartime decision-making could be markedly improved by harmonising different platforms and sensors to seamlessly work together.

The United States gradually adopted an approach framing digital attacks as an independent domain of warfare. This embodied the persistent desire by the American defence establishment to put everything in distinct, neat boxes. While awkwardly shoehorning cyber into a domain may appear initially frustrating to external observers noting that "everything is cyber", a sprawling, globe-spanning, multi-conflict force such as the US employs must often do so to remain effective. The conversation in the US military and government over the role of cyber-warfare has been an evolutionary one; from a minimal acknowledgement of information operations to a fully-budgeted Cyber Command, standalone doctrine,[111] and dedicated exercises. As defined in 2005 by then Air Force Secretary Michael W. Wynne, the new role of the air force would henceforth be to "to fly, and fight in Air, Space and Cyberspace".[112] The US perspective codified cyber-warfare as a distinct field, with its own set of rules and considerations.

The distinct domain approach is gaining traction in the West. After several Russian-attributed network attacks against Estonia and Georgia in 2007 and 2008, respectively, an alarmed NATO scrambled to come to grips with the vulnerability of its member nations' networks and its own incapacity to conduct effective wartime operations in cyberspace. Most immediately, NATO's Cooperative Cyber Defence Centre of Excellence (CCDCOE) solicited a group of experts led by Michael Schmitt to identify the circumstances under which network attacks constitute *jus ad bellum*, thereby qualifying above the threshold of warfare.[113] In 2014, NATO expanded its defensive ethos to incorporate a network attack as a legitimising catalyst for invoking Article 5, which covers NATO protocol in the case of an attack against a member state.[114] Eventually in July 2016, NATO conventional wisdom coalesced around a declaration mimicking their US member state doctrine; cyber was announced to be a distinct domain of warfare alongside air, sea and land.[115]

Military culture has a huge impact on how cyber is integrated. Different nations have adopted broadly different approaches to cyber-warfare depending on their existing culture and limitations. Some chose to more thoroughly adhere to the distinctions that make the intangible software space unique; the United States visibly fences operating within and against networks militarily. Conversely, Russia recognised the added value of cyberspace, but did not necessarily expose the seams between those who fight within the domain and those who fight outside. In Russian military thought, networks are a virtual medium through which existing doctrine and operating procedures are channelled and reflected. This more integrative approach to offensive network operations means that they often purposefully fly below the threshold of warfare, in a bid to avoid retaliation. China views network operations as an equaliser, an opportunity to shatter existing symmetries and turn the advantages embodied in modern networked warfare into vulnerabilities. Lastly, countries such as Israel retain ambiguity over the actual doctrine surrounding cyber-warfare, but appear to be employing it nonetheless as a combined arms package in support of kinetic operations.

Approaches to digital warfare may vary, yet they crucially intersect where operations meet wartime necessity—the need to forcefully erode the will of the enemy,[116] or alternatively seek to complement ongoing war efforts. Any suggested scope for cyber-warfare must encompass this diverse reality, as has been shown throughout this chapter. While the major wielders of offensive network capabilities vary in their perception of their utility and the terminology used to describe them, they intersect on doctrine and deployment more than they initially seem to.

To conclude, there is a common view of cyber-warfare: it is designed to contribute to joint efforts to affect information for military-strategic goals. The above analysis is complemented by several high-level observations. The first is that offensive network capabilities cannot single-handedly achieve strategic objectives; such capabilities are best used when complemented by others. For some such as the

United States it may entail incorporation into kinetic strikes, while others—namely Russia—place more significant weight on large-scale information operations that include psychological warfare. As a corollary, offensive network capabilities are consensually viewed as integrated into the spectrum of information operations. Finally, OCOs are a natural evolution of warfare and doctrine, as they gradually formed out of a modern-day dichotomy: the need to offset the strategic advantages of a highly networked force, while simultaneously capitalising on the strategic vulnerabilities of the highly networked nation which it defends.

2

CHARTING INTANGIBLE WARFARE

A century has passed since warfighting first dipped into the electromagnetic spectrum. While the study of war often focuses on the kinetic, it is the imperceptible mediums that have some of the greatest advancements in the modern battlefield. From the advent of radar in the years leading up to the Second World War, to radio and satellite-guided munitions, to vying for digital information superiority today, war waged through invisible means has been embedded in every nation's doctrine. Cyber is the first aspect of intangible warfare to be widely recognised as sufficiently distinct to merit its own domain of war. But how truly different is it from its predecessors? How unique are its operational characteristics and associated challenges?

There is a familial relationship between what we now call cyber-warfare and electronic warfare. Many of the attributes uniquely attributed to cyber-warfare have long been associated with combat operations in the electromagnetic spectrum. Examining similar concepts such as electronic warfare, electromagnetic warfare, command and control warfare, information operations and cyber operations, they become visibly related in their characteristics, and can be bundled together under the term *intangible warfare*. Efforts to separate them entirely are therefore often artificial, as are some observations on cyber as the first man-made domain of war.

OFFENSIVE CYBER OPERATIONS

Offensive cyber capabilities are an evolutionary step, representing seven decades of cumulative experience and development in other forms of intangible warfare. The fundamental necessity in tactical cyber operations to be intimately familiar with the particular equipment being targeted dates back to radar jamming efforts in the Second World War, as does the notion that employing such capabilities may lead to their detection and loss. The Cold War saw strategy evolve from deceiving operators to directly impacting the functionality of devices and equipment. Perhaps most importantly, the idea that situational awareness on all levels—tactical, operational and strategic—can be critically manipulated by non-conventional capabilities has coalesced over the span of the last five decades.

Whereas bullets, shells and missiles function as intended against a wide range of possible targets, intangible warfare is unique because it may require the development of tools to defeat the specific technology of a particular enemy.[1] The Second World War era of British attempts to jam German flight guidance radar and the intricate network operations against military platforms done today both share the undeniably crucial need for intelligence and familiarisation with the adversary. Aspects of intangible warfare both historic and modern represent the desire to increasingly weaponise the adversary against itself and erode the fundamental trust a battlefield commander places in their technological toolset.

In some respects, cyber-warfare is not a distinct domain of warfare. Much like electronic warfare, battlefield network capabilities may affect targets but are frequently wielded by operators in the classic domains of war—land, sea and air. Understanding this integrative notion analytically then provides researchers and practitioners with the ability to implement lessons learned from previous iterations of intangible warfare—in doctrine, strategy, operations and incorporation of new technologies and capabilities.

Already in 2001, then US Air Force officer Gregory Rattray railed against the supposed novelty of so-called cyber-warfare. Particularly, Rattray rejected "the assumption that strategic information warfare should be treated as a completely new phenomenon because of the 'virtual' or nonphysical nature of operating in the cyberspace environment".[2] While this chapter shares that premise at its core, Rattray then proceeded to assess offensive network capabilities as an evolution of strategic air warfare.[3] Indeed, the analogy to the early days of air warfare may be apt when discussing how network operations are described but it does little to explain how they differ from traditional warfare. So instead, this chapter will establish the claim that cyber is an evolution of other forms of intangible warfare.

The chapter unfolds by reviewing a century of literature and warfighting, dating to the inception of combat by electromagnetic means. The analysis is ordered chronologically, examining in sequence the two World Wars, the Cold War period, the dawn of network-centric warfare in the 1990s, and the subsequent global rise of information operations. With each such iteration, intangible warfare matured and accrued more of its modern characteristics. Today's cyber-warfare did not materialise suddenly with the onset of the internet. Offensive network operations are reflective of a century of doctrinal, operational, and technological bricklaying.

The rapid cycle of innovation and counter-innovation did not originate with intangible warfare, although it is a core commonality of all its aspects. The necessity for technological advancement to counter enemy advantages has always played some role on the battlefield. Personal armour was introduced to reduce the kinetic impact of blows and arrows but was later rendered less effective with the advent of gunpowder. Stone fortifications were countered by increasingly sophisticated siege equipment from trebuchets to catapults, mechanisms designed to return the offence–defence balance to an acceptable equilibrium. For centuries, such development cycles were glacially paced;[4] the scientific method and overall prevalence of battlefield technologies had limited significance to warfare. This is not a novel observation, as Clausewitz once remarked in his writings: "Fighting has determined everything appertaining to arms and equipment, and these in turn modify the mode of fighting; there is, therefore, a reciprocity of action between the two".[5] Yet at the time, strategy, manpower and logistics were vastly more important.

Technology increasingly became a significant strategic element during the First World War. While some of the main participants—such as the British military—adopted key elements of technology-enabled warfare even prior to the war,[6] it was only tested at scale when combat erupted. Uniquely for the time, the incorporation of technological advancement heavily influenced doctrine. The oft-recited perception of the First World War as one of trenches and attrition was in part due to the then-conventional wisdom that the introduction of machine guns meant open warfare doctrine was rendered obsolete; technology had now lent to its wielders an unstoppable killing power that could not be directly countered. Commanders and strategists suddenly had to become innovators.[7] Although some Allied commanders, such as the American Expeditionary Forces' commander in chief John Pershing, held to their refusal to alter doctrine to compensate for new threats,[8] the field of war inexorably changed.

So dawned the first of many cycles of modern military counter-innovation. The necessity to counter asymmetry-inducing technologies meant that existing tools had to be wielded better, and new tools developed. The lag in adjusting to new battlefield realities was in part the reason for the war's drawn-out campaigns, with battles ending in unfathomable death counts on all sides.[9] Doctrine gradually adjusted with the first large-scale use of joint operations combining infantry, creeping barrages of artillery and—to a degree—tanks. These new inventions—ushered in by British forces in 1916, late into the war—were seen as a stalemate-shattering technological advancement capable of once again upsetting the offence–defence balance.[10]

First Cycle—The Second World War

In 1939, British intelligence officers in Oslo received an anonymous tip. They were instructed to adjust the daily BBC World Service German-language news broadcast slightly, to signal to a (presumably) German turncoat to provide the information he offered. They did, and after calling out "*Hallo, hier ist London*," they subsequently received a staggeringly comprehensive report on cutting-edge German technological developments. These included details of large military radars being developed and deployed, of large-scale rockets and even unmanned aerial vehicles.[11] Successfully co-opting the use of radar had seemingly allowed the Germans to circumvent one of the greatest challenges that plagued both the German Luftwaffe and its allied counterparts—conducting precision night-time aerial bombing runs, when anti-aircraft guns were blind.[12]

As Reginald Larson—one of the fathers of modern Western intelligence—recounts, British agents soon confirmed the existence of a German blind-bombing radar-based system known as *Knickebein* ("Crooked Leg"), codenamed Headache by British forces.[13] The radar was an ingenious invention in which two radio beams were transmitted in slightly different patterns from two locations on the coast of occupied France, the beams calculated to intersect roughly over targets in the United Kingdom. Luftwaffe bombers would then fly along one beam until their receivers indicated they had reached the transmission intersection and then release their payloads. British intelligence launched an exploratory sortie in June 1940, intent on intercepting these German radar transmissions.[14] Their efforts bore fruit and resulted in the development of a crude jammer with the appropriate codename of Aspirin.[15]

The British countermeasure did not spell the end of what later came to be called "The Battle of the Beams." Aside from waging actual warfare, both parties were now thoroughly engaged, for the first time in history, in a brief but fast-paced battle of technological wit. The Germans had not yet given up on electronically transmitted guidance systems, deploying the successful X-Gerät (X-Device)[16] radar in late 1940 with improved jamming resistance.[17] This was countered again by the British, which led to the far less successful Y-Gerät—or Wotan II—in 1941. By that time, and aided by Enigma-decoded information and sheer chance, British intelligence had been able to pre-emptively counter Y-Gerät, rendering it entirely ineffectual.[18] Soon after, the Germans were forced to recall their efforts as attention was redirected towards the build-up on the Eastern front, where an invasion into Soviet territory was imminent.

The contest over radar dominance in the Second World War was the first large-scale instance of what shall henceforth be called *intangible warfare*; the adoption of intangible means of warfighting such as electromagnetic transmissions and later attacks against digital networks. Even the earliest forms of intangible warfare included many of the same characteristics as the data-based attacks often discussed today within cyber-warfare, including often imperceptible effects and caution over employing measures to avoid exposing their existence to the adversary. The Second World War—in this regard—would be the dawn of modern intangible warfare.

In the years leading up to the Second World War it became increasingly clear that technology now played a pivotal role in developing modern strategy. The Great War saw an entire generation decimated due to previously unseen firepower. In the war's aftermath, it soon grew apparent that new capabilities and techniques were needed to increase combatant survivability and once again produce a favourable flow of combat. Thus, the years between the two World Wars saw an explosion of new technology. Major developments included the formalisation of air as a strategically pivotal domain of combat, wide adoptions of armour corps, initial forms of rocketry and the genesis of computing and electromagnetic transmissions for communications, navigation, and remote object detection. The groundwork was laid for the next cycle of counter-innovation.

Need bred necessity. In particular, the mounting necessity to develop means for technology-assisted warfare meant that the reliance on these capabilities similarly increased. When war again erupted in Europe in 1939, all major belligerents had already identified both the potential of the new unseen aspects of war and the possibilities afforded in disrupting them. Radio communication had become standard issue for armoured, infantry, navy, and air forces seeking to operate massive forces jointly in relative harmony. Electronic navigational aids became crucial towards effectively directing large-scale assaults beyond enemy lines and prevent miscalculations. For the first time, impacting the adversary's freedom to operate in the electromagnetic spectrum became a priority.

The Second World War was a mass exercise in counter-innovation embodied by constant attempts to jam adversary systems; it was the dawn of what would later be known as electronic countermeasures (ECM). Technological innovation was often shrouded with secrecy, as any means of rebalancing existing symmetries were hoarded, guarded, and applied with care. As Robert Watson-Watt—one of the pioneers of Western radar technology—recounted, this new set of capabilities were shrouded in so much secrecy that they often simply failed; operators were not aware of the value of their own missions and could barely practice in advance.[19]

This caution to avoid publicising capabilities was not singularly exercised by British forces. Jammers operated by German forces

were left inactive until absolutely necessary to assist other operations.[20] In 1942, a German naval battle group spearheaded by the
battleships *Scharnhorst* and *Gneisenau* attempted to break the British
blockade at the French port city of Brest. Despite routine patrols by
British aerial reconnaissance employing active radar measures, the
battle group escaped initial detection. As an investigation would later
reveal, the German ships enacted large-scale jamming operations
once they set sail from port.[21] This allowed the battle group to gain
distance against its would-be pursuers, eventually allowing the ships
to reach German ports. The jamming itself was not even particularly
technically effective—it was by virtue of operational surprise that it
achieved its results.[22] As Watson-Watt later recounted on the failure
of British radar operators to discern German jamming from equipment failure: "If I am held to my reiterated statement that radar is not
merely an equipment or a group of equipments, but a system, then
the radar system did fail but the electronics held out; the men behind
the electronics were lamentably far behind."[23]

There is a trust relationship between machines and the humans that
operate them. Operators must trust that the equipment functions as
intended, and machines require that the operators use them correctly
and appropriately act on their output. Operators of electronic countermeasures quickly realised that they had little feedback to assure
them that their efforts were successful. Unlike the bullet or the bomb,
getting confidence in measures that have no physical presence was an
unintuitive process. Operators of electronic equipment would at
times lose faith in their capability to assess that it was indeed functioning as intended. Rudimentary radars and communications equipment
frequently failed due to a host of operator and mechanical reasons.
Adding jamming and adversary interference to the list of failures often
meant a core loss of trust in the capability itself.[24] It was difficult to
tell when a system was misbehaving, and if it was—why. Trust in
systems and the persistent desire to subvert an adversary's trust
became key aspects of intangible warfare, and remain so today.

Improvements in communications were pivotal to the war effort,
for both attackers and defenders. Whereas wireless communications
were gradually introduced in the First World War, by 1939 they had
become essential for joint force operation, artillery targeting and

range finding, and coordination of strategic effort. Reliance on radio signalling increased explosively. It was so meaningful that British research and development efforts resulted in an airborne communication jamming platform named Jostle II, that was subsequently used extensively in their Middle Eastern campaign. Jostle II was designed to disrupt communication networks employed by German tanks and armoured fighting vehicles (AFVs), which in turn affected their ability to operate in unison effectively.[25]

Another key aspect of intangible warfare would emerge around the same time; it is highly dependent on intimate familiarity with the adversary's equipment, systems and operational techniques. Put simply, a great deal of high-quality intelligence collection was required in order to successfully fashion jamming or disruptive electronic countermeasures.[26] While equipment familiarity had always helped tactics, it was a prerequisite for jamming. Each adversary device functioned and transmitted differently. Consequently, technology seeking to degrade the performance of these devices had to accommodate their unique characteristics. Intelligence operations to detect, investigate and map equipment characteristics and vulnerabilities were key to supporting the research and development processes of countermeasures and counter-countermeasures. It was a subtle and risky endeavour.

As a dangerous corollary, battlefield commanders and strategists realised that the dependency on jammers was a double-edged sword; to use an electronic countermeasure was to risk losing it.[27] A jammed signal would soon inform its users that their equipment had been compromised, prompting them to respond. In some such cases, even limited recalibration, frequency changes or modifying the transmission could eliminate the effectiveness of the jammer entirely. This created a palpable tension between operational commanders, who sought to increasingly wield these powerful new tools, and command staff, who had understandable concerns over compromising vital yet brittle capabilities. There was no easy solution to this issue and the uneasy balance remains to date with cyber operations. The Second World War thus marked the first conflict in which the considerations of intangible warfare grew dominant in the battlefield.

Finally, the opportunity to remotely manipulate adversary equipment for effect was the first indication of an underlying theme for all intangible warfare: it is at its core an exercise at deceptively weaponising the adversary against itself. Rather than seeking to directly impact the enemy, initial forms of jamming, spoofing and decoys sought to actively disrupt the operational decision-making process, in turn causing the adversary to act against its own stated goals. Borrowing from US military strategist John Boyd's OODA Loop,[28] intangible warfare graduated from disrupting the first observation phase to impacting the subsequent orientation, decision and actioning phases. While some disruptive effects were previously possible, influencing electronics has made the exercise possible at scale. Causing the myriad devices used in the battlefield to betray their operators quickly became a viable operational goal, a necessity to prevail in the modern war effort. The Second World War had affirmed the need for strategic investment in countering adversary technologies, not just forces and resources.

Second Cycle—Cold War and Electronic Warfare

In 1973, and in spite of concrete intelligence prompting Israel's preparedness, Arab forces led by Egypt and Syria successfully attained strategic surprise by concurrently launching attacks against Israel. In the first several days of what would later be known as the 1973 Arab–Israeli War, Israeli forces were beaten back from their strongholds in the Sinai desert and the Golan Heights. Existing Israeli Defense Forces deployments clashed with freshly furnished Soviet technology in great numbers. Even as they encountered numerous difficulties on the ground and in the air, the Israeli military had within several hours proactively launched naval forays intent on engaging Soviet-made missile boats in use by the Syrian Navy. While considered a marginal series of battles within the broader conflict, they would prove to be deeply prescient of a larger trend. The naval skirmishes included the first ever engagement to effectively use ECM (Electronic Countermeasures) to subvert guided missiles.[29]

On paper, the odds were stacked in Syria's favour. The Soviet SS-N-2 Styx missiles used by the Syrian and Egyptian navies had twice the range of their 20-mile indigenous Israeli counterpart, the Gabriel.[30] The Styx had previously been used in the region, when a volley fired from an Egyptian vessel sank the INS Eilat in late 1967, marking one of the Israeli Navy's greatest disasters.[31] This previous painful experience prompted Israeli missile boat captains to engage all forms of countermeasures upon establishing hostile contact, which included flares, anti-air gunfire and electronic radar spoofing. The aggressive posture, coupled with capably operated ECM and accurate guided missile fire, proved decidedly effective. The Israeli Sa'ar 4 missile corvettes evaded or otherwise destroyed all incoming threats, while sinking five Syrian ships.[32] The strategic effect was significant even for a war with minimal naval aspects; Syrian ships were thereafter confined to their ports, and Israeli corvettes continued to harass coastal targets with impunity. The naval Battle of Latakia had demonstrated that integrated intangible warfare can measurably augment operational efforts.

While the naval theatre proved a success for the Israelis, the airspace was not as easily dominated. A presumption of superiority by the Israeli Air Force (IAF) had resulted in sorties against some of the world's densest anti-aircraft defensive perimeter. In the first phases of the war, failure to employ ECM in order to contend with surface-to-air platforms such as the stationary SA-6 Gainful or the mobile SA-7 Grail[33] resulted in staggering losses. The IAF rushed to provide air support and suppress advancing Syrian and Egyptian ground forces,[34] and the lack of electromagnetic superiority proved costly. While the war itself had many consequences, one key aspect stood out prominently. It was a war in which a prevailing over-reliance on technologically superior military hardware was thoroughly probed for vulnerabilities.

A further interesting characteristic stemmed from the 1973 Arab–Israeli War and other Cold War era conflicts pitting US and Soviet technology against each other; intangible warfare has evolved in earnest beyond deceiving human operators to include deceiving the weapons themselves. The increasing miniaturisation of computers meant that the actual weapon platforms could be partially automated, thereby improving accuracy and response time while freeing operators to pursue mission-critical tasks. Communication networks to facilitate large-scale defensive operations and coordinate operations by disparate assets were becoming increasingly common. These networks served as a potent advantage while similarly resulting in a sizeable increase in potential targets for intangible warfare.[35] By falsifying signals, disrupting sensory input and affecting communication between devices, weapons could be manipulated beyond their original intent. And by attacking areas where electromagnetic defensive measures were concentrated, an adversary's capabilities could be reduced.

The coalescing of doctrine around the electromagnetic spectrum soon gave birth to the ancestor of the doctrine employed today in US cyber-operations. In 1969, the US Joint Chiefs of Staff issued policy intended to clarify terminology around the use of electronic warfare. It was defined as: "military action involving the use of electromagnetic energy to determine, exploit, reduce, or prevent hostile use of the electromagnetic spectrum and action which retains friendly use of the electromagnetic spectrum."[36]

Electronic warfare was initially divided into three key parts.[37] The first, electronic countermeasures (or ECM), was the prevailing term for measures designed to affect devices operating on the electromagnetic spectrum, mainly for communication and radar. As these measures proliferated, countries sought to defeat them by using the uninspired second term, electronic counter-countermeasures (or ECCM). These defensive measures included frequency hopping transmitters to defeat jamming, and techniques designed to identify actual targets within chaff clouds defensively released by aircraft. Lastly, the third component was electronic support measures (or ESM), a term which encompassed all operations and capabilities that enable ECM and ECCM. Otherwise framed, these were intelligence efforts on enemy electronics and electronic warfare capabilities. As previously mentioned, and much like its modern cyber counterpart, electronic warfare was and remains highly dependent on high-quality intelligence. Both offensive and defensive capabilities must inherently be tailored to the adversary technology they are designed to affect.

The Cold War also saw the expansion of terminology beyond countermeasures to proactive attacks. This initially included terms such as electronic attack (EA) and electronic warfare (EW). The increased integration of offensive electronic capabilities led to their recognition as not just reactive measures that attempt to defensively stymie an aggressor's momentum, but as a key component of joint warfare. As early as 1975, Soviet documents discussing large-scale military exercises began including a separate section on electronic warfare.[38] By 1977's Soviet Baltic Sea Exercise VAL-77, electronic warfare was identified as a key tenet of joint operations to be tested throughout combat trials.[39]

As noted by Western observers, Soviet forces adopted an intangible warfare doctrine that rather closely mirrored its US counterpart, calling it Radioelectronic Combat (REC).[40] Notably different, however, was the mathematical approach adopted towards its use. Recognising that it is far easier to recover and respond to REC efforts at degrading command and control, Soviet planners modelled adversary behaviour in order to identify its critical time windows. Those were reportedly defined as "the sum of times required to complete a sequence of steps in control,"[41] or alternatively as the decision-making window for operational command. The goal of REC would be to attack directly in that critical time window as to disrupt crucial decision-making and attain maximum impact.

Third Cycle—Command and Control

The strategic significance of networks increased at an explosive pace throughout the second half of the twentieth century. The growth was largely in step with wider trends in computing; namely the increase in available computational power and storage volume, miniaturisation, and declining hardware costs. Computers rapidly became more adept at handling additional aspects of the complex modern battlefield, and so uniquely capable of facilitating large-scale joint operations and precision targeting. Military networks for communication, logistics, command and control, targeting, and telemetry became ubiquitous and progressively more connected. In countries where the development of joint warfare lagged, military observers noted Western progress with concern. Such was the case of the Soviet Union, which in its final years correctly identified its own growing capability discrepancy with the West as partly a result of a woefully underdeveloped computer hardware industry.[42]

Saddam Hussein's refusal to vacate his forces from occupied Kuwait in 1991 had triggered one of the most prominent joint warfare campaigns to date. In Operation Desert Storm, relatively accurate and coordinated firepower was brought to bear across the warfighting domains, all overseen and enabled by an expansive surveillance grid comprised of satellites, monitoring aircraft, shipborne radars, allied facilities in the region, and tactical equipment. The result was an overwhelming strategic success that reverberated globally, later hailed by some as a "revolution in military affairs" (RMA).[43] Global observers from militaries in Russia, China and others all noted the arrival of precision joint warfare on an unprecedented scale, crucially enabled by a sprawling Command, Control, Communications, Computers and Intelligence (or C4I) mechanism deeply woven into all operations.[44]

This increasing reliance on networked warfare also signalled the deepening symbiosis between intelligence and combat operations. As former Director of National Intelligence James Clapper wrote in 1994,[45] Desert Storm revealed that the era of precision guided warfare created an insatiable need for intelligence.[46] To proceed effectively, operations now required additional coverage that was available consistently before, during and after conflict—and it had to be of higher quality.

In the 1990s, the United States—then the chief pioneer of intangible warfare under an increasingly networked reality—evolved its military doctrine beyond the electromagnetic spectrum in its perception of intangible warfare. Namely, it used an umbrella term for its kinetic and non-kinetic efforts against the flow of information: *command and control warfare* (C2W). Initially, the term was in fact coined as command and control countermeasures, reflecting the innate tendency of intangible warfare to embody counter-innovation.[47] Almost immediately, the term was reinvented employing warfare vernacular.[48] Most importantly, the concept represented a doctrine comprised of five interlocking components:

> [Command and control warfare is] the integrated use of operations security (OPSEC), military deception [(MILDEP)], psychological operations (PSYOP), electronic warfare (EW), and physical destruction, mutually supported by intelligence, to deny information to, influence, degrade or destroy adversary C2 capabilities, while protecting friendly C2 capabilities against such action.[49]

With C2W, intangible warfare had fully progressed beyond support capabilities woven into the existing domains. It instead became an operational goal of its own. C2W was a result of an increasingly ingrained understanding that modern command and control networks essentially formed new centres of gravity. As US Army Field Manual 100–6, dated 1996, explains: "C2W applies to all phases of operations, [and] offers the military commander lethal and non-lethal means to achieve the assigned mission...."[50] The very technology that was used to enable modern warfare had finally become a primary target instrumental in facilitating battlefield success. All five key pillars of C2W embodied a continuation of previous efforts at intangible warfare and a prescient look at modern-day cyber operations. Technological military deception (MILDEP) efforts woven into operations in order to create adversary false situational awareness were in place since the earliest days of radar-based ECM.[51]

The US perceived that intangible capabilities supported, enabled and empowered their physical counterparts in a larger doctrine. Of C2W's five pillars, only one (physical destruction) was inherently kinetic, while the others (EW, MILDEP, PSYOP, and OPSEC) were potentially both kinetic and non-kinetic. Offensive electronic capabilities, much like their subsequent cyber counterparts, were viewed as a particularly potent means when combined intelligently with other available assets. In some cases, such as the US Navy's former Space and Electronic Warfare specialty area assessed to be analogous to C2W, intangible warfare was pre-assumed to function alongside the traditional approaches.[52] The constant improvement of C2 and C2W systems alike tend to create a see-saw effect. As C2 systems are created with "anti-C2" fixes, C2W systems are developed to counter them. The lethality of Counter-C2 assets such as HARM [High-Speed Anti-Radiation Missiles] and sophisticated jamming modulations must continue to stay ahead of C2 systems upgrades.[53]

The brief era of C2W was also accompanied by the realisation that military intangible warfare is one component in a larger battle for information dominance. The notion of *information warfare* was gradually introduced as a contest for controlling the overall flow of facts and situational awareness before, during and after conflict. The concept later extended to operations conducted routinely even in neutral or allied territory—such as propaganda or influence operations—as part of the larger war of narratives. Intangible warfare had effectively spread beyond the military domain to aspects of grand strategy. As stated in the US Joint Doctrine for Command and Control Warfare dated 1996, "Command and Control Warfare (C2W) is an application of [Information Warfare] in military operations... to attack or protect a specific target set—command and control (C2)."[54]

Considering that the US itself viewed C2W as tactically and operationally subordinate to larger information warfare efforts,[55] it's no surprise that other countries do not make the distinction even today. Instead nations such as Russia, China, Iran and Israel view information warfare holistically, as a paradigm that envelops all operational levels including what is often called operations other than war (or OOTW). The holistic approach again indicates how artificial the distinction between cyber warfare and information warfare may be to some; one is contained within the other, the latter can often also be the former, and in many cases, they are implemented and operated by the same personnel. Thus, the notion of "cyber" as a distinct operational domain is similarly artificial, as cyberspace is merely the latest manifestation of the transfer and storage of information.

As everything today is data-based, information warfare extends to far greater reaches than it has before. Electronic warfare elicits tactical-level effects in an attempt to influence remote adversary hardware by way of the electromagnetic spectrum. Command and control warfare, now largely an abandoned term, indicates an operational-level doctrine originally designed to guide targeting against critical places where information flows, and is analysed and disseminated. Strategic information warfare is the attempt to impact the flow of information beyond the scope of military operations, in order to influence decision-makers and the population they are entrusted with securing.

Fourth Cycle—Cyber-Warfare and Information Operations

The rise of network centric warfare (NCW) in the 1990s aligned with the mounting realisation that intangible warfare is difficult to accommodate as long as it is not practised frequently and at scale. While the increased networking of military forces was already proceeding apace for several decades, doctrinal texts on warfighting as a networked force reliant upon information technology were still lacking. As in many cases, NCW terminology was far ahead of its actual implementation. This was reaffirmed by US Department of Defense researchers in their surprisingly candid introduction to the book *Network Centric Warfare*, dated 1999: "The truth is that we are not experts on NCW and far more importantly, in our opinion, no one is. In fact at the current time, NCW is far more a state of mind than a concrete reality."[56]

Network centric warfare also further reinforced the notion that integrating information networks into warfighting did not constitute a radical departure from the classic tenets of warfare. While digital networks and an increased reliance on computing presented both new opportunities and various challenges, the reality of combat remained largely the same.[57] The introduction of network centric warfare—and its equivalents as observed by global US adversaries and allies—is an intriguing mirror image of offensive cyber capabilities. Where NCW recognised the rising use of information gleaned from disparate sources as an asset used by commanders,[58] cyber-warfare commensurately recognised this new reality as an equally substantial vulnerability.

The trend of technology-dependent warfare was not unique to Western militaries. Through careful dissection of the coalition-based Desert Storm operation, the Chinese People's Liberation Army (PLA) soon diverted considerable efforts into attempts to both counter US NCW while also harnessing its advantages locally. This notion was one of the key contributors to the modern Chinese doctrine of local wars under conditions of informatisation. Informatisation was the modern holistic approach that stated that information must be harnessed, fused and wielded at scale to enable pursuit of all national objectives, both military and otherwise.[59] While military analysts previously deemed the PLA far behind their US counterparts at the time of Desert Storm, by 2013 US Department of Defense analysts observed an overwhelming emphasis in PLA drills on joint networked warfare.[60]

The 1990s also introduced a key final component; that intangible warfare fundamentally represents a contest for information superiority. Military doctrine—ever vigilant over the need to visibly achieve objectives—became increasingly broad in its references to the significance of information itself. Intangible warfare began affecting more than just individual systems and streams of data; it influenced the overall perception of conflict itself. The 1940s saw the introduction of electromagnetic capabilities to influence specific systems. The 1970s saw the maturation of these approaches towards influencing localised platforms working in unison. The 1980s witnessed the graduation into abstract regional and global data networks. The 1990s led to the critical evolutionary step of influencing psychological decision-making processes and global public perception. Intangible warfare now encompassed a wider range of activities than ever before.

In November 1992, the US Department of Defense issued Directive 3600.1 succinctly titled "Information Warfare," detailing definitions and responsibilities for the US information order of battle. Notably, the doctrinal document underpinned its policy section with the following opening text:

> U.S. Armed Forces shall be organized, trained, equipped and supported in such a manner as to be able to achieve a distinct information advantage over potential adversaries in order to win quickly, decisively, and with minimum losses and collateral damage.[61]

The transition in vernacular from information warfare (IW) to information operations (IO) was so rapid as to almost be unnoticeable. Within four short years, warfare was swallowed up as a subordinate element of a far more ambitious scope for "information operations."[62] In December 1996 the Department of Defense reissued Directive 3600.1 with a comprehensively altered opening policy statement suggesting a far loftier agenda:

> The Department of Defense must be prepared for missions from peace to war to include military operations other than war (MOOTW), such as peace-keeping and humanitarian operations, opposed by a wide range of adversaries including State and non-State actors. To meet this challenge, DoD activities shall be organized, trained, equipped, and supported to plan and execute [information operations]. The goal of IO is to secure peacetime national security objectives, deter conflict, protect DoD information and information systems, and to shape the information environment. If deterrence fails, IO seek to achieve US information superiority to attain specific objectives against potential adversaries in time of crisis and/or conflict. The goal of IO is to promote freedom of action for US forces while hindering adversary efforts.[63]

While information operations were at least superficially far broader in scope, their core parts bore a striking similarity to previous iterations of intangible warfare doctrine, namely Command and Control Warfare (C2W). Numerous official US documents on IO enumerate its five core competencies as Electronic Warfare, Psychological Operations, Military Deception, Operations Security and Computer Network Operations. These five elements are nearly identical to their predecessor C2W equivalents, with the exception of removing physical destruction and replacing it with computer network operations. Cyber had merely replaced the last kinetic crutch in the broader spectrum of intangible warfare. The transition was complete.

The latest iteration of intangible warfare was a shift to a doctrine more reliant on achieving non-physical objectives. A dependence on data served as the recognition that control of information has generated a completely new set of strategic, operational and tactical goals. Each of the five core competencies represented a conduit for shaping perception, situational awareness, and the decision-making process across all operational levels. Information had finally outstripped in importance the equipment and networks through which it was sent.

The transition to an information-led doctrine was not without its difficulties. The marrying of peacetime and wartime operations, the inclusion of computer network operations (CNO), the exclusion of physical destruction, and the inclusion of global perception management (PM) into scope all contributed to an awkward marring of boundaries. For example, the US Information Operations Roadmap from 2003 proceeded to call for aggressive proactive psychological operations (PSYOPs), advocating for vast resources invested into offensive information capabilities both during conflict and peacetime.[64] At the same time, the document recognised that due to the potentially global reach of propaganda in the internet age, the quality and coherence of messaging must be improved significantly and coordinated with numerous agencies.

The 1990s also saw the birth of "cyber" as acceptable terminology. Cyber-everything had seemingly sprung into existence within a few short years, thereby launching a trend that would accompany military affairs well into the twenty-first century. But the etymology of the word dates back to the 1940s, thereby charting a similar—albeit distinct—course through modern history. In essence, cyber represented the ever-increasing interaction and dependence of man and machine;[65] an apt phenomenon accompanying the realities of intangible warfare. This new family of terms came to roughly encompass the notion of targeting all manner of devices, the networks they comprised, and the software that powered them.

While it rapidly gained popularity, "cyber" as a term did not contribute any clarity. One of the earliest and oft cited cyber-warfare articles remains Arquilla and Ronfeldt's 1993 article, "Cyberwar is coming!"[66] Interestingly, while it is often looked upon with derision as alarmist and overreaching, considering the nascent state of publicly acknowledged intangible warfare it was rather prescient. Namely, the authors distinguished between two key terms: netwar and cyberwar. The former bears a striking resemblance to ongoing influence campaigns waged against Western governments, often attributed to Russia: "Netwar refers to information-related conflict at a grand level between nations or societies. It means trying to disrupt, damage, or modify what a target population 'knows' or thinks it knows about itself and the world around it."[67] Cyberwar, now a frequently abused term, was also defined fairly well by the authors and mimics somewhat the modern Chinese approach: "Cyberwar refers to conducting, and preparing to conduct, military operations according to information-related principles. It means disrupting if not destroying the information and communications systems... It means turning the 'balance of information and knowledge' in one's favor, especially if the balance of forces is not."[68]

Cyber-capabilities therefore epitomise the cyclical nature of intangible warfare. It is the counter-narrative to network centric warfare, a direct assault on the new centres of gravity created by the modern combatant's increased dependence on complex networks of sensors and data streams. As information became pivotal to waging warfare, it almost immediately spawned a reaction that sought to target it. This did not go unnoticed by the United States, which in 1998 formed a miniscule force tasked with defensive cyber-operations—the Joint Task Force—Computer Network Defense (JTF-CND). The task force increased in size and importance over several years, evolving to encompass additional responsibilities. In 2003, offensive cyber capabilities became more widely acknowledged.[69] By 2005, the US military established the Joint Functional Component Command—Network Warfare (JFCC-NW), tasked with coordinating offensive network operations.[70] The significance of targeting and defending information networks has grown so rapidly that, by 2007, then National Security Agency head General Keith Alexander claimed: "USSTRATCOM [United States Strategic Command] has also begun to develop tactics, techniques, and procedures and other concepts into cross-mission strike plans."[71] Network operations, both defensive and offensive, have increased in visibility and perception; enough to eventually merit the 2009 creation of US Cyber Command (USCYBERCOM), established to:"conduct full-spectrum military cyberspace operations in order to enable actions in all domains…."[72] Since the 1990s, there has been an incredibly dynamic growth process for network capabilities within the United States military.[73]

Cyber capabilities and information operations were codified into NATO and US doctrine as distinct but overlapping terms. "Cyber" was recognised as a catch-all term for the protection and manipulation of data and the digital systems that handled it. Separately, information operations encapsulated a far broader range of concepts and operational procedures, from wartime tactical operations to grand-strategy attempts at shaping adversary perception. Not all experts uniformly agree to this delineation. To wit, Dorothy Denning—a leading US information security researcher—published her seminal 1999 book *Information Warfare and Security*, in which she essentially treated information warfare and hacking operations as largely overlapping.[74] By doing so—and similarly to Russian information operations doctrine—Denning supported the idea that all operations against data are a part of the same spectrum.

A Revolution in Military Affairs?

Military cyber capabilities are neither new nor revolutionary. They are the latest incarnation of intangible warfare, the discipline of achieving objectives by targeting technologies. It is more than using technology as the conduit for deceptive content, as influence campaigns or psychological operations seek to do. Intangible warfare positions the technology as the instrumental piece through which success can occur. Since the dawn of the twentieth century, societies have embraced computer technology and so it became a viable target. Cycles of counter-innovation led to increasingly complex solutions, so networked equipment became both an asset and a crutch. When said crutch is successfully impaired, the combatant stumbles and is rendered ineffectual. Cyber-warfare is merely the latest method of targeting military vulnerabilities.

The supposedly unique circumstances around attacking computer networks are, as shown, not all that unique. The secrecy around capabilities and their use dates to the very genesis of electromagnetic warfare in the Second World War. The voracious dependency of computer attacks on high-quality, high-quantity intelligence can easily be rooted in the Cold War contest of ECM and ECCM. Evidence of intangible warfare as a contest for information dominance can be traced to the 1990s doctrine on command and control warfare.

Attacking networks is a combination and escalation of all characteristics of intangible warfare. As time passed and connectivity increased, both the circumstances and potential impact intensified. Where once jamming radar could generate a localised tactical effect, network attacks now potentially enable an adversary to cloud a regional defence grid. Where proximity to the transmission was once needed, attacks can now potentially be carried out by distant troops from the relative comfort of their remote facility. But in order to generate this magnitude of effect, the need for omniscient, ever-present, and extensive intelligence resources has sky-rocketed.

With the exponentially increased levels of complexity of modern communication networks, the challenges of intangible warfare have similarly scaled. Such abilities are more brittle than ever, and guarded with fervour. Where once their discovery would mean eventual countermeasures of uncertain value, discovering modern deployed offensive capabilities can potentially ruin both years-long operations and the potency of the very weapons themselves. The defensive task has similarly become seemingly insurmountable and includes defending everything from civilian critical infrastructure to tactical battlefield satellite uplinks. Each such piece of equipment incorporates numerous hardware and software components developed internationally by a convoluted and largely opaque supply chain, making it a daunting task to certify that a system is truly uncompromised.

Cyber has finally pushed intangible warfare over the brink of domainhood because all the existing parameters have been intensified to the point where they may impact warfare itself. Attacking devices and the data they communicate can have dramatic impact on achieving battlefield objectives. Where once they were peripheral, computers have now permeated through every facet of warfare. So, while cyber is an independent domain, it has only become so because of the unprecedented dependence of all other domains on data.

Offensive cyber capabilities are the digital network aspect of the larger spectrum of information operations. Even as NATO members scramble to assemble fresh doctrine and strategy to combat within this supposedly new space, other nations have already recognised that a wider approach is preferable. The insistence on differentiating cyber from both the other domains and other forms of information operations is intriguing; as shown, the United States itself has by now officially recognised the role of CNO (*Computer Network Operations*) as a component in the grander strategic literature on information operations. The role that network attack capabilities play in OOTW far outstrips their utility in combat.

This historical analysis is by no means an exhaustive look at all military attempts at shaping the information battlespace; it was intended as a sobering look at the evolutionary nature of intangible warfare. Similarly, this analysis is limited in scope as network-attack capabilities are still in their operational infancy. As Russia, China, the United States, North Korea, Iran, Israel and others rush to signal their willingness to operate militarily in networks, the efficacy of these operations remains to be gauged. The relative dearth of real-life, wartime military network attacks does not however prevent the critical examination of the historical processes that led to modern doctrine.

Cyber-warfare is the latest incarnation of the counter-innovation cycles characterising the modern battlefield. Countering radio with jamming has evolved into countering digital command and control with network attacks. Remediating the advantages of radar has advanced to compromising theatre-wide sensory awareness by targeting the network-centric mindset. The underlying logic is the same, but the modern reliance on networking has enabled its application on an ever-greater scale. The historical lessons of cyber-warfare are therefore that its true uniqueness stems from its unprecedented reach, sophistication and scope, not from it truly being a new domain of warfare.

3

TARGETING NETWORKS[1]

Offensive cyber operations (OCOs) epitomise the unrealistic desire for cleaner, less violent conflict. If a nation can be coerced through the targeting of its digital infrastructure, its resolve should theoretically diminish to the point of surrender. This approach to conflict would understandably be appealing, if it were accurate. The aspiration to preclude violence by achieving digital supremacy seems a goal worth aiming for, but time and time again reality has suggested otherwise. Instead, offensive network operations can assist both tactical and strategic efforts, if all their particular advantages and disadvantages are accounted for.

While nations occasionally advertise slivers of information on how they pursue network operations, the methods and objectives for doing so remain understandably undisclosed. We have so far covered the scope of cyber operations, their heritage, and their similarities to other forms of intangible warfare. Yet it is clear to both outside and inside observers that there is some uniqueness to targeting networks. What are the characteristics of offensive network operations carried out by military forces and their direct equivalents?

Grouping offensive cyber operations into presence-based operations and event-based operations is intuitive and helpful. Presence-based operations are all offensive network activities which include a lengthy intrusion phase meant to establish a persistent hold inside adversary networks. Event-based operations include direct attacks intended to cause immediate effect against targets, by compromising their integrity, available resources, or ability to operate as intended. Most publicly acknowledged network intrusions—including many activities for intelligence or criminal purposes—would be presence-based operations if they include an offensive outcome. Direct attacks against military hardware in the battlefield would often be event-based operations.

Event-based operations are roughly analogous to firing a weapon. When such an attack is initiated a digital payload—a stream of data—traverses one or more networks, during which it attempts to subvert the target's normal functionality to achieve an effect. Unlike classic electronic warfare, the attack may involve back and forth communication between attacker and target. Impact on the target, if successful, is immediate or near-immediate. Such capabilities are meant to be reusable. An event attack may be launched by a local fire team (such as an electronic warfare squad), a platform (such as an aircraft or surface ship), or from remote territory (for example, a command centre or intelligence facility). These types of attacks—like their kinetic counterparts—often have localised impact meant to augment or support other, conventional strikes. Such tactical network warfare can therefore work well in a combined arms package, jointly deployed alongside kinetic capabilities.

Presence-based operations are roughly analogous to clandestine sabotage operations. An initial intelligence operation results in access to the adversary's relevant networks. From that point, the intruder's malware foothold is used to enumerate enemy servers and endpoints, gathering information, infecting additional devices and networks, and identifying potential weak points that may subsequently be attacked for effect, all the while continuously evading defences. Specialised offensive modules are optionally deployed where needed, with the intent to activate when the command to do so arrives.[2] The potential risk to friendly weapons and capabilities from discovery is far greater due to the extended presence "behind enemy lines", as is the chance of failure. But the potential benefit is commensurately immense, possibly resulting in an advantage of strategic proportions. These operations may serve as the surprise prelude to an offensive campaign, a one-shot opportunity to hurt an adversary, or as a supporting means of exerting pressure on enemy governments in conflict.

Distinguishing between categories of offensive operations matters. When the two are researched together, they blur together and become difficult to translate to effective military doctrine. Examined separately, presence and event-based operations are shown to have distinctive characteristics with unique advantages and disadvantages. They require different manpower, resources, and operational approaches, and can be applied against different targets for varying effects. Some may be more easily relegated to battlefield use, while others are best kept for strategic manoeuvres.

As always, there are exceptions. Not all network operations would fit as neatly into these categories. Some intricate campaigns establish a foothold as a presence-based operation and then proceed to use it as a trusted conduit through which to launch event-based operations against other targets. Yet by using this framework it becomes easier for decision-makers to prioritise some activities over others, and determine which capabilities should remain owned by intelligence agencies and which should be wholly operated at the discretion of a commander. These are consequential differences; some of the biggest issues in effectively understanding and using offensive cyber capabilities are a result of their historic ownership by intelligence agencies. These agencies are often reluctant to release their prized assets only to see sensitive, painstakingly developed capabilities burned to support an infantry force while on mission. Agencies predominantly had a mandate for collection, and now increasingly find their access co-opted for attacks.

Military network operations are discussed often but abstractly. Many of the techniques employed by military forces are a product of industry standards, and publicly-available information security best practice. Similarly, we now have access to a wealth of publicly documented state-attributed operations of varying qualities and effects. While the overwhelming majority of those operations focused on intelligence collection, an ever-increasing corpus of documented network attacks now exists. However, as military networks often differ in some respects from their civilian counterparts, doctrine also differs. Despite these circumstances, a decade of widely publicised coverage on state-nexus hacking operations, when coupled with leaked documents from the US intelligence community, affords a partial yet meaningful glimpse into how offensive units can and do operate in cyberspace. Public disclosure has led to previously unfathomable levels of visibility by external observers on the workings of network operations and those who practice them.

Responsibility for counter-intelligence traditionally lies with domestic law enforcement and intelligence agencies. Historically, there was little that the average citizen could or would do. Private intelligence firms more commonly focused on corporate espionage, risk analysis, or criminal cases. Wary of retribution and unaware of political context, unaffiliated citizens would avoid wilfully involving themselves with the clandestine affairs of nations. In the internet age, that is demonstrably no longer the case. Private individuals and organisations frequently interact with or even strategically disrupt state-operated network intrusions. Daily friction between private entities and government adversaries has never been higher.

In early 2013, US cybersecurity company Mandiant published an extensive analysis of what they assessed as a state-operated cyber-espionage campaign.[3] They had dubbed the phenomenon of a lengthy, targeted network intrusion by a capable adversary an "advanced persistent threat" (APT). Unusually for the time, the private company proceeded to thoroughly identify the perpetrators, a process the information security community calls "attribution." Mandiant labelled the group "APT1" and pointed an accusatory finger squarely at the Chinese military. Incredibly, they went further and identified the responsible party to be the People's Liberation Army 3rd General Staff Department (3PLA), and even provided the specific unit indicator, 61398—along with the building in which it allegedly resided.[4] Several individuals were also directly discussed in the report.

Mandiant understandably felt compelled to explain the unusual decision to meddle in the affairs of nations. In their APT1 report, the company authors declared that "It is time to acknowledge the threat is originating in China, and we wanted to do our part to arm and prepare security professionals to combat that threat effectively".[5] While many other security companies have since shied from outwardly attributing intrusions to specific nations and organisations, detecting and discussing similar state operations has become a wildly successful and largely consensual practice.[6] In some cases, companies such as the Moscow-headquartered Kaspersky Lab have even taken to publicly outing malicious network activity undertaken by their host countries,[7] thereby exposing themselves to possible government retribution. For the first time in history, publicly interfering in intelligence operations had become a viable business model for an entire industry.

The information security industry's new interest in tampering with network operations provides unprecedented transparency into how these play out. The tempo of state-affiliated intrusions has increased as countries have ramped up operations, and private security companies have been remarkably successful at repeatedly unmasking many of them. As private companies assisted victims in mitigating and defeating network incursions by military units and intelligence agencies, the models, methodologies and techniques they have for doing so correspondingly improved. The information security community's expertise has gradually become one of the best sources of publicly-available knowledge for analysis of state network operations.

The inherently secretive nature of cyberspace operations does not mean that there are no official publications on the topic. Numerous official US Department of Defense documents are crucial in understanding official perspectives on how offensive network operations are conducted. The reports run the gamut from tactical accounts on how units operate on the field,[8] joint publications on doctrine,[9] strategic guidelines,[10] oversight reports,[11] and even operational integration roadmaps.[12] While parts of the content remain redacted within these publications, when carefully amalgamated they become incredibly telling.

By some estimates, dozens of countries are actively pursuing offensive cyber capabilities as part of a national security strategy.[13] Some nations—such as Israel—are notoriously tight-lipped when it comes to their plans. Others, such as the United Kingdom, and most of all the United States, discuss their broader strategies more frequently. Deterring nations in cyberspace is difficult, and some may say impossible. Whatever nations can do to signal one another that they too can strike through this supposedly new medium may theoretically prove useful, but it is unclear to what degree. So, publicly acknowledging at least some form of offensive network operations may be a viable deterrence play, even if the results are not easily measured.

Military Offensive Network Operations

The history of intangible warfare shows that investing time and resources into a weapon that may strike a single target once could still be a cost-effective decision. Others may be developed to be made routinely available to battlefield units. How would these processes look? What forces, procedures, and relationships would that entail from inception to deployment? By expanding on the existing industry models for assessing cyber operations, additional processes undertaken in a military context can be accounted for. These include research and development cycles, intelligence gathering, tasking, and infrastructure management. Employing effective cyber operations is a major undertaking; models assessing these processes should reflect that reality.

Broadly speaking, all network operations can be reduced to four critical parts: *intelligence, capabilities, operators*, and *infrastructure*. Intelligence represents all information gathering and assessment required to find, compromise, and succeed against a given target. This includes prioritising adversary networks for targeting, analysis of used technologies, and identification of exploitable vulnerabilities—human or otherwise. Capabilities include all software and hardware components employed by the attacker to breach the target and influence targeted systems. The operators are the actual individuals deploying the tools and manipulating them as necessary. The infrastructure is the sum of all virtual logistics necessary to effectively communicate with the target, maintain operational security, extract information, and control the tools being used. While some simple, limited operations can be carried out without one of the above components, any meaningful offensive manoeuvre will almost certainly have all four.

Operations do not occur in a vacuum. They do not truly begin with reconnaissance against the target and do not end after activating the offensive payload.[14] There are several strategic and tactical phases preceding the operation itself, and several that follow it. Similarly, there are processes that run concurrently to the network intrusion itself, in constant interaction with work carried out by operators to facilitate their success and draw from it. These additional efforts are not peripheral; they are instrumental to an operation's success and are an integral part of understanding offensive military capabilities in cyberspace.

The model used in this chapter expands on the Department of Defense's Cyber Threat Framework model to include all processes and stakeholders associated with a network operation's success. The goal is to reflect all relevant efforts by broadening the aperture to include intelligence providers, planners, research and development, software engineers, and kinetic combatants. Additional stakeholders and phases are assessed in each one of the model's four consecutive steps; *preparation, engagement, presence*, and *effect*.

The *preparation* step entails all the prerequisite processes to create viable operational capacity. This includes often ignored but substantial investments in research, development, initial targeting and strategic intelligence gathering. It is impossible to effectively attack any networked device without knowing what exactly to attack and how to attack it, and possessing the integrated capabilities to carry out the attack. The *engagement* phase is the initial contact with the targeted networks, in which forward defences are subverted and initial compromise established. This original intrusion vector serves as the basis for all subsequent activity against the target. The third phase—*presence*—involves gradually infecting other nodes within adversary networks, hunting for the objectives while collecting intelligence. This step's duration ranges from possibly months or even years to a matter of seconds for event attacks. In the fourth and final *effects* phase, offensive payloads are activated against the target, hopefully resulting in the intended effect—which may or may not be immediately discernible. This phase includes all post-attack procedures, such as folding back the attacker's infrastructure, covering up forensic evidence, and conducting attack damage assessments.

Certain operations particularly highlight the long build-up to the full OCO life cycle. In 2010, Iranian engineers submitted a request to the Belarusian information security company VirusBlokAda to assist them in investigating an incident.[15] It seemed that malicious software had infected some of the computers and servers at the

Figure 2: Presence & event-based operations mapped to the Cyber Threat Framework

	PRESENCE-BASED OPERATIONS	EVENT-BASED OPERATIONS
PREPARATION	– Extensive targeting cycle – Infrastructure set-up – Stealthy, agile, modular malware development – Vulnerability research	– Limited/field targeting cycle – Robust, aggressive, intuitive tools – Vulnerability research – Deployed integration
ENGAGEMENT	– Initial infection – Phishing, remote software compromise, supply chain, or insider – Potentially lengthy	– Device-to-device compromise – Payload selection by operator or automatically – Quick/instantaneous
PRESENCE	– Lengthy (up to years) – Lateral movement to objective(s) – Support intel collection – Support R&D	– Minimal – Auto lateral movement – Limited or no persistence
EFFECT	– High-impact visible cascading effect OR gradual clandestine impact	– Transient or lasting impact against targets

Natanz nuclear facility, part of the illicit Uranium enrichment program then operated by the Iranian government. They were not aware at the time, but they had stumbled on a clandestine offensive network operation intent on causing physical damage to the plant's centrifuges.[16] The malware was soon discovered to have specifically targeted the industrial control systems orchestrating the plant's operations,[17] subtly masking its own activities as to fool local technicians into believing that physical centrifuge failures caused by the malware were accidental.[18] The offensive toolset, named Stuxnet by the researchers who discovered it, soon became the analytical cornerstone for offensive operations.[19] For the first time, malicious software was employed by one state to kinetically target another; some hailed it as the dawn of cyber-warfare.[20] Though it may not have been warfare, it was by most accounts an OCO driven by military organisations and their equivalents.

Stuxnet was unique.[21] As researchers scrambled to dissect the malware and its many characteristics, it became evident that its uniqueness was both a blessing and a curse. On one hand, researchers were afforded unprecedented intimate access to a thoroughly engineered, complex, targeted instrument of offensive state network operations. On the other, it was one of a kind, which immediately raised the question of how representative it was as an attack with a physical outcome. Stuxnet was the result of a complex network operation, requiring extensive investment in research, development, adversary simulation and a sensitive targeting cycle. While not used within military conflict, it was perhaps used to preclude one. Therefore, it remains a highly instructive example of a large-scale presence-based operation.

Preparation

Preparation encompasses all efforts prior to contact with the enemy. The Cyber Threat Framework defines preparation as all collective efforts to identify targets, develop capabilities, assess victim vulnerability, and define the scope of the operation.[22] Each of these processes reflects months and perhaps even years of investment in both material and operational resources. So, while it is the least discussed, the preparation phase of any offensive network operation may also be its longest.

Before operators ever interact with adversary networks, planners must first initiate a *targeting* cycle. This may seem deceptively trivial; anyone seeking to target an adversary will simply go after its networks. In reality, finding, identifying and mapping relevant networks for attack can be difficult.[23] Modern military organisations and other large entities use dozens of distinct networks.[24] Identifying which one hosts the targets you are interested in attacking is no negligible feat. It requires in-depth intelligence and an understanding of the adversary. In many cases, sensitive or operational networks do not interface directly with the public internet or perhaps even with any other networks.[25] This makes the task of identifying them and attaining access that much harder.

Targeting has a role that does not begin in cyberspace. Rather, targeting is "the process of selecting and prioritizing targets and matching the appropriate response to them, considering operational requirements and capabilities."[26] But whereas in kinetic targeting it may be sufficient to know the location of the target, this is not the case for targeting virtual assets. The process entails analysing all relevant networks, identifying those which it would potentially be prudent to target, and then prioritising between those based on available and potential capabilities.[27]

Targeting cycles are decidedly different for presence and event-based operations. Targeting for presence-based operations is commonly conducted by the strategic intelligence entities that develop network intrusion capabilities. Traditionally, those were owned by signal intelligence (SIGINT) organisations, which in varying jurisdictions are either civilian or military entities.[28] As such, it is often a derivative component of those entities' prioritised intelligence requirements (PIRs). PIRs form a fundamental national security agenda which the agency is expected to action, whether by collecting intelligence or preparing for eventual network attacks.[29] Targeting is therefore a long-term process during which intelligence on the adversary is accumulated. Over time, targeting increasingly provides the insight required to properly prioritise between networks by balancing compromise feasibility and relevance to the objectives at hand. The result is a highly and continuously curated list of specific targets.

Some targeting cycles may require expert external assistance. In NSA documents leaked by Edward Snowden in 2013, cooperation efforts from 2004 between the NSA and the Defense Intelligence Agency (DIA) are described.[30] The DIA's Joint Warfare Support Office assisted in mapping and analysing materials about an alleged Russian military base buried deep in Mount Yamantau. As indicated in the document, the US intelligence community remained in the dark despite information on the facility first surfacing over a decade earlier.[31] The DIA asset helped targeting efforts to identify Russian entities associated with the project, marking them for subsequent signals intelligence collection[32]—and potentially network operation—efforts. Prioritisation and identification of the operational target was thus jointly determined by the two agencies.

Targeting for event-based operations would take place in proximity to the attack itself.[33] As a result, this cycle could commonly be conducted by the same unit carrying out the attack, or by friendly forces supporting it. This alongside the employment of pre-packaged network capabilities means that the decision-making process is faster and can be conducted with far less available resources. To facilitate effective targeting, local reconnaissance assets scanning for signals and automated network mapping procedures may identify adversary networks operating in the region, possibly even auto-assigning possible capabilities to use against them.

Different goals and opportunities mean that some targets may be chosen for both event and presence-based operations. Since the 1990s, the United States has gradually modernised battlefield connectivity for its deployed forces. A part of this process, titled Warfighter Information Network–Tactical, or WIN-T, is a prime example of how saturated the network landscape can be. Dedicated line-of-sight radios, satellite-communication terminals,[34] and other mediums form a combination that services a host of networks, including the general-purpose NIPRnet, the classified SIPRnet,[35] and local standalone data and voice networks.[36] Many of these networks enable unclassified support functions that are not mission critical. Others carry sensitive targeting information, communication, or intelligence data. Some of these networks may be inaccessible to an enemy as they are transmitted over a medium for which the attacker has fewer ways of gaining access. Others rely on commercial satellites and even the public internet as the transmission medium of choice. Completing the targeting process by successfully classifying which networks both matter and are pragmatically reachable is therefore a challenging commitment. In some cases, these networks may be subjected to a long-term compromise in the form of a presence-based operation. In other cases, locally accessible datalinks such as a regional network cell might be the target of an event-based attack. Interestingly, the US military has now officially terminated the WIN-T project with a plan to gradually phase it out, citing concerns that the project's architecture is indeed too vulnerable to a determined, well-resourced adversary.[37]

Acquiring and developing capabilities is a crucial and continuous process. Capabilities in network warfare include all hardware and software used to intrude, exploit, and affect enemy platforms. There is some limited merit in downplaying the complexities of this process in comparison to conventional military assets; unlike conventional weapons, effective network intrusion tools can be and are developed by many, from enterprising individuals to intelligence units bursting at the seams with funding. The effectiveness and robustness of capabilities varies wildly, depending on the skills and resourcing of the developers. Similarly, the development cycle for a potent offensive network capability is also classically deemed to be shorter,[38] easier, and cheaper than the development cycle for conventional weapons.[39] However, the unique circumstances in which these tools are used are well worth examining; network weapons are often fragile, specific, and difficult to reliably test.

Presence tools must be stealthy, agile, and modular. They must be stealthy as the majority of their life-cycle will be spent clandestinely embedded in adversary networks. They must be agile, to enable operators to creatively use them to traverse adversary networks, collect intelligence, and weaponise valuable targets. Finally, they often must be modular, to allow operators to only deploy necessary capabilities at any given moment, thereby reducing the footprint of the tool and improving operational security.[40] Each deployment of a highly engineered network attack tool should ideally be carefully managed as to only include the components currently needed to facilitate success. The expectation that presence tools must be stealthy introduces a significant weakness; operational capability can become quite fragile. The pervasive idea that offensive network tools are single-use stems from this very issue,[41] even if it is not entirely accurate. Malware found by defenders or researchers can result in defensive fingerprinting within days of its discovery.[42] If that happens, it is not just the discovered operation that is threatened. Detection of an offensive platform risks its compromise against all of its current targets, globally. On the other hand, compromised malware is not lost forever; tools can be modified, obfuscated, and redeployed anew in a gambit to avoid detection.[43] Regardless, compromise presents a momentous risk of capabilities, which explains in part why the more capable intelligence agencies guard them so religiously.

The McDonnell-Douglas F-15 Eagle aircraft first entered service in the United States in 1976.[44] Originally designed to counter the Soviet Mig-25 Foxbat, its role exponentially grew along with its exemplary service record to include additional missions. Despite significant Soviet developments in fielding newer aircraft and air defences, the F-15 continued to receive upgrades to its avionics, weapons, radars and targeting subsystems.[45] Adaptations and newer variants such as the F-15 Strike Eagle and the latest F-15X meant that the same platform could retain its utility even against modern threats. As a result, it remains a highly active combatant platform both domestically and internationally, and is expected to continue its active service worldwide well beyond 2025.[46] This means that that the F-15 will see at least fifty years of active service.

It is almost inconceivable to envision network attack tools enjoying the same operational longevity as their kinetic counterparts. Even if the vast majority of their components are updated over time, any malware is likely to be rewritten from scratch at some point. One of the longest known offensive network operations platforms—codenamed Regin by its private sector discoverers—was ostensibly operating since at least 2003[47] and has been attributed to either GCHQ,[48] the NSA,[49] or both. At the time of its discovery in 2014, security company Kaspersky claimed that it was "one of the most sophisticated attack platforms we have ever analyzed".[50] Once publicised, and with its various mechanisms for communication and stealth thoroughly mapped and defended against, NSA operators would have had to immediately cease all intrusion activity until sufficient changes could be made and new detection evasion mechanisms deployed. In the meantime, most if not all existing access to sensitive networks would have to be shut off as the tools would be erased from infected computers. Such an event is both an enormous investment of time and resources and potentially a major strategic compromise of access to valuable targets. It would be as if one successful loss of an F-15 to an enemy air defence battery would cause the immediate grounding of all F-15 aircraft globally—even against other adversaries—until such a time as countermeasures could be developed, tested and deployed for the entire fleet against that particular threat. This is evidently not the case.

There are mitigating factors unique to OCO capabilities. An entirely new suite of highly capable modular tools can be created in a handful of years before they are ready for operational deployment. The catastrophic loss of a platform and its associated infrastructure can be strategically damaging, but is more easily recoverable than its physical counterparts. As an example, the F-35 took well over a decade from inception to the production of its first plane, and a decade further until it reached initial operational capability. Production of each additional F-35 is a costly and time-intensive affair. While the F-35 is perhaps an unusually hefty example in comparison to other warplanes, it highlights the disparity between even the most intricate offensive software project and its hardware counterparts. Analogies have their limits.

Conversely, event-based tools must be robust, aggressive, foolproof and intuitive to operate. As they would likely be used by front-line units, such tools should require no expertise to wield them effectively. They should be able to operate against a wide range of targets in a slew of contingencies, while generating similarly predictable and consistent effects. Battlefield operators will not have time to dynamically redeploy modules or carefully orchestrate network traversal. The weapon must therefore be capable of completing its objectives without further assistance. As a result, resource exhaustion attacks such as the commonly seen denial of service, or generic destructive payloads as seen in wiper malware,[51] are good examples of preparing event capabilities.

Both presence and event capabilities require investment in vulnerability research. Hunting for vulnerabilities includes all efforts to locate exploitable flaws in software and hardware used by the adversary. These flaws can then be exploited for compromising the target and getting it to either behave unexpectedly, or preferably to run arbitrary code. Vulnerability research targets a broad range of targets from generic-use software such as Microsoft Windows or Google's mobile Android operating system, to dedicated software for military hardware and other niche platforms. It is a valued component in most network attack campaigns.

Some vulnerabilities are more useful than others. In August 2016, an entity calling itself The Shadow Brokers began releasing information seemingly pilfered from the NSA's network intrusion unit, at the time still named Tailored Access Operations (TAO).[52] Assessing the potential fallout from the significant compromise of some of their most compartmentalised capabilities, the NSA decided to attempt to pre-empt any damages that might occur from having their most sensitive attack tools repurposed against others. They disclosed to Microsoft that they had previously discovered several vulnerabilities affecting multiple versions of their Windows operating system.[53] At least one of them, codenamed EternalBlue in the leaked files, was a coveted asset for operators. It enabled zero-interaction remote code execution (RCE) on Windows, and it was "wormable". It could remotely infect vulnerable computers without any engagement by the user, which meant that when used correctly after an initial compromise, it could be used to infect additional computers on the same internal network with ease. It was a highly virulent vulnerability.

Duly alarmed, Microsoft quickly released a patch for the vulnerability. Coding it MS17–010, the company strongly advised all its customers to quickly patch their systems or risk extreme susceptibility to external compromise.[54] Many did so, and a great many more did not. Two months later, a tidal wave of malware infections swept across the world. severely impacting users both large and small.

The malware proceeded to encrypt files on infected computers, locking them and generating a pop-up message. The ransom note informed the victims that they were to pay a sum of the popular cryptocurrency Bitcoin to the malware authors, after which they would receive the decryption key to their files. The ransomware—dubbed WannaCry by researchers—had weaponised the MS17–010 vulnerability by way of the Shadow Brokers leak.[55] The effects of the malware were staggering; some organisations such as the UK's National Health Service suffered greatly from the malware, resulting in the costly, albeit temporary, loss of operational capabilities.[56] It seemingly could have all been prevented if they had patched their ageing inventory of computers and servers.

WannaCry was later attributed both officially by the US and the UK government, and unofficially by private sector researchers, to the North Korean government. Though it certainly qualified as an attack through its impact, the assessed goals of the operation prevent WannaCry being truly treated as an act of warfare between nations. Regardless, it was a stark reminder that even disclosed vulnerabilities may retain their potency, particularly against targets that are difficult to routinely patch and defend. As a sad underline to this bleak message, even WannaCry itself was insufficiently severe to get all Windows users to inoculate against EternalBlue. Less than six months later, a second wave of infections from the NotPetya malware once again used the same vulnerability—albeit as a secondary means of self propagation.[57]

Are WannaCry and NotPetya representative of the modern networked battlefield? Likely not, or at least not without additional complexity. Most military networks are segmented in some form from the internet and are comparatively more resilient. There are dedicated defenders and support staff tasked with keeping internal software up-to-date with the latest available patches. Yet WannaCry and those like it are instructive in one key area; effective vulnerability research is paramount to the success of a reliable network attack platform. This requires the expert attention of vulnerability researchers who comb through the source code of products used in adversary networks and systems. If an offensive tool is detected and made public, losing a powerful vulnerability can be a crippling loss as inoculation becomes widespread. Similarly, public disclosure of a vulnerability used in weapons means it cannot reliably be used in sensitive operations again, even with different tools being deployed. The risk of discovery becomes too great.

Software vulnerabilities are difficult to find, for both attackers and defenders. From the offensive perspective, effectively exploiting critical software in a manner conducive to intrusions is increasingly difficult, especially for some products with a sizeable investment in security. At least from the purview of some private threat intelligence companies, there is a diminished reliance on zero-day exploits in favour of simpler tools that mimic authentic user behaviour.[58] At the same time, there is no shortage of vulnerabilities, as data indicates that publicly disclosed, high severity submissions nearly doubled in 2017 and have maintained those levels throughout 2020.[59] From the defender's perspective—as a controversial RAND report indicated in 2017—unless the tool weaponising them is discovered. vulnerabilities last an average of almost seven years without being exposed.[60] Thus, maintaining an expert workforce entrusted with continuously hunting for new useful vulnerabilities is paramount.

For event-based operations, the final component of preparation is integrating capabilities for use with forward-deployed platforms, such as an installable module on an aircraft. Presence-based operations are most commonly handled by remote operators, much like drones. However, in some cases, especially those involving segregated networks used to communicate sensitive data, proximity or line-of-sight access is required for some steps. In these cases, often called "close access operations", forces may find themselves conducting network operations directly in the field, be it by aircraft, naval vessel, ground vehicle, a friendly insider, or actual boots on the ground.

There are recent examples of event-based attacks in which network capabilities were supposedly integrated into battlefield platforms. The US military operates infantry cyber teams alongside electronic warfare assets to map out enemy networks, identify targets, and action against them.[61] The Russian military has allegedly disrupted Royal Air Force sorties over Syria by way of an event-based attack launched from a deployed electronic warfare vehicle.[62] Developing a reliable, robust, battlefield-deployable offensive cyber capability is becoming increasingly viable—although expensive. So, while attacking networks may seem to be low cost, achieving battlefield readiness and conducting event-based operations may include hefty development, targeting, and intelligence cycles.

Engagement

Network operations are not carried out only by operators. Even once attack capabilities are fully developed, vulnerabilities weaponised, and targets identified, operations need extensive support from other functions to succeed. Those who facilitate the attack will likely need subject-matter assistance from intelligence analysts, reverse engineers, software developers, and decision-makers.

The Cyber Threat Framework defines the initial engagement phase as: "Threat actor activities taken prior to gaining access but with the intent to gain unauthorised access to the intended victim's physical or virtual computer or information system(s), network(s), and/or data stores."[63] Put simply, this phase embodies the attempts to intrude on the enemy; it is the first active contact with its networks, intent on establishing a digital beachhead. However, this framework obfuscates the characteristics of the engagement phase, which may occur months in advance for presence-based operations, or in adjacency to the desired effect for event attacks.[64] Not all cases are created equal, but all share one notable commonality—the engagement phase starts the operational clock.

In presence-based operations, operators or officers during the engagement phase are often conductors. They oversee the weaving of other orchestra members' capabilities into the operation as they are needed. Intelligence analysts are most familiar with the targeted organisation and would likely be best positioned to identify the valuable people to target. Technical staff may be required to assist in matching the appropriate payload to defeat enemy defences and evade detection. Senior staff may be necessary to prioritise goals and resources as the operation proceeds. They may also need translators if encountering materials or charts in unfamiliar languages, or specialists that understand the targeted equipment better.

A common approach to network intrusion is compromising an internet-facing server or device. Identifying and compromising these may be easier than directly penetrating segregated networks, but not all such targets are inherently useful. Operations frequently start by sending a malware-laced email to an individual within the organisation. Those intent on gaining entry to sensitive networks may first need to compromise individuals who routinely use them and hold trusted access to their assets. The reason for this is two-fold; first, there may not be a viable technological intrusion vector, as many sensitive networks are cut off from external inputs. Second, users are often the most vulnerable element in an otherwise secure network environment.[65] They are prime targets for social engineering as an intrusion vector, but that does not mean it is always a trivial endeavour. Successfully getting individuals to compromise their own security without arousing suspicion often requires its own brand of expertise, preferably provided by dedicated personnel.

Even supposedly aware defence contractors can be compromised by social engineering. In March 2011, the RSA security division of US technology company EMC reported a breach of its networks.[66] RSA soon admitted that the intrusion vector consisted of a phishing email containing a weaponised spreadsheet. According to the company's then head of security, "the email was crafted well enough to trick one of the employees,"[67] thus admitting the attackers into the network. Yet the effects had a far wider reach, as RSA was also the producer of the SecurID encryption tokens used for authenticating access in many government contractors and agencies. It was revealed two months later that defence giant Lockheed Martin—prolific designer and manufacturer of US military equipment—was targeted in part through counterfeit SecurID tokens created as a result of the RSA breach.[68] Securing presence on both RSA and Lockheed Martin networks was therefore a fairly long and involved process.

In event-based operations, the engagement phase can occur in seconds. As the targeting cycle is short, there are no phishing emails with human targets in mind. Instead, the engagement phase will focus on compromising accessible targets by exploiting remote software and hardware vulnerabilities. When using automated capabilities to target devices, it is sometimes possible to directly subvert software's intended functionality to gain entry. The engagement phase for event-based operations may not always result in persistent access to the target. Depending on what the desired effect is, it simply may not be necessary. For example, it may be possible to exhaust available networking or processing resources without ever being able to execute code directly on the target. If the goal is to prevent the target from functioning as intended, that may be sufficient. Such scenarios are more easily placed within a military context; see for example denial of service attacks, which bear similarities to conventional electromagnetic jamming.[69]

The potential perpetrators for event-based operations are more varied than their presence counterparts. In many cases, these could be forward deployed cyber units, as both the US and the UK are increasingly using.[70] In other instances, fielded assets such as human intelligence agents or specialised equipment may be required. As Edward Snowden revealed in a leaked document in 2013, the NSA's GENIE program to facilitate semi-automated network operations would at times rely on such assets. When necessary, field operators would physically infect adversary devices, plant hardware, or conduct short-range offensive SIGINT.[71] Reporting from 2020 even suggests AI-supported network operations, closing the gaps between event and presence-based operations.[72]

Presence

The presence phase is when most of the friction occurs between intruder and target. Malicious software is continuously employed to understand, dissect, and maintain a hold within the targeted network or networks, gradually extending the intruder's access until it locates servers or devices fitting whatever harm is intended.[73] It is essentially the process of extending and cementing the reach into the adversary's networks, two processes called lateral movement and persistence, respectively. As defined by the Cyber Threat Framework:

> [The presence phase includes] actions taken by the threat actor once unauthorized access to victim(s)' physical or virtual computer or information system has been achieved that establishes and maintains conditions or allows the threat actor to perform intended actions or operate at will against the host physical or virtual computer or information system, network and/or data stores.[74]

The presence phase embodies the biggest discrepancy between event and presence-based operations—time spent inside the targeted networks. Where presence-based operations unsurprisingly spend most of their life cycle in the presence phase, event-based operations may either have an inconsequential or perhaps even non-existent presence phase. When state-initiated intrusion campaigns are publicly reported to go on for months prior to detection, it most commonly refers to the presence phase. The key difference in time span reflects two wholly different operational tempos. For presence-based operations, the presence phase is essentially a cyclical process of expanding micro-intrusions; additional nodes in the network are identified, assessed, breached and subsequently mined for relevant intelligence. Each intrusion must be handled with care to avoid tripping any alarms or inadvertently informing network defenders of an active intrusion against them.

The attacker's point of entry is rarely the intended target. This is no different than a physical circumstance, in which operatives would need to infiltrate an installation in order to cause harm to someone or something within. In order to truly inflict meaningful damage, it is simply not enough to plant a bomb in the first room they access. Instead, operatives would need to quietly move between rooms, often gated behind locked doors and security measures, leaving behind explosives whenever they reach a structural weakness. For maximum effect, they would then need to remotely detonate their charges, bringing down the building. Networks are the digital equivalent of this process, stretched across thousands of servers and possibly months of slinking around.

Networks are comprised of many servers, computers, and other devices. On gaining entry, the intruder would proceed to perform what is often called "lateral movement". The process is intended to establish a deeper presence on the network, infect additional nodes, and locate targets of worth. While operating the software to facilitate these efforts is squarely within the purview of the operators, they require a tremendous amount of support from intelligence staff[75] to assess content pulled from devices and servers. Research and development resources may also be required to create dedicated modules to subvert specific technologies encountered.

Offensive presence-based operations are always intelligence operations first. Until such a time as a more active measure is needed, malicious software is tasked with either remaining dormant or collecting information, identical to its behaviour in an intelligence mission.[76]

Operators at the presence phase must then rely extensively on the assistance of intelligence analysts to assist in the further targeting and dissection of materials exfiltrated from the target.[77] In some cases, the offensive operation is carried out wholly by an intelligence agency.[78] The presence phase therefore comprises both assessing the independent intelligence value of the target, and simultaneously gathering information needed to help steer the operators towards the server or servers where attacking would result in the desired outcome.

When Russian operators initially infiltrated the Ukrainian power grid in 2015, they did not immediately wreak havoc on all they encountered. Instead, earlier intrusion efforts cleverly used the specialised protocols unique to these industrial networks to traverse the network, map its layout and glean information required to develop robust offensive capabilities.[79] In a subsequent operation, the presence phase included pivoting from the power company's corporate network onto its industrial network, leveraging an attack against both to simultaneously cripple the grid and prevent-based operators from fixing it.[80] Finally, advancements eventually allowed the operators to "de-energize a transmission substation on December 17, 2016"[81] by way of the CRASHOVERRIDE malware,[82] which was tailored to impact even relatively well defended energy grids. The Russians had achieved a malware-induced blackout, but they had done so after a considerable amount of time from the initial engagement phase. Success would not have been possible without topical expertise and accrued experience.

As reported, the operation revealed substantial command of the industrial communication protocols at use, a rapidly evolving sophisticated malware platform with highly specific modules, and politically relevant targeting, all the while dodging detection by wary defenders. These circumstances embody the intricacy and difficulties of successfully navigating the presence phase. It would not have been possible without technical expertise in industrial control systems, developers to generate the unique attack code, and intelligence support to guide the efforts.

The physical consequences of failure are rarely comparable to a botched kinetic attack. Much like the motivations behind operating drones over contested airspace, this is a key advantage of offensive network operations. Indeed, even at the absolute worst scenario in which the offensive measure both fails and results in a catastrophic compromise of the attacker's capabilities, the actual operators are certainly to remain free from physical harm. They are distant, often operating from the safety of their homeland, and as such will be untethered to physical harm concerns. They will be able to rebuild and attack another day.[83]

For event-based operations, the presence phase is nearly imperceptible. Capabilities employed in an event attack are meant to impact the target directly and then disappear, leaving as few lingering artefacts as possible. If tell-tale indicators remained—such as residual code left running, or malicious files left behind on targeted systems—it would simplify follow-on efforts by the adversary to develop future countermeasures. It is therefore significant for an event capability to be only minimally present within enemy assets, if possible.

A cascading effect—intentional or otherwise—may result in an event attack having a limited period of network presence. For example, an automated network attack tool designed to propagate through networks and rapidly destroy all infected endpoints and servers would require a limited presence as to ensure subsequent infections to additional targets. A good example of such an attack is the previously mentioned NotPetya destructive malware, which in 2017 heavily affected Ukrainian networks before breaching its scope to adversely impact various other entities globally.[84] The attack, which resulted in extensive damage to victims worldwide, was unusually publicly attributed by numerous Western intelligence agencies to the Russian military.[85]

The potential cost of discovery is arguably the most meaningful deterrent to attacking networks. In recent years, a growing trend amongst large vendors in the information security market has been to uncover massive state intelligence and surveillance efforts, often facilitated by elaborate malware campaigns. For an agile intruder, the immediate result of this compromise an effort to roll back all deployed tools, to attempt damage control. The product of this is a partial collapse of the attacker's intrusion infrastructure, but more importantly—the defender's eventual inoculation from future attempts to use the same tool in an offensive capacity.

Inoculation is neither instantaneous nor guaranteed. While the majority of established organisations may invest in the ability to rapidly mitigate vulnerabilities through patching and security updates, many others do not. Full population defence against a given malware variant or vulnerability may take time, especially in low-budget or high-complexity environments, such as critical national infrastructure or deployed military equipment. Detection equals exposure, and exposure means a hamstrung flexibility to operate—but it is not automatic defeat. This is especially true if the enemy relies on throw-away commodity malware or simply has a risk model where exposure of capabilities, targets, and infrastructure is internally viewed as acceptable cost. This is true for both Iran and Russia to a degree, as we will observe later in the book.

The presence phase is therefore the most sensitive component in many offensive network operations. The continuous friction with different adversary networks and the need to collect intelligence means that discovery and eventual inoculation are a big risk to attackers. Presence operators must continuously work to conceal their moves, clean up forensic evidence, and establish stable, covert communication channels that would reliably enable decision-makers to activate positioned offensive payloads once they are needed.[86]

Effect

The final effect phase is where triggers are pulled. Payloads are activated—disabling, disrupting or manipulating targets. Effects either translate into objectives, fizzle uselessly, or have unintended and potentially disastrous collateral impact. For presence-based operations, the effect phase is the culmination of possibly months of planning, targeting, intelligence collection, infection attempts, and dedicated engineering.[87] For event-based operations, the effect phase represents the primary thrust of the attack. When Richard Clarke declared in 2009 that "strikes in cyber war move at a rate approaching the speed of light,"[88] he referenced not the entire span of an operation but rather the span of time between the activation of the payload to its actual detonation against the target. Even so, payloads may be instantly triggered but may still take time to achieve their intended impact.

Distilling various definitions, there are three "attack" types when targeting networks—*disruptive*, *manipulative*, and *destructive*.[89]

Disruptive—or suppressing—attacks incur "temporary or transient degradation by an opposing force of the performance of a [target] below the level needed to fulfil its mission objectives".[90] Their utility increased with the rise of electronic warfare, where electromagnetic transmissions could be jammed to produce a temporary but potent effect.[91] Disruptive attacks have made a natural transition to cyberspace, where temporarily degrading available resources can adversely impact the efficacy of an adversary force.[92]

Disruptive network attacks are commonplace outside of military scenarios. Denial of service attacks capable of levying massive throughput of network traffic routinely disrupt the functionality of online services big and small. The targets range from global gaming communities such as Sony PlayStation Network[93] to major banks.[94] Typically, these attacks either exploit an implementation flaw in the targeted technology or simply attempt to overwhelm its available resources. By doing so, no legitimate connections can interact with the platform as intended, rendering it temporarily disabled for its original purpose. Similar approaches may be applied to military technology, platforms, and protocols.

Manipulation effects are about altering information or functionality in the adversary networks, thereby deceiving operators or preventing intended functionality. Such attacks attempt to alter perception, preventing an adversary from acting properly to further its own objectives. A scenario could include introducing a nearly imperceptible deviation to a weapon's targeting process, causing strikes to miss by what could appear to be a technical glitch. Kinetically, this is hard to accomplish with finesse, but it can be roughly analogised to physically tampering with a missile's warhead to covertly render it inert. When the missile fires, it seemingly behaves as normal until impact, during which the warhead does not detonate. During the heat of conflict—and until it happens repeatedly and consistently—it would be difficult to identify the fault as an attack. By the time it is discovered, it would likely already be too late. As the Stuxnet campaign demonstrated,[95] masking a manipulative effect to increase its longevity can cause an effect to be repeatedly successful over time. Hiding an effect does, however, require gradually introducing it; an immediate and blunt change of circumstance can vastly increase the odds of detection.

Masking an effect may be more applicable to presence-based operations than to event attacks. As previously discussed, event tools tend to be more geared towards instantaneous effects and are thereby less compatible with the subtle machinations of a presence effect. On the other hand, as event capabilities are designed to be more resilient if detected, they may not need to be so hidden. It would be surely preferable, but not a requirement per se of the attack.

Destructive attacks are aimed at inflicting damage on adversary networks, either on hardware, software, or both. These types of attacks are more rooted in conventional warfare, where destruction of enemy assets and personnel is often seen as the primary method of reducing its coercive combat effectiveness.[96] When applied to network operations, a destructive attack could cause "permanent" software damage—such as in the case of malware which completely erases all critical files on target servers,[97] or even permanent hardware damage, such as the previously mentioned Stuxnet worm targeting the Iranian nuclear project.[98]

One of the key challenges in the effects phase of a presence-based operation is in trusting the tools to work as intended. A presence-based operation is about painstakingly embedding offensive tools into enemy assets, infecting endpoints and servers as relevant networks are identified and penetrated. However, when the time comes to activate the payload, there is no guarantee that the attack will be correctly registered and actioned by the infected machine meant to do so. Unlike kinetic or even event attacks, the tools are wholly behind enemy lines when activated. The offensive payload is already pre-placed within adversary networks, increasing the chance that it might fail to be triggered due to some extrinsic circumstance outside of the attacker's control. In complex presence-based operations, several smaller effects must coincide in order for the overall objective to be achieved. Imagine rewiring an entire building to explode in a cascading set of dependant detonations, where a single charge not going off at the right time endangers the entire operation. Now consider doing so clandestinely, behind enemy lines. Without any actual bombs.

Even if the operation is successful, odds are there will be no visible indication of it. The other side of the reliability issue is in conducting effective battle damage assessment (BDA); the process of estimating the effects of an attack. This difficulty applies to both categories of operations, with some key differences. Intangible warfare normally has no physical aspects in its effects phase. Software is manipulated, hardware may be fooled or tampered with, but physical safeguards designed to protect equipment and operators from faults may often prohibit such drastic effects. This is of course not always the case. Presence incidents such as the centrifuge-crashing Stuxnet,[99] the generator-rupturing Aurora experiment,[100] and the German steel mill attack[101] remain among the few instances of significant physical damage to equipment directly caused by a software attack. Real world effects—especially lasting physical damage—are harder to accomplish. They are often also unnecessary.

A destructive event attack with no viable BDA means that it is often challenging to assess whether the target has in fact been degraded, either permanently or temporarily. This ambiguity is similar to electronic warfare, and reduces possible trust in using these tools to support other combat efforts. So, if as previously suggested the goal of the network attack was to facilitate a subsequent kinetic manoeuvre, operators may be forced to make a leap of faith that the attack has indeed been successful. In some cases, this leap may entail risking lives. Gambling on whether a software attack successfully degraded an anti-aircraft targeting radar's sensors to the point where manned fighters are safe to engage is a high-stakes endeavour.

Network attacks can be oddly harmful to their own BDA efforts. Kinetic attacks are often accompanied by sensors. Cruise missiles take out a target; the damage is then verified by satellites. The verification process provides crucial telemetry as to the success of an attack, and can include manned solutions such as special forces, submarines, or aircraft, or unmanned solutions such as drones or satellites. For presence-based operations, the attack tools are often their own sensors. As the offensive tools were originally implements for intelligence collection, they may be the only active source within the network. Indication as to the status of the targeted system or network may singularly come from that very system or network. Take down the network and crucial observation channels similarly disappear. For networks, it is usually only possible to conduct effective surveillance by observing the data coming in, out, and through the network. The activation of an earnestly crafted and destructive presence attack may cut off all such communication, taking with it the ability to observe whether the desired effect had actually been accomplished. In other cases, more subtle effects may subsequently be monitored by the very tools that deployed them.

Challenges and Opportunities

Comparing event and presence-based operations allows a discussion of how nations are integrating these capabilities into military doctrine and strategy. Both are markedly different in characteristics, duration, challenges, and opportunities and therefore should not be lumped together. Fundamental similarities exist between the two categories and are certainly helpful towards understanding networks as a medium for warfare; but useful observation of military capabilities will remain limited unless we recognise that offensive cyber capabilities must not all be treated the same.

Event-based operations represent the instances in which network attacks are somewhat analogous to the kinetic. Like firing a weapon, an event-based operation entails sending a payload from attacker to target in the hope of immediately reducing its integrity or capacity to operate. As a result, these capabilities are often more tactical in nature, easier to integrate with existing military OODA loops,[102] and are promising candidates for joint warfare. They are however limited in scope, may require extensive research and development, and could be limited to a specific subset of adversary equipment. A weapon suitable for disabling the missile defence radar onboard a US Navy destroyer may exploit hardware-specific vulnerabilities,[103] rendering it unsuitable against other targets. Consequently, battlefield operators deploying such weapons must have immaculate understanding of their adversary and a firm control of their own options.

Presence-based operations are intelligence missions with an offensive finisher; a form of digital sabotage. They may initially appear indistinguishable from intelligence campaigns as operators infect networks and gather information necessary to craft an attack. In these phases, even if the target detects the malware present in its assets, it may be difficult to assess motive and intent. Only once offensive modules are deployed can confidence in hostile intent increase. This adds an unfortunate layer of political nuance, as overly successful network intrusions may be misconstrued by the target as unduly aggressive. The risk of potentially undesired escalation has been aptly covered by Buchanan when discussing the "cybersecurity dilemma,"[104] an application of the classic security dilemma to network intrusions between nations.

Presence-based operations can potentially be high-risk, high-reward capabilities. Successfully pre-positioning assets in military or otherwise critical networks may potentially have meaningful impact on the course of conflict if used to facilitate strategic surprise or large-scale reduction in enemy capacity to operate. At the same time, presence-based operations are notoriously brittle, and their discovery can undo years of focused labour. By nature, such operations require the tight, intensive, unyielding support of friendly intelligence assets to map the threat, generate initial persistent access, and successfully manoeuvre through inscrutable tangles of military networks until the relevant targets are found. It is therefore understandable why these campaigns are often spearheaded by intelligence with core expertise on network intrusions, rather than deployed military forces.

The Lockheed-Martin F-35 Lightning II fighter aircraft is a fascinating example of a platform potentially vulnerable to both presence-based and event-based attacks. After two decades of development, the aircraft had started active deployment accompanied by a litany of issues with its onboard software. These included major in-flight failures of the radar system,[105] issues with its onboard avionics,[106] and "276 deficiencies in combat performance [designated] as 'critical to correct'...."[107] Additionally, both the onboard systems and the logistics software used to manage the F-35 have demonstrated numerous flaws and vulnerabilities during security testing procedures, many still yet to be addressed with new ones found as of 2020.[108] While onboard systems are unlikely to be directly connected to the internet,[109] targeting one or more of the F-35's prized array of sensory inputs and communication methods is viable by a knowledgeable adversary. To that end, evidence suggests that the F-35's most recent software version still presents a sizeable attack surface.

An event-based attack might try to overwhelm or otherwise compromise some of the F-35's tactical data links, used to share data with allied assets in the air and on the ground. For compatibility purposes, this communication may occur via the Link-16 protocol, an encrypted legacy protocol used by NATO forces since 1975. While it has undoubtedly undergone improvements over its life cycle, the limitations in encrypting reliable airborne tactical traffic and the vast array of opportunities for US adversaries to intercept, analyse and exploit Link-16 protocol vulnerabilities raise the possibility that it may be compromised during an attack. Link-16 includes targeting information, location of friendly forces, and directives from command forces.[110] Interestingly, even oversight reports have indicated some issues with the Link-16 data that forced pilots to revert to voice communication.[111] Others have indicated intermittent problems with the Multifunction Advanced Data Link (MADL) system used to communicate between fifth generation stealth aircraft,[112] causing pilots to "lose tactical battlefield awareness".[113] Successfully compromising the F-35's data links is therefore not unfeasible and may severely degrade aircraft battlefield performance.

The effects phase in this particular instance could include one of several options. As an example, a manipulation attack could alter the pilot's perception of the battlefield by adding, removing, or moving specific targeting points fed to the radar subsystem by external channels. Blips on the radar may disappear and reappear, representing false targets to the pilot. A disruptive attack could alternatively try to overwhelm sensory input, or prevent awareness that the aircraft is being acquired by a ground-based air defence battery. The effects would be nearly instantaneous, limited in scope to the targeted aircraft, and tactical in nature.

A presence attack against the F-35 could take months to prepare, culminating in an elaborate effects phase saved for evoking strategic surprise or a dire need. Rather than targeting a single aircraft or sortie, attackers would instead target the peripheral networks that interface with the F-35 during its operational life cycle. These could be on-base networks, maintenance forces, or third-party software providers. By doing so, an adversary may temporarily degrade or otherwise completely disable a large number of aircraft.

One supposed innovation in the F-35's software is the Autonomic Logistic Information System, or ALIS. With one ALIS station present at each unit operating F-35s, it allows semi-automated fleet management, mission management, logistics and maintenance.[114] As described by Diana Maurer, director of defense capabilities and management at the Department of Defense, "ALIS is integral to the more than 3,300 F-35 aircraft that the US military services and foreign nations plan to purchase. A fully functional logistics system is critical to the operational success of the F-35".[115] As with other parts of the Joint Strike Fighter program, ALIS has been plagued with critical faults, with official oversight analysis continuously warning that "cybersecurity testing to date identified vulnerabilities that must be addressed to ensure secure ALIS... operations".[116] These faults do well to instruct on two relevant aspects; how ALIS might be vulnerable to presence-based operations, and how exploiting these vulnerabilities could lead to a strategic advantage when triggered in the effects phase.

The issues in ALIS are varied. Attempts to deploy it in test environments forced support personnel to lower network security settings to allow users to log on.[117] Incorrectly handled maintenance data resulted in one instance in "major damage to a weapons bay door"[118] from an incorrectly loaded bomb that got loose and struck the aircraft. In June 2017, a software error in ALIS grounded an entire F-35 unit until the issue was addressed.[119] A Government Accountability Office report from 2020 highlighted claims that ALIS had consistent issues with data reliability, falsely grounding operationally fit aircraft.[120] A separate report reiterated that "F-35 testers stated that since 2016, they have identified a number of cyber-related ALIS deficiencies, most of which remain open today".[121] The issues are so severe and pervasive, that a Department of Defense consensus reported in 2019: "because most of the ALIS source code has not been updated in years and contains numerous security vulnerabilities, the software should be completely re-designed".[122] The situation had not improved meaningfully by the end of 2020.[123] It would therefore seem that the system can both be a boon to aircraft operators and an attack vector for offensive network operators. A single platform now presents a diverse, varied attack surface that can potentially be exploited during wartime.

Using the common cyber threat framework is useful to establish the key differences between the two high-level operational categories. Event-based operations have a far lengthier preparation phase as offensive technology is researched, developed, deployed and integrated to combatants. Conversely, presence-based operations obviously skew more heavily towards the intelligence-heavy presence phase. In each of the four phases—preparation, engagement, presence, and effect—circumstance and characteristics differ between the categories. But even a relative dearth of empirical evidence does not mean they are inscrutable to outsiders. Analysis of previous operations, industry practices, military doctrine, and vulnerability assessments creates a rich tapestry of possibilities and challenges for military network operations.

4

VIRTUAL VICTORY

APPLIED CYBER-STRATEGY

Developing military offensive cyber capabilities is a challenge; using them well, perhaps even more so. Across all levels of activity militaries are now attempting to accommodate a set of tools that offer unique possibilities while creating fresh issues. Operational capability does not exist in a void. Used in harmony with other approaches it can create opportunities for kinetic attack, and weaken, surprise or confuse an adversary. Misapplied capabilities can burn away years of technical, intelligence, and operational labour after activation. So how can military forces get the best out of tactical and strategic options available to them?

Both presence and event-based operations can aim for lofty objectives or tactical effects, depending on the scenario. Yet, by their cautious, ponderous, and time-consuming nature, presence-based operations tend to skew more strategic and event-based operations achieve immediate, visible effects and are easier to package neatly, making them more robust as tactical instruments. This differentiation can profoundly impact the expectation of results from network attacks. The distinction between types also supports the difficult process of identifying which offensive capabilities should be relegated to battlefield commanders and which should remain within the remit of intelligence agencies or other rear-active forces.

OFFENSIVE CYBER OPERATIONS

Network operations can also impact grand-strategy, if used for systemic compromise of adversary civilian infrastructure. A comprehensive set of strikes against economic, industrial, and critical infrastructure targets is theoretically possible, but would require an immense commitment of resources. Attempting to whittle away national resolve or resilience by repeatedly and aggressively targeting civilian infrastructure is controversial but certainly not unheard of, even if success is tenuous at best. It is less clear whether a cyber-based approach that attempts the same is feasible. Unlike strategic bombings, which ostensibly use a handful of weapon types to achieve their goal, each and every target pursued through such an aspirational network campaign would need to be pursued individually and over time, both developing and maintaining presence for when activation is required. A single destructive, virulent event attack could not take out a power grid, economic infrastructure, or a country's industrial base. It is never truly that easy.

The frequent ambition to try and assess offensive network capabilities through existing strategies is understandable. As Arquilla and Ronfeldt pithily wrote, "People try to fit the new technology into established ways of doing things."[1] Analogies employed by strategists and historians equating elements of cyber-warfare to electronic warfare, strategic air power, or even Cold-War era nuclear standoffs can sometimes be helpful. Some of these analogies have been carried out within this work as part of its conceptual framework. But as shown in previous chapters, these analogies do not produce a holistic grasp of what role offensive network operations holds in military thought. There are key differences in learning from existing strategies and attempting to shoehorn them directly onto the considerations of network operations. Direct analogies between kinetic and cyber have severe limitations impacting their usefulness beyond a thought exercise.

Conventional strategy still matters. Simply put, without weaving network operations into conventional strategy and doctrine, they may be used incorrectly, or at best used sub-optimally. In order to assure that the capability to attack networks is properly integrated, we must first recognise that elements of existing strategy already incorporate the prerequisite building blocks. In some cases, historic strategists commented on warfare in a fashion inherently hostile to the basic characteristics of cyber operations. In other cases, history represents a powerful foundation upon which modern network strategy can construct.

There are other valuable taxonomies for offensive network activities. Buchanan has similarly identified two different primary types of operations. The first category is similar to presence-based operations, as it includes time-consuming highly targeted network intrusions initially similar to intelligence operations.[2] The second category offered by Buchanan is markedly different and includes indiscriminate wide-scale operations that are not necessarily targeted beyond their original point of entry.[3] These cascading operations often begin with a single point of infection and continue to spread and infect additional endpoints within the targets, causing widespread damage. The first difficulty in this taxonomy is that it excludes most event-based operations viable in a military context, such as attacking specific warfighting platforms with a reusable but ultimately tailored capability. The second difficulty is that while less discriminating offensive capabilities are both easier to develop and likely to exist, they immeasurably raise the odds of incurring collateral damage, an often-dangerous proposition for any but the most total of conflicts.

Not all theorists uniformly agree that cyber-warfare is a discipline of consequence. In an impassioned call against the supposedly over-zealous attention to military OCOs, Libicki claimed that the very risk around network security is transient. As he claimed, cyber-warfare can only exist as long as human developers continue to introduce vulnerabilities into code.[4] As methods improve, these vulnerabilities would likely decrease in quantity and significance, and with it so would the prevalence of offensive network operations for strategic gain. More cautious voices such as that of strategist and historian Lawrence Freedman have viewed cyber-warfare as an instrument of potential operational, although not necessarily strategic, significance.[5] Considering that cyber-warfare has yet to prove its lethality, doubting its strategic contribution is understandable. Instead, by invoking well-known strategic elements, forward-thinking observers can identify areas to which offensive network operations can meaningfully contribute in a multitude of ways.

Applying Force to Networks

Should network attacks be applied against deployed forces or command structure? When should sensitive, brittle, presence-based assets be activated? What are the guidelines for pacing the use of recurring event-based capabilities in a tactical scenario? These questions can adversely impact long-term operational success. Misapplying force may mean permanently losing a capability that could otherwise inflict significant costs upon an enemy. As always, these are not uniquely cyber-related issues. Strategists have long concerned themselves with such aspects including intelligent use of capabilities via the *economy of force* principle, as well as examining how force asymmetry and geography matter.

Commanders and strategists should aim to maximise the economy of force, a key element of modern joint operations. At the heart of the principle resides the understanding that available resources are limited, and in order to achieve victory they must be applied intelligently.[6] Seemingly trivial to understand but difficult to implement at scale, efficiency is at the heart of all good strategies.[7] Expending overwhelming resources where they are unnecessary means that they may not be subsequently available where the odds are less favourable. Similarly, it is the economy of force that has historically allowed numerically disadvantaged forces to prevail against unfavourable circumstances.[8] By judiciously employing available assets, even asymmetrically weaker adversaries can achieve objectives.

Some offensive network operations are difficult to reconcile with an economy of force. As previously discussed, presence-based capabilities often inflict limited visible effects, require immense prepositioning, and are immediately expended upon first use. Network capabilities may almost seem anathema to a commander seeking reliable resources that would be made available when and where required. Aircraft are reusable after refuelling and maintenance; stand-off missile stores can be steadily replenished; special forces are flexible and re-deployable;[9] these circumstances do not easily align with the characteristics of cyber-warfare.

Expending network force well requires forethought, but it is possible. The possibilities run the gamut from providing options to national leaders, to enabling joint operations, and even facilitating gradual psychological operations.[10] From a different angle, presence-based operations can offer either sustained, low-yield stealthy force, or a burst of visible strategic effect. Event-based attacks—more similar to their conventional counterparts—may offer recurring, gradually diminishing levels of effect when deployed intelligently alongside other, conventional capabilities. Efficient use of cyber is a worthwhile aim, but it is easier to do when accounting for the differences between the two categories.

Presence-based operations do not conform to conventional perceptions of economy of force. With non-negligible odds of failure, they lack the visceral reliability of many kinetic capabilities. Premature detection of the intrusion prior to the effects phase—sometimes days or weeks in advance—could result in a complete loss of hard-earned operational capability.[11] However, successfully activating an attack at an opportune moment can result in a high-yield effect that could alter the operational environment. As such, presence-based capabilities should not be ignored; they simply must be properly accounted for.[12] Indeed, as previously suggested, the economic consideration is vastly different if presence-based capabilities are deployed stealthily or overtly.

Stealthily using presence-based capabilities means diffusing the effect over time. A promising case study demonstrating a stealthy gradual effect is the Stuxnet campaign. Rather than attempting to achieve a single destructive effect against the centrifuge cascades within the Natanz facility, the malware deceptively influenced both software and hardware over a lengthy period of time.[13] The strategic effect was achieved by maintaining a clandestine presence phase over an extended period, dispensing the effect in a controlled, subtle fashion. It is a useful example of how operations can represent a highly economic use of limited force; comparable in-conflict scenarios could have significant utility. It does, however, require a remarkable level of operational and technical rigour.

Alternatively, an overt effects phase for a presence-based operation can result in a single-shot, high-yield effect. Expending such a capability may represent the culmination of months or years of operational effort to pursue a single objective. One such plausible scenario would be a wide-scale sudden disruption of military satellites servicing an operational theatre. Such an attack could impair navigation, targeting, and intelligence efforts, all crucial in combat operations. However, activating such a capability would nearly always ensure its detection, thereby allowing defenders to eventually expunge the malicious presence out of their networks. The attack would therefore best be employed alongside a complementary kinetic effort, preferably one that capitalises on the new weakness exposed as a result of the offensive.

An event-based capability should be able to consistently achieve a similar effect against different targets. For example, a protocol denial of service attack targeting tactical communication networks by flooding them with superfluous control messages could degrade the operational capacity of local units.[14] While such a capability is inherently more robust and reusable, diminishing returns are expected as local forces eventually realign, adopt evasive or mitigating procedures and attempt to deflect the attack in various ways. Thus, while repeat uses are available, judicious dispensing of the capability is advised as to avoid the eventual inoculation by defending forces.

Opposing forces are rarely at parity; some form of asymmetry is assured. Even in the Cold War, the quintessential example of military balance, the United States and the Soviet Union were never equal in their capabilities. Soviet technology proved immensely capable at ballistic missile technology, while the United States was unparalleled in its adoption of computerisation. Adversarial asymmetry is fundamental to warfare in both historical and contemporary circumstances, and can manifest equally across equipment, experience, troop quantities and even will.[15] However, as history instructs, conventional overall weakness does not necessarily mean defeat.

Cyber-warfare feeds on asymmetries. Offensive network operations can be used by a conventionally weaker adversary to reduce technology-centric power discrepancies. Yet this is not necessarily the most likely scenario. As previously covered, nations with a history of success in signals intelligence and electronic warfare often flourish in devising a potent network offensive. As cyber operations require domain expertise in networks and intelligence, they may eventually favour the already strong. An asymmetrically stronger force may opt to capitalise on its technological superiority to shatter already weaker network defences, cushioning the way to a kinetic attack. Contrary to some conventional wisdom, cyber-warfare is not necessarily the toolset of the weak. It is simply a versatile toolset.

Highly networked adversaries do not lack weaknesses. Countries heavily reliant on digitisation often exhibit such reliance both in the civilian and military spheres. The more networked a country is, the more critical functions within a country are dependent on the proper functioning of computer systems for every-day and crucial services. As a result, the attack surface for targeting civilians may be vastly larger in such cases. A high quantity of networks certainly does not mean they are all equally well-defended. Many such systems run antiquated software, exhibit fundamental lapses in security good practices, and are generally vulnerable to well-established offensive techniques available even to low or mid-level adversaries. Even supposedly secure military networks within nations topping the charts in offensive cyber may have such critical weaknesses.

In 2014, a 25-year-old British man called Sean Caffrey successfully compromised a US Department of Defense satellite network used to communicate with global partners and personnel.[16] The Enhanced Mobile Satellite Services (EMSS) allowed globally dispersed personnel to securely communicate over commercial infrastructure. At the time, Caffrey satiated his curiosity by publicly releasing a list of 30,000 phone numbers and the details of 800 network users. Were such a compromise to have taken place by a hostile, relatively well-resourced threat actor, the result could range from assisting in gathering targeting intelligence on local personnel to compromising details on planned operations. In times of conflict, the network and its neighbours could have found themselves targeted by an operation seeking to disable or influence them. It clearly did not require meaningful expertise or dedicated resources to successfully target the network. Attacking it for effect would have been a different matter altogether.

Networks may transcend geography, but they ultimately remain subordinate to it. While information is carried globally at tremendous speed and volume, traffic is always facilitated by a physical infrastructure.[17] Impair the infrastructure sufficiently, and the virtual component will be similarly impaired. Thus, while the US and its NATO allies have deemed cyberspace an independent domain of warfare, it is instead the penultimate one which permeates through all others. Land, sea, air and space all facilitate the transit of networks, and consequently one can attack the latter through any of the former. US doctrine recognises this approach, with diagrams of the modern operational environment visualising cyberspace as awkwardly woven into all physical domains.[18]

It is generally difficult, though not altogether impossible, to fully sever a nation's connection to the global internet.[19] In many cases, nations rely on several high-volume fibre-optic cables to carry the bulk of their internet traffic beyond their borders. Consequently, degrading these crucial data arteries could effectively cripple a nation's access to the internet, as providers would be forced to fall back to limited-bandwidth, high-cost solutions such as satellite connections. One such notable incident occurred in 2008, in which two distant undersea cables carrying much of the Middle East's internet traffic were severed, supposedly by a combination of inclement weather and wayward ship anchors.[20] The result was staggering; Egypt reportedly lost seventy per cent of its internet traffic, while other nations similarly suffered. A decade later, United States officials reported with alarm increased activity by Russian submarines as they prowled known undersea cable routes.[21] Similar effects can be achieved at the global routing level by targeting the internet's underlying protocols,[22] but such attacks are far more likely to be transient. Regardless, the risk to internet infrastructure remains substantial.

The geographic aspect of network warfare creates new complications. Beyond the overall risk of collateral effect, attacks against networks risk unintended damage against third parties. This is a plausible prospect; militaries often rely on civilian telecommunications infrastructure. Public global infrastructure used for military purposes is often shared by multiple other countries for civilian use.[23] Attacking infrastructure, or even information conduits such as local satellites or optical fibre trunks, can cause significant damage that extends far beyond the parties at conflict.

Intentionally co-opting third parties is also a significant risk. If military equipment is stationed on foreign soil it would be highly visible, and impossible to carry out at scale without overt or coerced permission from the hosting country. Conversely, offensive cyber capabilities could easily use third-party servers without the knowledge of the nations in which they physically reside. This creates an issue of ambiguous complicity when one party to a conflict is attacked through an intermediary country.[24] It can be challenging for a victim to assess the role of the third country, or even the degree of control that the attacker has over the third-party assets.[25] There is therefore a substantial risk of drawing additional participants into conflict.

A small preview of the dangers of network geography can be seen in the 2008 Russo-Georgian War. The conflict offers a rudimentary example of offensive network operations due to the then-infancy of Russian doctrine and capabilities, and the technological limitations of Georgian network infrastructure.[26] Yet one incident stands out. After a denial of service attack knocked out Georgian government websites, they were relocated to civilian US jurisdiction to avoid follow-up attacks. The websites, however, were not hosted by the US government, but rather by a small privately-held company named Tulip Systems.[27] By any kinetic equivalent, this could potentially then expose the company, its assets, and peripheral providers to attack by both the Russian government and its supposedly unaffiliated hacktivist supporters. Were subsequent attacks to occur with damage incurred to US infrastructure, it could have compelled the US to respond more aggressively.

Manoeuvres

Achieving surprise in conflict is an ambitious but worthy goal. Liddell-Hart labelled attaining surprise as one of the most vital elements in war,[28] and Clausewitz placed it "at the foundation of all undertakings."[29] It is a quintessentially timeless aspect of warfare, referring to the art of attacking an unsuspecting target at an unexpected time.[30] Tactical surprise entails forcing an adversary into a localised battle under conditions suboptimal for the enemy. Strategic surprise is achieved if the actual state of war has been unexpectedly forced upon an enemy, thereby causing disadvantageous resource allocation, disposition of forces, state of readiness, and overall capacity for defence. Attaining tactical surprise is commonplace, whereas strategic surprise is a far more difficult ruse—yet it potentially carries a far more significant pay-off.[31]

In presence-based operations, surprise may manifest as a prelude to the onset of conflict. To mask large-scale movement of forces or the intent to attack, network operations may be used to disable or otherwise degrade early warning and situational awareness systems. Alternatively, targeting military equipment may reduce available adversary forces, while targeting network infrastructure could sow useful chaos. This would allow attacking kinetic forces to operate with increased freedom, improving the odds that a gambit at strategic surprise may succeed. As a result of their extensive clandestine pre-positioning, presence-based operations can complement a strategy of surprise.

In event-based operations, surprise may be limited to a tactical play. An attack against local equipment or networks could be used to reduce situational awareness, thereby facilitating manoeuvres that would otherwise result in direct engagement with an adversary. As such attacks do not necessarily generate visible effects, they could even avoid tipping victims off to having been attacked, thereby allowing kinetic units to surprise an already-degraded enemy. By targeting networks, it becomes viable to initiate hostile contact while one party remains unaware.

Deception in cyber-warfare entails fooling both man and machine by targeting the latter. Devices may be deceived in similar fashion to people by tampering with sensory inputs, thus causing them to reach incorrect conclusions and generate false outputs.[32] A machine inherently has no reason to doubt its trusted pipelines of information unless explicitly instructed to do so, and may accept false input as real as long as it is constructed and authorised correctly. Much like deploying inflatable tanks to fool spotters, creating a digital equivalent of an optical illusion can be useful. However, if the deceptive component of a network attack is detected, it becomes more difficult to achieve recurring success through same malicious component.[33] Much like with malware itself, detection of deception means eventual inoculation. Where deployed tank dummies may continuously foil adversary reconnaissance efforts until sensory technology is improved, network deception is far more difficult to maintain over time.

Destruction in its classic sense is simply not an essential component of offensive network operations. Network attacks offer an interesting dichotomy of destruction; they are simultaneously capable of immense collateral harm and unprecedented pinpoint accuracy. Applied correctly, modern intangible warfare can be the "ultimate precision weapon", as labelled by Rattray.[34] The gradual increase in precision targeting is a fulfilment of the underlying premise of technology-assisted warfare;[35] cyber is merely the continuation of this trend. As the previous chapter detailed, the elaborate targeting cycles of presence-based operations support this notion. Operations include extensive periods of lingering within adversary networks and conducting repeat micro-targeting cycles against specific servers. Offensive payloads must be specifically crafted and configured to work against the target. As a result, the surgical fitting of payload to effect[36] implies peerless specificity and control over the intended impact.

Collateral damage is assumed as an inherent by-product of network warfare.[37] Even a seemingly localised event-based effect can escape its intended boundaries and potentially wreak havoc on a wide scale if it is improperly developed and constrained.[38] As US doctrine acknowledges, "collateral damage from this type of attack is not always predictable".[39] For event-based attacks, a destructive attack might permeate well beyond the intended local network if operators are not cautious. Considering the modern battlefield often relies on multi-use infrastructure shared between military forces and civilian population, the result can be a catastrophic loss of access, data, or services that may be difficult to quickly recover. The 2017 NotPetya worm—originally targeting Ukrainian servers and computers—quickly led to a blaze of infections that wiped many thousands of devices and networks worldwide.[40] For presence-based attacks, engaging destructive payload against an identified network can similarly have unexpected consequences. In one example, an overly-successful attack against military aviation could have ramifications on civilian aviation, as a result of data-sharing conduits, overlapping networks, shared personnel, or collaborative air traffic monitoring services.

Network attacks are often disregarded as they mostly result in transient effects.[41] Weapons seemingly have limited utility if their effects are ephemeral, making them potentially unreliable when pursuing objectives. This argument, which holds some merit but can be challenged, represents three contestable claims; (1) that the effect means less because it is disruptive rather than overtly destructive; (2) that damage applied to software is easily recoverable; and (3) that the weapon viability itself is transient.

Disruptive attacks may still be valuable. On the campaign scale, a well-positioned presence-based attack can enable strategic surprise. Tactically, disruption can enable retreat, limit casualties, or be followed by a more permanent conventional attack. A failure to scramble defenders due to disrupted situational awareness can result in a hefty cost, redistributing relative advantages and potentially knocking out assets that would otherwise come into play. Event-based attacks can similarly be used to create a temporary breakdown in command and control or communications, thereby facilitating conventional strikes.

Damage sustained from network-warfare is indeed often easier to recover from. Physical effects from such attacks are a rarity—although still possible. The destructive value of software attacks can be reduced through principles of network resilience, which include the use of redundancies, segmentation, backups, and emergency recovery procedures. However, military conflict unfolds at a rapid pace. Once an attack has taken place, recovery would still take precious time, during which the impacted system will be diminished, misleading, or inactive. In many cases, deployed assets such as warships or aircraft do not have the means to completely restore their own software. Consider a navy destroyer that suffers from a catastrophic corruption of its shipboard systems. As shipboard staff are mostly operators of weapon platforms developed in the defence industry, specialty personnel may be required to restore the ship to full functionality. In such cases, this would require a costly trip back to a friendly port, where an extensive repair and readiness cycle would be undertaken to recover ship functionality. By the time the ship is ready to re-enter the conflict, it may have already missed it entirely.

The notion that offensive network capabilities are transitory stems from the idea that detection means inoculation. As previously discussed, this is true in some cases but not all. As Max Smeets notes, well-resourced forces can mitigate the perishability of their offensive capabilities by investing in robust, modular, and difficult to detect platforms.[42] Event-based capabilities would rely on hard-to-patch or systemic vulnerabilities that are not easily mitigated. Presence-based operations would attempt to ensure operational security so that capabilities are not wholly compromised when used.

A key element in manoeuvres is agility. In warfare, agility commonly relates to the effectiveness in which a force can shift from one circumstance to another; it is a measure of dynamics.[43] Circumstances include both purposeful and unexpected changes between states. The former can occur simply when trying to adapt from one operational scenario to another, while the latter occurs when entities are forced to adapt to external change. Agility is thus a measure of speed, adaptability, and resilience.[44]

Agility is heavily addressed in modern warfare doctrine as a prerequisite for conducting effective joint operations.[45] The rise of network-centric warfare has emphasised the deep fusion of technology with the decision-making process and increased the operational tempo, requiring all participating forces to be increasingly agile to keep up.[46] Concurrent sensory input from dozens of sensors both local and remote can assist in engaging numerous targets in rapid succession or even simultaneously. Forces must be able to respond, redeploy, and in some cases switch payloads to be able to fulfil an entirely different mission package. Agility may also occur on a strategic scale; the sudden need to respond militarily to an escalating situation may require disparate forces to suddenly and wholly realign priorities.

Agility is particularly fascinating as it is one of the key areas in which event and presence-based operations differ. Agility for event-based operations reflects one of the greatest difficulties in offensive network operations; they have to be robust. Much like its electronic-warfare ancestors, whenever an operator "fires" an event-based capability at an adversary, it must work, no matter the circumstance. This is particularly difficult due to the inherent sensitivity of software-based attacks to minute changes in the adversary environment. If the vulnerability it exploits is no longer present, or if the attacked platform is configured differently from what the attack tool needs in order to function, it may cause an unexpected effect or altogether fail. Even the most specific event-based attack capability has to be agile in order to foster commander trust and be properly integrated into operational planning. An unreliable capability is one that will not be chosen when the need arises.

Presence-based operations initially appear almost antithetical to agility. Requiring extensive pre-positioning, bespoke attack tools, focused intelligence, and limited reusability, such capabilities seem intrinsically hostile to a high-paced operational environment, in which "forces must be prepared to transition rapidly from one type of operation to another".[47] This is indeed one of the key reasons why they are best suited for strategic or theatre-level effects rather than localised operations which experience more variable changes in circumstance.

Agility manifests differently for presence-based operations. Rather than flexibility at the operational level, presence-based agility can manifest as robustness and modularity of the attacking infrastructure itself. Often, high-quality government malware is modular in nature,[48] indicating a strategic understanding that a flexible framework can be adapted to specific needs as an operation unfolds. The intrusion and lateral movement mechanisms may stay the same, with attack modules developed and deployed as required for specific target types. By strategically investing in such capabilities, military planners can shorten the time it takes to develop an operational solution against a high-value presence-based target once it is compromised. Agility remains relevant, but relates to an entirely different scale of time.

Conclusions

Network operations can fit well into modern military strategy. Introducing a distinction between presence and event-based operations helps distinguish between the unique considerations of each. Where the former is flexible, surprise-enabling, and strategic, the latter is robust and ideally predictable. These differences can help decision-makers identify when various capabilities may be best served and when they should be held in reserve. The considerations may be somewhat unique, but the process is not.

Presence-based operations are naturally incongruent with a battlefield tempo. Their targeting phase is ponderous, and they require significant prepositioning. Establishing even a single presence-based capability against a hardened target of military value can take many months. However, once established, high-level command staff must be aware of the potential strategic utility of such capabilities. If a deceptive element can be incorporated into the activation phase, its effects could possibly stretch on for the duration of the conflict. Subtle malicious manipulation of command and control telemetry, or minute disturbances in targeting latency could wreak havoc across an entire operational theatre. Conversely, if protracting the engagement is not feasible, a single activation burst could similarly prove lethal in the earlier stages of conflict. Fully blinding satellite communications with a network attack could significantly degrade operational capacity until it is restored, which could possibly take several days. Such a stretch of time is critical at the onset of combat operations.

Event-based operations can and should be delivered to deployed units. Where offensive cyber cells operate, they should have pre-packaged, resilient tools for their use. Such capabilities could allow them to temporarily degrade tactical communication networks, wipe local adversary networks, or blind vehicle-borne systems used by aircraft, maritime vessels, and ground forces. As adversaries become increasingly networked themselves, their cyber-attack surface commensurately grows. It is increasingly becoming possible to weaponise the adversary against itself; one only needs to target the systems that have become its operational crutches.

The infatuation of modern militaries with technology resulted in capability-based strategies rather than strategy-contributing capabilities. Originally a twentieth century phenomenon, the desire to offset adversary advantages by winning technologically is untenable in network operations. Over-reliance on air superiority arguably resulted in degradation of capacity to effectively hold territory. Adoption of drones and remote strikes similarly increased the distance between decision-makers and the battlefield, resulting in protracted, gainless conflict. We must not repeat these mistakes with network capabilities.

Strategic coercion of an adversary will not likely occur as a result of overwhelming its networks. Instead, understanding the centrality of networks to modern life and combat operations means identifying how centres of gravity have now uniquely become targetable. Where forces shunned concentrations of military mass, network operators seek the convergence of command and control. These hubs of activity are prime targets, potentially presenting an enormous strategic benefit with relatively minimal risk to materiel.

Cyber is also not the full answer to resolving conventional asymmetries. A substantially weaker nation is not substantially more likely to achieve victory by simply applying force against a stronger enemy's networks. It would require vast efforts, often beyond the reach of the weaker nation, to effectively minimise adversary asymmetries by degrading their ability to conduct joint warfare. At the same time, traditionally potent actors such as the United States, Russia, and the People's Republic of China are all global leaders in the maturity of their offensive cyber doctrines. Instead of decreasing asymmetries, powerful actors can use offensive network capabilities to increase them. Weaker parties to conflict also have less resource to spend on network defence and secure development of military resources. That means they must rely on commercially available solutions, imported military equipment, and ageing hardware. As a result, such parties may find themselves on the receiving end of persistent network attacks, rather than effectively delivering them.

This trinity of traditional military concepts—surprise, deception, and destruction—is enlightening. Tracing their origins back to the earliest days of warfare, network operations are intrinsically geared towards surprise and deception. Extensive prepositioning of operational assets and the ability to subtly manipulate software and sensors are conducive to the same principles of subterfuge offered by Sun Tzu. As hardware destruction by way of software attacks is difficult, the metrics must simply be calibrated to account for digital destruction and physical disruption. These can be accomplished at an unprecedented scale, yielding effects ranging from small tactical disturbances to wide-scale strategic disruption of operational capacity.

Offensive network operations are immensely useful to all manner of operations, and to both disadvantaged and dominant aggressors. It is the discourse that counts; to create effective doctrine, the involvement of intelligence, network operators, weapon developers, military command, battlefield staff, and policy makers is absolutely necessary. The separation of event and presence-based capabilities can assist in overcoming several existing issues, yet others will remain. Overcoming challenges is a long and arduous process, but one that can potentially result in a force multiplier effect across all aspects of conflict.

5

AMERICAN CYBER SUPERIORITY

In the twentieth century, the United States led the world in terms of military technological superiority. The quintessential display of this was the 1991 Gulf War, viewed with surprise and consternation by global adversaries who realised the discrepancy between US technological prowess and their own. In the years that followed and as other global powers acquired increasingly advanced technologies, the capability to target information networks at scale still seemed a distinctly US advantage. Information leaked from several high-profile incidents shone a crucial light at how developed the US technological capacity to wage network warfare truly was; developing offensive platforms, encroaching on adversary networks, researching equipment used by enemies, and creating subtle yet significant effects. Yet as before, the US had led by technology rather than by strategy; the former continued to precede the latter, often creating mismatched capabilities and a lack of coherence on how to achieve goals with offensive cyber operations. The US now realises this, however, and is adapting to a new reality.

Documentation on the US approach to "cyber" as an operational space is vast. It includes policy directives, national strategy documents, doctrinal publications both general and service-specific, and significant coverage of various programs advancing OCOs as means to generate effects. Alongside these publications, private sector research has uncovered several network intrusion campaigns commonly associated with US agencies, including toolsets presumably employed by the NSA, CIA, and elements of the military. Overall, US coverage of all matters cyber offers a rare, unique glimpse into the evolution of a discipline within the US military. Despite the inherent sensitivities of cyber operations, the US offers ample evidence to assess its capabilities and approach.

There is value in established discipline seniority. Network operations being a likely US practice since the earliest days of the modern internet means that US-based units accrued decades of hard-earned experience in penetrating and influencing adversary networks of all kinds. We can turn to the 1998–99 Kosovo War as a suggestive precursor to the "cyber era" US approach, in which "a recurring theme in our discussions with military operators is, well, if we can drop a bomb on it, why can't we take it out by a computer network attack".[1] Other snippets of evidence suggest network operations stretching even further back to the early 1990s.

The US de facto leads the school of thought envisioning "cyber" as an independent combatant domain. The domain approach entails observing all efforts to attack and defend networks as doctrinally and operationally distinct from the physical domains. While US doctrine clearly and loudly identifies the interdependence between networks and the physical domains, it maintains that the oversight of the former requires separate command from the latter. These efforts have resulted in an intricate amalgamation between cyber, electronic warfare, and information operations; due to the numerous overlaps between the three disciplines, it is often unclear where one begins and the other ends. The US approach has both advantages and disadvantages, but could nevertheless benefit from the distinction between event and presence capabilities to provide commanders with options, while retaining the strategic sensitivity of presence-based operations.

A 2015 Defense Science Board (DSB) report highlighted the dangers of lumping all OCOs under the same framework. The report definitively stated that "the United States must maintain—and be seen to maintain—an array of scalable offensive cyber capabilities—including high-impact strategic cyber attack options—as an integral part of its cyber deterrence posture."[2] The DSB conflated the robustness of partly-visible event-based attack capabilities and the necessity for strategic presence-based capabilities to remain covert. The report then claimed that "Unlike precision-guided munitions, cyber weapons cannot be bought and deployed on a delivery system… with confidence that they will work when needed. A highly talented cadre of cyber warriors must work together closely with intelligence specialists and technologists in a highly classified environment."[3] This remark disregarded the potential utility of deployable event-based capabilities which already existed within the US arsenal, while assuming that all OCOs must remain under the purview of remote operators within intelligence units.

We can look to space for evidence. Reliance on satellites as means for communication, geolocation, intelligence, and research has skyrocketed since their inception in the twentieth century. Scores of satellites now dot Earth's orbit, providing essential services for both civilian and military use. Their significance makes satellites prioritised targets. Disruption in space-based services can have immediate and widespread effects, including errors in navigation, or inability to communicate globally. Corrupting an adversary's navigation systems can cost precious hours and days in how quickly they are able to deploy forces. In the early stages of conflict, this can make a strategic difference. Several countries now visibly invest in kinetic anti-satellite (ASAT) solutions, both as a deterrent and as a capacity. But satellites, at the end of the day, are orbital transmitters wrapped around a lightweight computer. Everything they do begins and ends with data transfer and processing. This makes them prime targets for offensive cyber operations.

Enter the US military's Counter Communication System, or CCS. The military has declared initial operational capacity (IOC) on the system in March 2020.[4] A project initiated in 2004 and upgraded many times over, CCS is described in budget reports as a platform which allows for "expeditionary, deployable, reversible offensive space control effects applicable across the full spectrum of conflict".[5] It is a packaged, robust, configurable weapon for creating transient effects in data-based systems, and is continuously upgraded to reflect advancements in the satellites it seeks to disrupt. This perfectly describes an offensive cyber capability for event-based operations.

OFFENSIVE CYBER OPERATIONS

This chapter will present the argument that the US is technologically best positioned to conduct OCOs, but a rigid approach to cyber as a domain slows effective integration. The United States has the operational experience, technical expertise, and high-quality intelligence required to be consistently successful in employing OCOs. At the same time, a technology-first strategy means that the creation of capabilities is often detached from considerations of need or the requirements of the forces who may eventually employ them. The United States' well-developed, co-opted national defence industry is fully capable of crafting packaged event capabilities that could then be delivered to deployed forces. At the same time, The National Security Agency has provable experience both operationally, in penetrating hard-to-reach network targets, and in developing advanced presence capabilities that could then create effects against these targets. Yet both have traditionally struggled in transparently delivering capabilities to the parties who need them based on a thorough understanding of the threats and opportunities. Bridging these challenges by shattering some existing boundaries between cyber and the other domains could position OCOs as valuable across the spectrum of military operations.

Even more so, the relatively advanced US approach to OCOs means that second-order integrations may be a viable reality. These include using offensive cyber capabilities to enable or deliver other network operations rather than just support kinetic forces and attacks. Event capabilities and their presence counterparts could work in tandem by having one facilitate the other, though this requires an intimate familiarity with the characteristics of each, and an established trust between forces and their available capabilities.

Permeation is preferable to segmentation. US efforts to foster collaboration between the physical domains and their new cyber sibling are admirable, but it is preferable to simply acknowledge that cyber is an inseparable facet of all domains. Rather than fitting OCOs in a neat domain-shaped box, Cyber Command should act as a beacon of expertise facilitating the integration of network capabilities into every unit, along with the strategies and doctrine that support them. With vast experience in the intricacies of conducting effective presence-based operations, it is now essential to understand how these operations fit across the services. A tight-knit relationship with a highly capable defence industrial base and the ability to roll out modular capabilities to combat platforms means that the US could lead the charge in truly integrative event-based OCOs. While these principles may already exist on paper, their implementation despite historical barriers is at the heart of future US success. Technological and capability superiority will not suffice; their manner of use is key.

Segmentation by Doctrine

The US approach to network warfare is one of the most publicly accessible. With experience in targeting digital communication that spans decades, US strategists have increasingly recognised the utility of pursuing networks to accomplish a broad spectrum of objectives ranging from limited tactical effects to broad strategic success. Mounting US attention to the operational significance of networks famously resulted in the 2009 creation of Cyber Command, and later in its mid-2018 elevation to a full "unified combatant command", thus recognising its importance towards US strategic success.[6] Each of the military services has its own cyber command and these openly declare offensive cyber mandate. Even as the Pentagon dramatically reflected at the time of US Cyber Command's elevation that "the cyber domain will define the next century of warfare", the specifics of how the US seeks to accomplish this are worth examining in depth.

US thoroughness has resulted in meticulously crafted doctrinal maturity, enabling operating forces to potentially integrate a vast array of capabilities. The linchpin of published US network warfare doctrine is Joint Publication 3–12: Cyberspace Operations, a core document providing some of the standard definitions, objectives, and approaches all US forces are expected to implement and adhere to.[7] Yet Joint Publication 3–12 is not enough on its own for a complete reconstruction of US doctrine and strategy for OCOs. A full view can only be accomplished by reviewing other relevant documents, policy directives, committee hearings, technical specifications of capabilities, and leaked classified materials.

Operating against networks is formulated in the broader context of a new US strategy for victory in modern conflict. This approach is perhaps most commonly presented as "Multi-Domain Operations", an integrative strategy heavily favouring joint operations across multiple warfighting domains, incorporating asymmetric capabilities and focusing on near-peer threats such as Russia and China.[8] Modern US doctrine has accepted that technological advancements uniquely allow adversaries to challenge US forces where they were traditionally perceived as vastly superior.[9] Similarly, military strategy acknowledges the difficulties presented in overwhelming modern defences with directly applied force, a traditional strength of US forces in the last several decades. As the US Training and Doctrine Command (TRADOC) states, "The cost of penetrating prepared enemy defenses is now too great for current conceptions of forward positioning and expeditionary maneuvers to effectively deter adversaries and prevail in armed conflict."[10] This is precisely the type of challenge that military OCOs may help ameliorate.

US strategies consider offensive network capabilities to be crucial for achieving objectives in modern conflicts. Contributions may be direct, by actioning against a target or otherwise enabling conventional forces to accomplish their missions. The Department of Defense's previous 2015 Cyber Strategy already articulated this as the aspiration to "build and maintain viable cyber options and plan to use those options to control conflict escalation and to shape the conflict environment at all levels."[11] This in turn supports the notion that OCOs can make measurable contributions to commanders in various circumstances. A similar understanding is etched within US Cyber Command's mandate: they must "rapidly transfer technologies with military utility to scalable operational capabilities."[12]

The rich tapestry of US official documentation on OCOs indicates a comparatively firm grasp of the value that such capabilities may lend. Many of these were not originally intended to be publicly available. In 2013, NSA leaker Edward Snowden included several policy documents among the trove of materials he released to media outlets. Most relevant among those is Presidential Policy Directive 20 (PPD-20) on offensive cyberspace operations. Cloaked in its intended classification, the document provided a candid view of some crucial aspects of operating in and against networks.[13] On the risk of uncontrollable cascading effects caused by the misuse of OCOs, the document warns against "cyber effects in locations other than the intended target, with potential unintended or collateral consequences that may affect US national interests in many locations."[14] On the need to selectively employ OCOs to avoid unduly risking brittle capabilities, the policy claims that an effort must be made to "identify potential targets of national importance where [Offensive Cyber Effects Operations] can offer a favourable balance of effectiveness and risk as compared with other instruments of national power."[15] In respect to the possibility of attaining various levels of surprise and the spectrum of possibilities, the document provides a lucid articulation: "[Offensive Cyber Effects Operations] can offer unique and unconventional capabilities to advance U.S. national objectives around the world with little or no warning to the adversary or target and with potential effects ranging from the subtle to severely damaging."[16]

There is a good measure of strategic wisdom to unpack in PPD-20. While it lacks detail on how OCOs concretely may be employed, the document directly articulates several of the strategic contributions that they may have, as reviewed in previous chapters. Within three pages of content, the directive refers to the unique economy of force consideration offered by military OCOs alongside the dangers of detrimental collateral effects that may occur by incorrectly employing them. Similarly, the above quotes indicate a desire to subvert conventional centres of gravity that are increasingly proving resilient to US technological prowess, instead opting for an indirect approach that may carve a path towards coercing an adversary.

PPD-20 turned concerns over the impact of OCOs into presidential shackles. As the document stipulates, approval for operations required an extensive escalation chain reaching the top echelons of decision-making, including the president. Understandably, this subdued the appetite for pursuing and integrating such operational capabilities to any scalable degree. This frustration with the inability to wield the considerable resources available to them eventually resulted in the 2018 National Security Presidential Memorandum 13 (NSPM-13). The Trump Administration's NSPM-13 streamlined the security forces' ability to conduct offensive operations, thereby encouraging agency and military use when required. There is some suggestion that the increased discretion pertains specifically to "actions that fall below the 'use of force'", which includes many of the activities previously described in this book. If that is the case, then this new directive would do well to empower tactical and operational use of offensive cyber capabilities, while maintaining presidential oversight for strategic capabilities that may purposefully or inadvertently cause more substantial harm.

US literature on doctrine and strategy is cascading by design. Documentation exists at the national level, which in turn leads to overarching integrated military doctrine, finally resulting in service-specific doctrine and strategy. Service-specific doctrine entails translating the broad criteria set at the higher levels and extracting potential utility for service objectives. The Army does not operate under the exact same conditions as the Air Force, Navy, or Marines. Each have their existing concepts of operations (CONOPS), platforms, and procedures. It is in the service-specific doctrines that we observe the US at its best; the multiple layers of integrations are indicative of a comprehensive push towards having more OCOs made available for combat missions.

The aforementioned JP 3–12 doctrinal document exemplifies both the strengths and relative weaknesses of the US position on OCOs. The odd relationship between "cyberspace", the electromagnetic spectrum (EMS), and information hampers overall clarity. As US capabilities deepen across these three categories, the lines between them appear increasingly blurry and the attempts at distinguishing between them become laboured. While the document stipulates that "Cyberspace is wholly contained within the information environment",[17] only "cyberspace" is defined as a distinct domain, while the "information environment" is relegated to a separate publication.[18] Electronic warfare is shunted even further away as a sub-publication of information operations.[19] This is particularly puzzling, as Joint Publication 3–13.1 on Electronic Warfare contains the following accurate qualification:

> Since cyberspace requires both wires and wireless links to transport information, both offensive and defensive cyberspace operations may require use of the [electromagnetic spectrum] for the enabling of effects in cyberspace. Due to the complementary nature and potential synergistic effects of [electronic warfare] and computer network

operations, they must be coordinated to ensure they are applied to maximise effectiveness.[20]

The complete dependence between the EMS, networks, and information is unbreakable, because they merely represent different layers of the same communication model. In a positive indication of progress, US doctrine does indeed reflect the multi-layered approach to networking, though rather awkwardly. Joint Publication 3–12 outlines the "cyberspace layer model" as comprised of three layers; a physical network layer encompassing physical hardware, terrain, and transmission medium; a logical network layer that refers to the links and networks that make up "cyberspace"; and the cyber-persona layer which reflects the actual use of information. This is useful to a degree and echoed in official NATO cyber doctrine,[21] but does not serve to fully address the depth of the relationship between networks, the EMS, and information.[22]

Within the services, the Army's implementation of network warfare is novel. Grouping its tactical approach under "Cyberspace electromagnetic activities" (CEMA),[23] the Army has recognised the natural relationship between the electromagnetic spectrum as the medium, cyberspace as the networks, and information as the payload. The Army doctrine manual on the topic acknowledges both opportunity and risk, advantages and disadvantages, and how these broadly manifest. It thereby offers a cautious but optimistic vision as to what could be done with such capabilities, both against US adversaries and against US assets.[24]

The rigid adherence to cyber as a domain is perhaps the key element which separates the US from its closest adversaries. Even elements within the US chafe against the restrictions that "cyber domainhood" imposes. US Army Cyber has indicated its desire to change its title to Army Information Warfare Command to reflect its broader responsibilities.[25] The US Air Force empowered the re-established 16th Air Force to own "information warfare", and delegated cyber, electronic warfare, and signals intelligence to the organisation.[26]

It is unclear when an EW platform that defeats an adversary system by transmitting a digital payload qualifies as a cyberattack in US military thought. It is similarly unclear if an OCO manipulating adversary propaganda in a war zone qualifies as a cyber operation or an information operation. It is unclear because these distinctions are often artificial. All OCOs are dependent on the electromagnetic infrastructure carrying data. The network effects employed almost always directly impact information in some form—either by corrupting it, manipulating it, preventing access to it, or otherwise occluding its intended use. The three-way interplay between the electromagnetic spectrum, networks, and information must be a core aspect of the fundamental approach.

This fuzziness in boundaries has not gone unnoticed. Over the course of several years, the US defence establishment has done its best to work within the confines of their cyber domainhood. One such prominent example is Joint Force Headquarters—Cyber (JFHQ-C), a structure designed to unify representatives from each of the military's services as they conduct defensive and offensive cyber operations.[27] Despite all efforts to the contrary, the US cannot avoid the simple reality of ubiquity; computers and networks permeate everything, and cannot be contained within a separate domain of warfare.

The US is gradually adopting an information operations doctrine broader than cyber, as it had in its previous iterations. It is the unavoidable response to the realisation that the vast majority of network operations occur between conflicts and under the threshold of warfare. If China seeks to continuously pilfer US intellectual property from the private sector, this must be countered consistently. If Russia is poised to foment public sentiment with disinformation and otherwise disrupt sovereign elections, the US must identify, repel, and ideally deter further such attempts.[28] None of these enemy actions would be considered "cyberwar", or even warfare-threshold OCOs individually. They are dangerous in aggregate, when hundreds of small information operations accrue to a dangerous strategic campaign. This evolved notion of a consistent state of information friction is what eventually crystallised into the "persistent engagement" approach within US Cyber Command.[29]

Persistent engagement broadly assumes two key realities. The first is that friction between adversaries is a constant, both in and out of conflict. Targeting networks, as previously discussed, does not occur strictly within the spectrum of warfare, which means that defensive efforts entail a constant hunt for adversary operations within your own networks, even as you seek to do the same. A unique circumstance of network operations is that they are constantly occurring, even if they may not initially be intended for attack. Counter-cyber operations are brought to the forefront of strategic planning.

The second assumed reality is that once you are defeating adversaries within your own networks, it is already too late. As with conventional warfare, it is classically better to fight away from your own territory. Rather than always defending against network intrusions by detecting them as they occur, the new reality is to consistently seek out adversaries and disrupt their capabilities before they ever attempt contact or pursue their objectives. The notion of persistent engagement aims to consistently push the point of friction beyond friendly virtual territory and into neutral or adversary space. If done effectively alongside classic network defence efforts, it can be a relatively proactive way of conducting offensive cyber operations in service of a greater strategic agenda. That is what the United States' "Defend Forward" approach sought to establish.

"Defend Forward" was codified in the 2018 Department of Defense Cyber Strategy. The document summary highlights a more assertive, proactive approach to defending and targeting networks. Interestingly, the document acknowledges the role of both presence and event-based operations. Firstly, for the role of long-standing strategic presence ops, "[The US DoD] will conduct cyberspace operations to collect intelligence and prepare military cyber capabilities to be used in the event of crisis or conflict. We will defend forward to disrupt or halt malicious cyber activity at its source, including activity that falls below the level of armed conflict".[30] And soon after, on the battlefield role of robust event capabilities: "During wartime, U.S. cyber forces will be prepared to operate alongside our air, land, sea, and space forces to target adversary weaknesses, offset adversary strengths, and amplify the effectiveness of other elements of the Joint Force".[31]

While a strategy document does not immediately change the entire organisational culture of a behemoth such as the US defence establishment, it suggests a gradual evolution towards a nimbler approach to OCOs, which places them on both a broader scope of competition and a more complete spectrum of information operations both during and before conflict. As explained by Secretary of Defense Esper in

2019, the overall goal is to "maximize the effectiveness of the department's cyber warriors."[32] Practicalities aside, there is a deep understanding that OCOs have a fundamental role to play in facilitating success against modern adversaries of all kinds.

Event-Based CEMA

The US military holds all the pieces required to successfully use event-based operations. It trains and forward-deploys operators meant to augment kinetic capabilities with OCOs. It develops advanced platforms intended to both deliver network payloads and simplify their use. The US enjoys a rich technological landscape alongside a highly capable intelligence apparatus, thereby supporting the elaborate operational life cycle required. It is likely that some high-profile operations—particularly ones involving special forces—incorporate such capabilities to a degree. Yet, the systemic incorporation of offensive cyber operations across the spectrum of joint operations is still an ongoing process.

Efforts to combat the Islamic State (IS) and similar extremist parallels lean more heavily on OCOs. In 2016, the US coalesced its network efforts against the Islamic State around Joint Task Force Ares. The task force was uniquely empowered to conduct offensive cyber activities against the organisation and its assets in support of the international coalition's efforts to combat its regional capabilities. One responsibility of Ares was "Operational Glowing Symphony",[33] primarily focused on limiting the ability of IS to spread propaganda and communicate globally. JTF Ares has reportedly carried out both event and presence-based operations against the Islamic State, and continued to do so into 2020, before reportedly shifting its resources to state adversaries, with a focus on the Pacific.[34] Some likely presence-based operations within Glowing Symphony included crippling networks and locking down accounts used by IS. Reported event-based operations included slowing down IS networks, disrupting downloads, and adversely impacting their ability to utilise networking effectively.[35]

The Army is currently exploring how to best extract operational value from network operations. Efforts have ratcheted up largely due to the realisation that after a year of counter-insurgency operations against poorly equipped adversaries, the US is woefully underprepared to tackle advanced adversaries in electronic warfare and battlefield network operations.[36] This perception emanates from sensitive presence-based operations and lumps all military OCOs together, limiting the Army's capacity to own operational capabilities that could then be integrated. More nuance and distinction between event-based and presence-based capabilities at the doctrine level could afford the Army firmer boundaries to seek capabilities within, thereby generating opportunities that do not require top-echelon approvals. Even as they struggle to map out the various possibilities of OCOs, the Army has identified networks as an adjacent and dependent space to the electromagnetic spectrum. At the deployed force level, this could mean event-based capabilities targeting adversary infrastructure, weapon platforms, and communication networks.

Despite doctrinal difficulties, the multi-decade prominence of the NSA as a top provider of signals intelligence uniquely positioned US forces to both acquire a deep understanding of adversary systems and provide potential reach where it would otherwise be difficult to acquire. Persistent tapping of data arteries within the global internet grid and sensitive adversary networks means that event-based operations may be carried out against these without ever introducing a presence component. Injection of traffic to block, manipulate or otherwise influence data streams could be a crucial vector towards impacting adversary networks. This may manifest in programs similar in approach to the now-exposed NSA QUANTUMTHEORY project, which "dynamically injects packets into a target's network session to achieve CNE/CND/CNA network effects."[37] Packet injection allows an adversary to externally splice data into networks that may otherwise be inaccessible, thereby manipulating or compromising them. It would, in most modern cases, require the ability to subvert whatever encryption is employed.

These capabilities often have steep intelligence requirements. Crucially, the documentation for QUANTUMTHEORY specifically called out the incorporation of passive signals intelligence (SIGINT) in targeting,[38] thereby indicating the capacity to conduct high-resolution targeting as required for such capabilities to be employed judiciously. The need for tight support from operational intelligence for effective OCOs was also referenced in the NSA's Sentry Eagle program, in which the NSA "provides SIGINT that supports the planning, deployment/emplacement and employment of [Computer Network Attack] combat capabilities."[39]

Abilities may only be valuable if their use is streamlined. Communication is key; battlefield commanders must be made aware of a capability's existence and the circumstances around its optimal use. One such example may be in event-based capabilities such as QUANTUMSKY—which facilitated disruption of web access—or QUANTUMCHOPPER—which enabled disruption of file transfers.[40] Seemingly tactical capabilities, they could still prove advantageous to deployed forces in certain scenarios. Yet, the classified documentation around these belies their internal sensitivity, thereby suggesting that they were reserved for strategic operations or compartmentalised special activities. It is unclear if comparable capabilities were or are available to support regular deployed forces.

Both 2016 and 2017 were painful years for the US intelligence community. A series of intrusions against institutions embodying national sovereignty, such as the Democratic National Committee (DNC), reflected the vulnerability of the United States to outside intervention by way of information operations. These breaches were accompanied by data stolen from the NSA, CIA, and other leading agencies, weighing down public trust in the institutions and inflicting damage on the US capacity to conduct OCOs effectively. The grand theft of NSA data presumably originated from a group calling itself The Shadow Brokers,[41] while the CIA content was leaked directly to Wikileaks where it received the codename Vault 7.[42] An additional leak from the NSA—the contents of which were not publicly disclosed—was traced to NSA contractor Harold Martin.[43] The aggregate leaks shine a partial spotlight on many US presence and event-based capabilities.

The capacity to remotely compromise systems is instrumental to all OCOs. Specifically, in event-based operations where the attacker is presumed to not have the time nor the capacity to conduct a long, cautious targeting campaign, the solution needs to rapidly succeed, with a high probability rate. Within the Shadow Brokers leak, the

EternalBlue tool was a coveted exploit enabling wormable remote code execution, thereby allowing automated compromise of vulnerable Windows-based systems.[44] The exploit—one of several in the same family and numerous others of differing quality in the Shadow Brokers leak—soon became a prominent component in several high-visibility malware campaigns such as WannaCry and NotPetya,[45] notorious for their unusual virulence and destructive impact on affected systems and networks. While both self-identified as financially motivated ransomware, they were soon attributed to the North Korean[46] and Russian[47] governments respectively, thereby becoming de facto rampant event-based operations.

Incorporating OCOs into military hardware for battlefield use is a difficult process. It requires significant research and development resources; intimate collaboration with intelligence agencies providing required telemetry and defence industries to facilitate integration; a trust relationship with combatant forces resulting in coherent, realistic enumeration of requirements; and a deep understanding of the operational life cycle into which these capabilities may eventually integrate. For this purpose, the US military–industrial complex is uniquely qualified. Companies such as Lockheed Martin, Raytheon, and others enjoy close relationships with their in-service peers and the wherewithal required to develop platforms over numerous years. While the vast majority of technologies remain understandably classified, a glimpse into some publicly filed patents reveals just how prominently they feature throughout the OCO life cycle for event-based capabilities.

Perhaps the best-known instance of field-worthy event-based capabilities is the EC-130 Compass Call aircraft. Originally deployed in 1981, later evolutions of the aircraft have also demonstrated the capacity to directly target networks. As cheerfully explained in 2015 by Major General Burke Wilson of the US 24th Air Force, "Lo and behold! Yes, we're able to touch a target and manipulate a target, [i.e.] a network, from an air[craft]".[48] The Compass Call's onboard equipment facilitating these new capabilities is provided by British defence contractor BAE. BAE is purportedly also behind the Suter event-based operations platform previously referenced. As such, this indicates a long-term dedication from BAE towards providing the US and its allies with robust platforms to conduct battlefield cyber operations. A similar commitment to equip the EA-18G Growler electronic warfare aircraft with event-based capabilities has been strongly suggested.[49]

Autonomous platforms are likely to feature heavily in US event-based capabilities. One such example is a 2016 patent submitted by Selex Galileo, an offshoot of Italian defence company Leonardo, dedicated to providing solutions to the US Department of Defense.[50] The patent details a dedicated unmanned aerial system (UAS) small enough to evade detection by air defence systems and designed specifically to deliver electronic warfare and event attacks, reducing the risk to manned assets and paving the way for kinetic attacks. The drone documentation reveals the desire to deliver "[radio frequency] based cyber effects".[51] This is the quintessential manifestation of an event-based capability as an evolved electronic warfare attack vector, one that simply seeks to target a platform by attacking its software rather than its sensors.

Other technologies encapsulate the spirit of event-based operations, even if they do not recognise it as such. Various elements within the US military have increasingly begun to adopt the use of Digital Radio Frequency Memory (DRFM).[52] The technology allows the rapid recording, digitisation, manipulation and retransmission of electromagnetic signals so that they could be weaponised against an adversary in a form of evolved jamming. The transmission itself can be modified at the digital level, potentially generating different effects. This form of jamming thoroughly blends electronic warfare with event-based OCOs, as transmissions affect adversary platforms on the software level rather than on the sensory level. This natural escalation results in substantially more flexibility and the ability to manipulate targeted systems at a high level of accuracy, creating effects previously impossible. Manipulation of the digital payload within a signal can result in the creation of altogether new information, increasing the range of options available to the commander using the ability.

Raytheon prominently features as a provider of OCO-related technologies. In one 2014 patent partially titled "Digital weapons factory",[53] Raytheon claims to be able to dynamically match an offensive network payload by assessing the targeted adversary equipment[54] in near-real time. This pairing is done transparently to the operator, thus saving them the otherwise steep requirement for familiarisation with offensive techniques normally beyond fielded forces. Presumably, the technology calculates a probability of success by analysing the hardware and software used by the target, and an offensive tool would only be created for use when a specific threshold is passed. A similar yet more specific 2014 Raytheon patent attempts to tackle ballistic missile defence (BMD) by outlining a system that can assess the vulnerabilities of a launched missile and attempt to pair a viable event-based capability to defeat it.[55] This is mirrored by a subsequent 2017 Raytheon patent detailing an integrated kinetic, electronic warfare and OCO-delivering system.[56] These technologies—if functional—can afford tactical agility which may then be realised in field scenarios. Bridging the gap between OCO subject matter expertise and combat forces could make event-based operations far more likely to be used routinely in the field.

Other patents attempt to simplify the overall battlefield awareness and direction of OCOs. Recognising that it is often difficult to grasp aspects of intangible warfare—and doubly so with battlefields becoming increasingly saturated with various networks—these technologies seek to streamline the process. In one patent from Boeing, a technology is offered to orchestrate both "cyber" and electronic warfare missions.[57] An older Raytheon patent tries to similarly assist by directly offering "command and control systems for cyber warfare".[58] These technologies prove how an investment in the preparation phase of the operational life cycle can then reduce overhead in the engagement, presence, and effects stages. By instrumenting all of those through a single unified platform, commanders can focus on how best to use the ability rather than the intricate characteristics of doing so.

Presence-Based Operations

The United States is arguably the best globally positioned actor to conduct high-quality presence-based operations at scale. Access to a wide range of adversary networks is enabled by supply-chain compromise, infiltration of third-party providers, cooperation from popular global service providers, proven niche subject matter expertise when required by operations, and a range of zero-day exploits against widely used products. As before, available evidence points to a high capacity to conduct operations in service of an overarching national security agenda, yet it remains unclear how thoroughly these capacities are made available to military planners. When trying to reconstruct the unusually rich tapestry of US offensive capabilities, a review of leaked, disclosed, and publicly researched evidence is useful.

The majority of available data pertaining to US network intrusion originates with the National Security Agency.[59] Through its historic role as the premier provider of signals intelligence (SIGINT), the NSA has organically grown into its mandate as the caretaker-in-chief of US computer network operations (CNO), an internal term which encapsulates defensive, intelligence, and offensive efforts against adversary networks. Under that mandate, it has become incredibly prolific at creating capabilities and compromising key US adversaries, ostensibly for the purposes of answering critical intelligence requirements within its area of responsibility. Unfortunately for the agency, its allies, and the US at at large, series of leaks and compromises of the infrastructure, tools, and documentation it uses to facilitate CNO resulted in an inordinate amount of public scrutiny. Rather uniquely, the world was given a partial yet surprisingly deep look at how a clandestine intelligence agency sought to weaponise networks and information.

It is crucial to note that the NSA does not operate solely under a direct military mandate and is primarily subordinate to its designation as an intelligence agency. As such, it is not uniquely tasked with accomplishing military goals, and is consequently often at odds when opportunities to conduct OCOs arise. As indicated by Ashton Carter, despite the creation of Cyber Command—which was designed to remediate some of this tension and concentrate military network operation efforts—the NSA has traditionally done relatively little to furnish the US military with OCOs at scale.[60] Instead, secrecy, bureaucracy and compartmentalisation mean that cyber capabilities are heavily classified, known by relatively few, and deployable only for very specific pre-approved intentions approved by the highest tiers of US decision-making. Yet irrespective of this and the limited signs of military offensive presence-based operations, previous network attacks carried out by the NSA are often an indication of overall US readiness to conduct them. Attacks that were carried out under the auspices of the NSA could theoretically be ported over to Cyber Command, where they would be employed according to military objectives and priorities. Perhaps the most well-discussed and visible of these sabotage operations is Stuxnet.

The Stuxnet malware targeting the Iranian nuclear project has been thoroughly scrutinised since its 2012 discovery. Even its purported operational designation as "Olympic Games" has been revealed in a *New York Times* article.[61] While the operation occurred outside of direct conflict, it is certainly revealing as a presence-based operation that could be mirrored for military purposes.[62] A tightly managed operational life cycle is indicated by a stealthy, modular capability incorporating dedicated attack components, and by Stuxnet's targeting of a specific set of software and hardware, wrapped in a self-replicating infection vector and several zero-day exploits. While a subsequent US capability in the vein of Stuxnet has not been publicly disclosed since, the tool remains a viable indication of US—and possibly Israeli—presence capabilities. Considering the vast increase in US resources and strategic focus, it is likely that similar capabilities against adversaries are routinely pursued to line up against potential US conflict contingencies, with some variations.

As previously indicated, deception is crucial towards maintaining the viability of a presence-based operation. Detection means mitigation, and subsequent decommissioning or at least revision of the attack tools. Importantly, Stuxnet maintained its deception even after commencing its effects phase,[63] a unique feature that enabled sustained—albeit lower intensity—impact against the target. Consequently, it was able to maintain continuous viability in a volatile environment, requiring a measure of operational nuance rarely found in other attack tools. Similarly, the self-propagation component of Stuxnet was coupled with code to detect specific hardware and software, thereby both identifying potential targets of interest while similarly avoiding harm against incidental infections This attention towards limiting collateral effects is operationally significant both to defend the tool itself, but also prevent uncontrollable cascading effects that threaten to harm to unrelated networks or devices and risk the overall mission. As former senior official Richard Clarke once said, "it very much had the feel to it of having been written by or governed by a team of Washington lawyers,[64]" emblematic of a highly disciplined and intricate security bureaucracy. This manner of discipline is difficult to develop and—doubly so—to maintain, as evident in the eventual breakout of Stuxnet, resulting in peripheral infections of unrelated targets around the world.[65]

Stuxnet was not enough. During the early years of the Obama administration—as Stuxnet was operating—the risk of a significant regional conflict in the Middle East peaked. Bolstered by conservative backing and a perceived sense of urgency, Israeli Prime Minister Binyamin Netanyahu reportedly became increasingly nervous. The illicit Iranian nuclear program, long suspected to have a military dimension, was rapidly advancing. Concern was growing in the US that additional viable solutions were necessary to deal with the Iran's nuclear aspirations that did not involve open warfare, in an attempt to reduce the likelihood of all-out war. A multi-target network attack seemed a potentially promising solution, one that could theoretically deliver success with minimal risk.

The plan for comprehensive strategic OCOs against Iran was reportedly folded into a program called "Nitro Zeus". In controversial coverage by *New York Times* journalist David Sanger, Nitro Zeus was described as essentially a series of presence-based operations against numerous critical targets, which Sanger claimed "would have required piercing and maintaining a presence in a vast number of Iranian networks, including the country's air defenses and its transportation and command and control centers."[66] The operational plan behind Nitro Zeus suggested a schism in US thinking on network operations at the time; it was both ambitious enough to suggest that strategic coercion might be singularly possible through cyber means, while at the same time too narrow as to thoroughly integrate the capabilities into a broader military capacity. In this sense, presence-based OCOs were almost viewed as a strategic extension of special forces, capable of surgical attacks on an unprecedented magnitude.

Offensive cyber operations are now a prime method of attacking Iranian interests while limiting escalation. During May and June 2019, several attacks took place in which explosive charges tore open holes in commercial naval vessels in the Gulf of Oman. Even as the United States accused Iran of orchestrating the attacks, it simultaneously mobilised an offensive response. This manifested several weeks later when US forces reportedly conducted an offensive cyber operation targeting "Iranian computer systems used to plan attacks on oil tankers in the Persian Gulf".[67] In September 2019, a combination of cruise missiles and drones slammed into the Abqaiq oil processing facility and the Khurais oil field, causing substantial damages and taking a chunk of Saudi crude oil production offline, accounting for 6% of global production at the time. Both the Saudi and US governments were quick to point the finger at Iran, which often provides material support to the Yemeni Houthis who claimed responsibility for the attack.[68] In a bid to retaliate while limiting a possible escalations spiral, the US eventually responded by way of an offensive cyber operation against Iranian information operations capabilities.[69]

An additional possible OCO capability can be found in a network espionage campaign dubbed "Slingshot" by researchers from Russian security company Kaspersky, who unmasked it. With forensic evidence indicating operational activity since at least 2012, the malware predominantly targeted Middle Eastern and North African nations. Defined by the researchers as a high-quality versatile platform, its truly unique differentiator was in the engagement phase of its operational life cycle; the malware appeared to spread in part by exploiting vulnerabilities in network routers. Once it had successfully done so, Slingshot's operators appeared predominantly interested in harvesting intelligence from affected endpoints. Yet, with a capable, modular platform, and having fully compromised its victims, it seems that a subsequent re-tasking could have turned Slingshot into a potent presence-based capability delivering a variety of offensive objectives.

Slingshot's most significant aspect is not the infection vector or its targeting; it is the context. State-sponsored network operations routinely get exposed by private sector research companies. More commonly than not, private sector analysis focuses on three aspects; (1) technical analysis of the malware and its components; (2) operational analysis of the threat actor's efforts; and (3), a victimology analysis of affected targets. Rather unusually, Slingshot's early 2018 reveal was rapidly followed up by a CyberScoop article indicating that Slingshot was "an active, U.S.-led counterterrorism cyber-espionage operation… used to target ISIS and al-Qaeda members."[70] The same article claimed the operation was under the purview of Joint Special Operations Command (JSOC), a subordinate part of US Special Operations Command (USSOCOM). The access and intelligence afforded by Slingshot was reportedly used to facilitate accurate targeting for kinetic operations. If supporting targeting was indeed the goal, it bolsters the notion that network operations are treated as part of the unique toolset provided mostly to US special forces, perhaps in tandem with the work conducted by Joint Task Force Ares.

On the technological side, a valuable case study is the NSA's Advanced Network Technologies (ANT) catalogue. Publicly released in 2013 by *Der Spiegel*, the catalogue purportedly originated from a subdivision of the NSA tasked with creating deployable hardware-software solutions for compromising networks.[71] Capabilities include access acquisition technology for firewalls, transmittable signals, mobile phones, servers, and personal computers. The now ageing catalogue includes a highly classified bevy of solutions befitting different scenarios, thereby allowing the NSA's customers and partners to directly request a capability for operational use. The comparatively high level of technical details and manner of explanations suggest that the target audience was not strategic planners, but rather operational planners familiar with the tactical details of the adversary. Despite the restrictive classification, the very existence of a formalised network capability catalogue in 2009 suggests an evolved approach to incorporating these capabilities across various branches of the US defence establishment. If an equivalent catalogue is available to strategic planners, using familiar military vernacular and detailing only the opportunities lent by such capabilities, it would likely encourage the co-optation of presence-based OCOs into the military operational life cycle.

Evidence also points to the publicly analysed "Equation Group" threat group as being synonymous with the NSA's Tailored Access Operations (TAO) unit. In a 2015 report from Russian security company Kaspersky, the researchers discovering the group's activity and tools claimed it was "probably one of the most sophisticated cyber attack groups in the world."[72] Within the same report, Kaspersky's analysis indicated evidence directly tracing the group's malware activity to 2001, with other anecdotal evidence suggesting some operational activity as early as 1996. If accurate, this positions the NSA as one of the most capable, persistent, and historically significant network operators in the world. The analysis similarly indicates that the malware in question employed several unique device persistence mechanisms,[73] was remarkably stealthy, and sufficiently modular to enable a wide array of payloads and capabilities. Some of the modules and exploits used by Equation Group were subsequently revealed in the aforementioned Shadow Brokers leaks, confirming the dazzling range of capabilities available to the NSA.[74] As such, the TAO's malware family was uniquely suitable for presence-based operations against a broad spectrum of global targets.

Occurrences of an integrated approach to cyber-warfare do exist. A relevant modern case for a joint kinetic–OCO strategy is the North Korean ballistic missile threat to the US. While volumes have been written on the missile threat itself, consensus broadly paints it as intricate and difficult to reliably overcome. Some of the parameters include wild fluctuations in the diplomatic relationship, a rapidly advancing ballistic missile program, and nuclear payloads arguably ready to be coupled onto warheads. The North Koreans have limited, tightly controlled internet access, reducing their online footprint and thus the available attack surface for OCOs. Conversely, a significant reliance on Chinese expertise, infrastructure and material support may also present opportunities. Increasing the chances of preventing a North Korean nuclear attack would require an integrated effort of all domains across the spectrum of operations. This entails a combination of kinetic ballistic interceptors, naval and air assets, electronic warfare, and presence-based operational capabilities. The layered approach can help ensure that even if multiple attempts fail, additional measures would be deployed in turn in a cascading fashion, until they deplete or the threat has been mitigated.

The ballistic missile threat may be mitigated prior to an actual attack in an approach labelled "left of launch".[75] The approach includes all efforts to defeat missiles by targeting the systems and components that make up their operational ecosystems. Rather than defeating the missile once launched, the goal is to prevent the launch or otherwise pre-emptively thwart its odds of success. These efforts do not uniquely need to be network-based, as put forth by the Atlantic Council's Herbert Kemp:

> It is time to change the game from a purely defensive battle to one in which the fight is taken to the source—to attack the [Theatre Ballistic Missile] launch systems and their supporting infrastructures before missiles could be launched. All parts of the chain leading up to the launch event are potentially vulnerable to disruption or destruction, and the time is right to undertake a serious effort to engage the TBM threat "left of launch".[76]

Left of launch operations present a unique opportunity for a collaborative, full-spectrum US approach that incorporates pre-emptive attacks, supply chain sabotage, and presence-based OCOs. The US currently pursues each of these capabilities separately and it is becoming increasingly clear that it is determined to integrate them into a tiered defensive network. The Department of Defense outlined its approach in a 2017 declaratory memorandum: "The concept of operations for employing left-of-launch capabilities is set within the broader context of integrated offensive and defensive operations for countering offensive missiles."[77] The department envisioned joint efforts to defeat missile threats by relying on varying capabilities, including both kinetic and non-kinetic options.

Pursuing presence-based capabilities against the North Korean missile threat is compelling but challenging. While some reporters claim that such efforts have already manifested as an increased failure rate for North Korea's missile tests,[78] proliferation researcher Jeffrey Lewis claims no credible evidence has been offered to support this notion.[79] Conversely, a 2015 panel of former US military staff officers on comprehensive missile defence[80]—later iterated by Andrew Futter in 2016[81]—offered that targeting ballistic capabilities with network operations risks undermining the certainty of classic nuclear deterrence models, thereby reducing the overall security of all parties involved. The core characteristics of presence capabilities—that they are clandestine, difficult to track, and inconclusively effective—mean that all involved actors cannot guarantee that either their defensive or offensive measures would be successful. This lack of transparency damages the clearly communicated notion of mutually assured destruction, or really any deterrence model. Yet this lack of clarity similarly also means that nations may already be pursuing such presence-based capabilities, and arguably none are better poised to acquire them than the US.

Integrated Warfare

The US challenges in integrating "cyber" do not stem from a dearth of capabilities. A strategic investment in military OCOs and network espionage tools over the last two decades have resulted in perhaps the broadest range of both presence-and event-based capabilities globally. The possibilities are seemingly dazzling, including targeting critical infrastructure, dedicated military equipment, encrypted communications, industrial systems, and air-gapped networks. Targeting efforts span from entire countries to individuals in a collaborative effort that potentially includes thousands of staff, across numerous agencies, units, and companies. The scope of US network operations is gargantuan.

Rather, the challenges for the US stem from a lack of focused offensive strategy, and a disconnect between the available capabilities and those who may use them most effectively. Deployed forces have limited support from event-based capabilities that could augment their operational life cycles, and therefore cannot incorporate them into planning or rely on their availability when needed. Similarly, strategic planners are often disconnected from the scale and specifics of US network penetrations and capabilities. Only the highest of echelons have visibility into presence-based opportunities to inflict harm and facilitate success. US operators and units are likely capable of succeeding in both presence and event-based OCOs; it is the scaling aspect that remains lacking. Where many nations struggle to create capabilities, the US has them, but needs to do better at deploying them. It needs better doctrinal clarity.

The cycle of capacity acquisition could be reversed. A separation into event and presence-based operations can assist in facilitating part of this process. If strategic planners were clear on the scope and possibilities presence capabilities lend, they could then task relevant capacity creators such as Cyber Command to create and maintain them. Persistent network intrusions are primarily facilitated by agencies such as the NSA for intelligence objectives, and subsequently weaponised if the necessity to do so arises. Introducing a presence-based operational mentality into the calculus would mean initiating intrusions with an offensive intent already in mind, thereby perhaps changing the approach, intrusion vectors, or even the toolsets used. Doing so may then allow a smoother offloading of access from the intelligence agency that facilitated the intrusion to the operational agency that seeks to attack the underlying targets. In a similar vein, a coherent plan to create and integrate event-based capabilities across all existing operational domains could result in force multipliers that deployed forces could rely on. These capabilities would need to be prioritised, as due to differing technologies and networks, they would vary based on the region. A concerted, strategy-oriented push from acquisition planners in the Navy, Air Force, Marines, and Army could then solicit the sprawling US industrial base for event-based capabilities that would suit their requirements, rather than being approached by these companies with technologies as they come up with them. Technological innovation is crucial to success in modern conflict, but only if it is channelled to meet requirements.

Beyond acquisition, the US is uniquely poised to use event and presence-based capabilities as force multipliers for each other, effectively creating hybrid OCO opportunities. While it is becoming increasingly clear that OCOs can enable kinetic operators, if the benefits of networks operations are sufficiently well developed and understood they could also be used in support of each other. Opportunities in this space include using deployed event-based assets to breach networks and deliver presence-based malware that would subsequently be handed off to remote operators. Alternatively, networks compromised for presence-based operations may allow remote access to "air-gapped" networks, thereby facilitating follow-up event-based attacks against them. This would require painstakingly crafted bureaucracies and deconfliction channels that can rapidly decide how to jointly exercise multiple types of OCOs in mutual support. While this direction is implied in US doctrinal literature,[82] the reality is that such cooperation requires an intimate familiarity with the considerations on operating offensively in and against networks that still eludes the US military as a whole.

The notion that "cyberspace" is a domain remains a significant hurdle for this. Rather than viewing it as a distinct set of opportunities and capabilities, US military planners could benefit from increased integration, transparency, and seamless co-optation of networks into all aspects of operations. Networks are already integral to all other domains and will increasingly become so as technology progress and automation increases. The staggering US dependence on technology-led strategy would only deepen if cyber is siloed in its own domain, relegating innovation and creativity in using OCOs to Cyber Command and its subordinate force structures.

Perhaps the US's greatest current challenge is integrating its implementation of OCOs across all its different services, agencies, manoeuvring forces, and defence companies. Each contributes a crucial facet that arguably embodies desirable capabilities, were they to be integrated with one another. Both for event-based and presence-based capabilities, the US currently represents a reality where the whole is less than the sum of its parts, rather than the opposite. Even as the Army pursues CEMA it remains shackled to an overly broad doctrine that views all OCOs as a single, sensitive operational approach. The Air Force does well to incorporate cyber as a notion but limits itself by relegating it to support roles as a domain of unique effects distinct from the airspace which it naturally commands.[83] The NSA has spent immense resources on developing both access and capabilities to action against critical targets that could prove strategic to future mission planning. And the US defence industry has proven innovative in fashioning event-based capabilities that could then be delivered from various platforms, integrated into the operational planning process, and translated to jargon and concepts familiar to military users.

The combination of these disparate approaches could result in a strategic advantage that rebalances the US in an increasingly contested geopolitical climate. Where near-peer adversaries such as Russia and China pursue advances in technologies and OCOs of their own, the US already possesses them in a disjointed ecosystem that may fail to deliver effects when and where they are most needed. Recognising that the electromagnetic spectrum, networks, and information are all different layers of the same man-made construct could help the US in distinguishing which entities should "own" which part of the overall effort. Whether or not it will do so remains to be seen, though the efforts towards an integrated conception of cyberspace continues across the US armed forces.

6

THE RUSSIAN SPECTRUM OF CONFLICT

What was old is new again. Observing Russia's geopolitical disposition teases familiar themes reminiscent of Soviet thought. These include an increased belligerence characterised by friction with Western interests; diminishing cooperation with NATO and its contingent members; emphasis on austere national loyalty, and an ever-present concern with perceived NATO encroachment on its borders with the West.[1] Even as it is yet premature to claim a fully-fledged return to the dismal days of the Cold War, where global stability seemed on the brink, global tensions between Russia and its historic rivals are on a notable incline.

The Russian geopolitical mentality is that of continuous strategic contest. Rather than envisioning clearly defined bouts of warfare capped at both ends by periods of peace, armed conflict is viewed as a deterioration of existing relationships, either due to a perceived imminent threat or the pragmatic realisation of potential value to gain. This is a key idea within Russian strategic theory; envisioning warfare as a component part in a larger contest of will impacts their perception of conflict itself and the tools employed within it. Some tools that are deemed overly aggressive or even classically categorised as warfare by other nations may be designated as legitimate activities in Russian grand-strategy.

This holistic perception of conflict has intensified in the twenty-first century and is well articulated in Russian military thought. As

Chief of General Staff Valery Gerasimov claimed in his seminal 2013 article, "In the twenty-first century we have seen a tendency toward blurring the lines between the states of war and peace. Wars are no longer declared and, having begun, proceed according to an unfamiliar template".[2] Russian theorists such as Chekinov and Bogdanov describe what they call "a new-generation war", referring repeatedly to non-military actions before, during, and after armed hostilities ensue.[3] This conceptualisation of a new-generation war is echoed repeatedly both by Russian military theory and those who observe it,[4] and offers some parallels to the controversial Western concept of "hybrid warfare".[5] Adversarial actions bleed into the civilian sphere and draw on mass-media, psychological operations, academia, outreach, and diplomacy, months before armed conflict visibly erupts.[6] In this reality of diffused contest, discerning concrete elements of the Russian way of war may be challenging.

This chapter will argue that Russia has thoroughly integrated OCOs into a broader spectrum of information operations, but often fails to achieve objectives. Drawing on decades of established doctrine, Russia has recognised and enshrined the usefulness of affecting information as a means of limiting or avoiding conflict altogether. Yet technical and operational limitations have caused many operations to be exposed, while others were sub-optimally used or prematurely executed. Despite superlative coverage, Russia has enjoyed relatively limited advantages from the use of OCOs, though it is comparatively poised to gain significantly from a more thoughtful approach.

There is a schism within Western analysis of modern Russian doctrine. While some evoke the notion of a new "Gerasimov Doctrine" which suggests a novel form of hybrid warfare, others are fiercely sceptical that these terms introduce meaningful value.[7] The label "hybrid warfare" is often used when describing the complexities of modern Russian strategy. Originally coined by Frank Hoffman in 2009,[8] the underlying notion is that, within the context of conflict, Russia relies on a combination of multi-domain forces alongside non-military and irregular means of coercion. While this does indeed appear to be the case, there are noteworthy reservations preventing the term from being useful as a descriptor of Russian military behaviour. First, that the term is mainly applied to modern Russian doctrine implies novelty where it does not exist.[9] The observed "hybrid" approach to conflict is mostly a modern manifestation of the Soviet-era strategy for geopolitical competition. Second, it implies an intentional labelling from Russian military theory where there is no such effort. Much like when discussing cyber, the notion of hybrid warfare mostly exists within Russian literature when referencing Western commentary about it.[10]

Russian grand-strategy often implements the idea of "reflexive control". Harkening to Soviet-era military thought,[11] the concept calls for a gradual manipulation of an adversary's perception so that it organically begins to act against its own stated objectives.[12] This approach is thereby a subtler stand-in for classic coercive behaviour, which instead overtly seeks to compel an adversary to behave favourably. Reflexive control could manifest as either reshaping the information pipelines used by the adversary in the decision-making process, or manipulating actors of influence to generate a more favourable setting.[13] The idea may seem familiar from information operations such as those targeting the US 2016 national election, but in fact has a far broader scope. The flexibility of the term embodies both the dynamic Russian approach to coercion but also their deep-set aversion to armed conflict where it is unnecessary.[14]

Reflexive control entails a grand-strategy reliant on shaping truth to exert influence over a political adversary. While manipulating the flow of information has been an integral part of reflexive control since the Soviet era, it has flourished in modern Russian grand-strategy. There are numerous similarities and overlapping areas between Russia's modern information operations doctrine and Soviet reflexive control,[15] where both seek to manipulate perception and upset decision-making processes across all political and strategic levels. Thus, as previously examined with cyber-operations,[16] there are well-established historical roots for the use of information campaigns to achieve objectives while pre-empting conflict.

There are several key characteristics in the overarching Russian approach to information and its uses. The first is that the entire information space is viewed as a fundamental aspect of modern geostrategic competition; this in turn explains the scale and investment in capabilities that target information in various ways. In the absence of any Russian framing of cyber, they instead consider "information confrontation" as the holistic concept of competing with an adversary through targeting information and its underlying systems and processes.[17] Second, there is little concrete distinction between information operations and actual computer network operations. As Akimenko and Giles adeptly define it, "information is the most important object of operations, independent of the channel through which it is transmitted[18]". Third, electronic warfare—a traditional strength of the Soviets—is often viewed on the same spectrum as its information counterpart. As a result, the potential to integrate OCOs across all disciplines is well-present in modern Russian thought.[19]

Due to the holistic approach, network operations are often an indistinct component within the larger Russian information operations doctrine. Where Western doctrine often makes a comparatively clearer distinction between cyber-operations directly targeting networks, and information operations that primarily tackle human perception, Russian doctrine views the information space as a continuous spectrum of operational capabilities.[20] These capabilities in turn serve a wide variety of purposes, many exceeding the military-strategic.

Within Russian thought, influence operations would be in-scope for achieving strategic confrontation objectives. While campaigns such as the US 2016 election interference or its counterparts in Europe are a key aspect of Russian information strategy, they are beyond the scope of this book. They may compromise public perception, they may instigate unrest, and even encroach upon the sovereignty of countries by targeting their democratic processes. Yet in the absence of an attack they are soft—albeit potentially impactful measures—that require a different approach to assess.

Russia does not just pursue reflexive control; it engages in reflective strategy. Many of its core principles stem from its own perception of threat, which in turn leads to its adaptation of Western techniques and advantages.[21] Simply put, they often do to others as they fear will be done to them.[22] Indeed, Russian theorists often view the Western and principally the US agenda as bent on dismantling residual Russian influence. As Colonel Maruyev claimed to this effect, "Obviously, the U.S. and its Atlantic allies are Russia's principal geopolitical enemy",[23] later adding that "as previously, the Americans will continue actively to foist their values on the rest of the world relying on all the force and assets available to them."[24]

While often blamed for engaging in prolific sub-warfare behaviour, Russian strategy has implemented this approach in part due to concerns about the United States aggressively pursuing this same behaviour.[25] To a degree, the renewed vigour in targeting the information space stems from the identification of a relative US technological advantage in the wake of the Gulf War,[26] in which US network centric warfare doctrine was fully on display. While some may argue that mirroring doctrine is a sign of strategic weakness,[27] it can perhaps also be construed as a rare case in which a country proactively realigned its strategy to account for asymmetries. Russian military thought adapted Soviet concepts of dominating enemy perception to a modern, data-driven technological environment.[28]

The Russian aspiration for information superiority cannot be stressed enough. Both in and out of conflict, Russia views perception across all levels and the information that shapes it to be a core tenet of its strategy. In the warfighting space, the military acknowledged that it lags behind its Western counterparts and has commensurately sought to revitalise its position. This gradual yet enthusiastic investment included both a commitment by the military to invest in networked information systems,[29] while simultaneously committing to engaging in offensive and defensive operations in the information space.[30] As with other aspects of its strategy, information operations do not have a branch of their own, and instead incorporate various elements from across disciplines and units.[31] As Giles notes, Russia's approach to information operations "combines tried and tested tools of influence with a new embrace of modern technology and capabilities."[32]

The focus on information operations reflects an overall Russian aversion to overt armed conflict. As will be reviewed throughout the chapter, Russian grand-strategy aims to wield all available information tools to soften political adversaries, ideally to the point where conflict is altogether obviated. However, even if it remains necessary, such capabilities are meant to weaken resolve, sow discord, fragment alliances, and damage military readiness as to increase the chances of an offensive success and reduce its required duration. As

Chekinov and Bogdanov colourfully write, "A new-generation war will be dominated by information and psychological warfare that will seek to achieve superiority in troops and weapons control and depress the opponent's armed forces personnel and population morally and psychologically".[33]

The result of the holistic, war-averse Russian approach is that most of their offensive activities in cyberspace may not separately meet the threshold of warfare. Russian activity may appear sporadic, low-intensity, half-baked or perhaps even crude when observed at the individual level. It is only when they are viewed in aggregate across the spectrum of activities that grand-strategy emerges. Considering the Russian methodology and its Soviet precursor, this does not seem accidental; this is an application of reflexive control to strategic scale, and is possibly meant to appear non-malicious or insignificant on cursory inspection.

Network operations are difficult to untangle from the wide breadth of information operations. Military officers Kuznetsov, Donskov and Nikitin tackled this distinction by indicating that "cyberspace is a component and tangible framework of another, and more extensive, space commonly known as information environment".[34] As such, OCOs are subordinate to grand-strategy and thoroughly woven into the larger power dynamic. This is similarly reflected in those agencies that carry them out. The Russian Federal Security Service (FSB), Foreign Intelligence Service (SVR) and Russian Military Intelligence (GRU, or GU) have all proven to be prolific network operators. Over the last decade, these agencies and others have engaged in a wide variety of international information activity, including reconnaissance, intelligence collection, political and personal active measures, and direct sabotage of information systems and even critical infrastructure.

When examining specific event and presence-based activities, it becomes easier to chart the Russian potential for integrating offensive network operations into warfighting. As the following analysis will indicate, most of the building blocks required to achieve the desired effects already exist, contributing invaluable operational experience. Were Russian military forces to similarly invest in permeating network operations through their forces as they have with electronic warfare, it would likely prove highly advantageous.

Applied Strategy

Modern Russia aggressively pursues its geopolitical agenda. Regionally and globally, it asserts its interests through means both overt and covert. Its presumptive subservience to a monopolar Western order is no longer applicable; Russia is eager to contend over resources, land, and strategic advantages. It is also intent to aggressively pursue Russian detractors worldwide, including through violent targeting of its own citizens.[35] Yet this eagerness should not be immediately interpreted as an offensive slant. Russian military thought insists upon a clear and present geopolitical danger from an encircling NATO, bent on its eventual destruction.[36] The key motivation appears to therefore be a highly developed—perhaps overdeveloped—threat perception.

Geopolitical concern is accompanied by a realisation that the Russian armed forces cannot currently contend symmetrically with their NATO adversaries, at least not if those adversaries are fully committed to—as unlikely as that may be.[37] As its conventional military still lags in capacity, Russia seeks to leverage all available advantages to either accomplish strategic objectives without conflict, or at least limit conflict significantly. As their doctrine states, this includes "the intensification of the role of information warfare", including pre-emptively, "in order to achieve political objective without the utilization of military force".[38]

In 2008, separatist sentiment within Georgian South Ossetia and Abkhazia eventually culminated in an invasion by Russian military forces.[39] Lauding the supposed right of self-determination, the concern seemingly stemmed from the notion that pro-Western Georgia directly threatened the ethnic Russian population within the disputed territories.[40] Deeming it an unacceptable encroachment by the West, Russian forces proceeded to pummel the Georgian military into inevitable submission, both within the separatist territories and along Georgian cities. While there was never any serious concern that Georgian forces would withstand the determined Russian onslaught, the campaign exposed a slew of deficiencies in the Russian order of battle. Disorganised forces, mismatched capabilities, lack of joint operational cohesion and an overall strategic inefficiency led to some unnecessary losses—often due to negligence—before victory was achieved.[41]

The Georgia campaign spurred substantial reforms within the Russian military.[42] These can broadly be abstracted to three primary approaches; investment in personnel, advancements of capabilities, and changes in doctrine. Between the three and over the span of the following decade, the Russian armed forces have made substantive steps towards improving their capacity to respond to threats and deploy offensively and defensively. While many issues still plague their military forces, subsequent campaigns indicate a rapid learning rate alongside a determination to improve. Whether this was motivated yet again by a perception of threat or expansionist desires does not matter within this context; the end results were the same. Russia had acknowledged that it must revisit its strategic priorities and migrate away from its classic emphasis on enemy force destruction, in favour of more irregular, non-kinetic means of coercion.[43]

A key element within Russia's military overhaul was the recognition that their use of information was lacking. While Soviet-era strategies continuously advised altering adversary perception, Russian strategy was relatively slow to catch up in the implementation of these principles for the modern, internet era. As information operations continued, operations against the modern conduits of data—the very networks that became instrumental to modern life in the West—lagged noticeably. Indeed, up until the early twenty-first century, the internet itself was viewed by Russia with suspicion. Late adoption resulted in delayed adaptation.[44]

Russia now extensively relies on several strategic approaches that align well with offensive network operations. Among these; a reliance on asymmetry, the indirect approach, and targeting of perceived centres of gravity.[45] These principles have not only retained their Soviet usefulness, but saw new life breathed into them in the information era. Fuelled by wariness of NATO and emboldened by the lack of repercussions to their aggressive campaigns against Estonia, Georgia and Ukraine, Russian strategists increasingly deployed more aggressive network attacks against their adversaries.[46] Repeat experiences in different environments globally allowed Russian network operators to accumulate priceless experience in both event and presence-based operations.

The Russian approach to network conflict is the embodiment of the indirect approach. Far more so than its Western equivalents, who use such capabilities to operationally bypass defences, Russia seeks to employ network attacks across its entire spectrum of conflict to subvert and diminish an enemy.[47] Tactically, network attacks are used to support and augment kinetic effects. Operationally, they are employed to elicit reflexive control in various regions. Most importantly, network operations are strategically used to altogether obviate conventional conflict and coerce adversaries into favourable conditions.[48]

A helpful analogy can be found in the spread-spectrum communication technique. Its principle is straightforward; rather than transmit across a narrow frequency band, one chooses to transmit the same energy output across a wider range of frequencies. As a result, power output is diffused, resulting in a transmission that is less observable to eavesdroppers, more resistant to interference, and more resilient overall. The recipient can then reconstruct the original communication by correlating the received transmissions across all frequencies. This technique is now a mainstay in numerous modern platforms.

The Russian strategic approach to network operations is similar. Individual OCOs are not meant to be decisive. Instead, they rely on the indirect approach to target adversary centres of gravity in barrages of low-yield attrition attacks. Rather than attacking a single target with a high-impact effect, operations are carried out frequently and across a swathe of globally dispersed targets.[49] This is done in recognition of the symmetric limitations of conventional conflict that still plague Russian military forces, and the resulting desire to achieve success with minimal actual violence. The spread-spectrum approach to offensive network activities presents a useful way to integrate such capabilities into military doctrine.

This approach also emphasises the Russian perception that information is a crucial conduit towards attacking modern centres of gravity—the collective will of the people. Considering the Soviet view of modern Western democracies as socially fractured and incapable of sustained resistance to strife, applying consistent, diffused coercive pressure against these weak points could erode the adversary's capacity to put up a meaningful defence. Clausewitz once discussed applying destructive coercion up to where the enemy's will shatters;[50] the Russian approach applies much of this principle and upgrades it to eroding the adversary's civilian, instead of military, will.

OFFENSIVE CYBER OPERATIONS

Deception also plays a significant role in Russian network operations. Although its quality varies greatly between operations, there is a significant reliance on *maskirovka*—the Russian military term for deception—in all information operations, offensive or otherwise.[51] It is understandable why deception is imperative to achieving reflexive control in which adversary actions are seemingly organic rather than coerced. The spread-spectrum approach to stringing together a mass of network operations can appear as virtual chaff, masking true intent and the full extent of effects. Obfuscating the true magnitude of operations leaves adversaries affected yet not wholly aware of the true scope of Russian strategic intent, manoeuvres and wider political goals.

This does not mean that the Russian approach is optimal. Examining the strategic principles outlined in previous chapters, several are noticeably lacking. Among those is the absence of surprise, a comparative dearth of agility, and unsustainably high collateral damage. When these deficiencies accrue, Russia's apparent potency in cyberspace is somewhat diminished. The Russian approach to information operations may shine when they go unnoticed and are deployed against adversaries it has not engaged in active hostilities, but it may prove lacklustre against a determined, defensively inclined, actively engaged enemy.

The use of surprise is not strictly necessary for all forms of network warfare, but it can certainly help. An adversary unaware of an intrusion against its critical networks may find itself the victim of a highly impactful presence-based attack. Instead, Russian threat groups and their corresponding offensive infrastructure are frequently exposed, either due to activation of effect, or operational security mishaps that lead to premature detection.[52] These premature detections reduce the available range of capabilities available to Russia, if and when it chooses to commit to active hostilities with its adversaries. Whether this sacrifice is an intentional risk or not, it may impact Russia's chance at strategic success in achieving its longer-term goals.

Agility is integral for achieving long-term success. Having the operational capacity to pivot to different challenges, adversaries and circumstances ensures that a force can respond appropriately and win across different theatres. While Russia has made vast strides in modernising its military forces, agility in its network forces remains lacking. Electronic warfare has been thoroughly integrated throughout the Russian order of battle, in recognition of a Western reliance on networked command and control to facilitate joint warfare;[53] a similar process must occur for OCOs in order to achieve military success against dispersed enemy networks. Instead, most offensive military activities in cyberspace are still pursued independently by military intelligence (GRU), intelligence agencies such as the FSB,[54] and partially affiliated external actors. These often operate under a wider doctrinal umbrella and pursue multiple concurrent goals, and are thus not always capable of agile responses to crises and combat.

Finally, Russian network attacks often lack finesse. The NotPetya worm—widely attributed to Russian military intelligence[55]—spread to thousands of endpoints worldwide in a blaze of destructive data corruption, yet it was not seemingly intended to do so. Instead, the attack vectors targeting a popular Ukrainian software provider,[56] and the geopolitical context suggest that a regional effect was intended. Ending up thoroughly shattering its original operational scope, NotPetya proved to be one of the costliest network attacks in history.[57] Such a brazen attack unduly risks attracting the ire of previously uncommitted parties, either by increasing their support to Russian adversaries or perhaps even directly engaging in countermeasures. Tighter control over OCOs could have resulted in a more localised effect, that subsequently may have sent a more potent coercive signal to the Ukrainian government. Aggressive collateral impact at least partially resulting from poorly developed malicious software merely indicates that the Russian threat may not always be as severe as they aim to reflect. There is a constant tension between high technical capability and variable operational maturity in all Russian operations. There is meaningful variance in capability, agenda, and types of operations between different Russian network operations groups. The SVR appears more heavily geared towards thoughtful, clandestine intelligence collection operations in service of Russian grandstrategy. The GRU's Unit 26165[58] is an aggressive, prolific threat actor likely supporting a range of objectives, including information operations. Unit 74455 appears more focused on OCOs, to include critical infrastructure.[59] Those are just several of the key groups all operating within the Russian ecosystem of network operations.

Interestingly, they are not always complementary in approach, in part due to an institutional competitive culture of intelligence agencies vying for influence and prestige. Increased collaboration between agencies, units, and other actors leveraging their unique strengths would likely contribute to a far more coherent and effective overall Russian capacity to pursue OCOs to successful conclusion.

Russian operations predominantly and consciously skirt the threshold of warfare. Information operations which are largely non-kinetic are attractive towards such aims, allowing significant operational freedom with relatively little risk and high deniability. Such capabilities can be harnessed both in peace and wartime, and prey upon existing weak points within adversary societies. Russian information operations include fomenting anti-liberal sentiment and conservative nationalism, evoking sectarianism, and undermining the democratic institutions and the agencies set forth to defend them. There is no simple metric for success, as national sentiment and its results are difficult to measure. At the very least, however, it appears that Russian efforts are intended to weaken national resolve and shape the political landscape towards a more pro-Russian, favourable disposition.

To describe Russian operational intensity, it may be useful to observe it as a diffusion gradient; the further out an adversary is from Russia's territory, the more dispersed and less overt the measures tend to be, on average. Near-border, former USSR-bloc nations often receive the brunt of aggressive Russian measures, while western European nations are primarily the subject of passive propaganda, disinformation campaigns and intervention in political processes. As such, Georgia, Estonia and Ukraine have found themselves on the receiving end of significant Russian offensive measures meant to support concrete strategic goals. Conversely, the United States, United Kingdom, and other Western nations are often primarily targeted in measures meant to offset democratic resolve as a whole.

It is mostly within the band of territorial conflict that many of Russia's information operations can actually qualify as attacks, or OCOs. Many of Russia's presence-based capabilities are alternatively used for intelligence collection, which is then either weaponised in disinformation campaigns or leaked to damage adversary capabilities. In early 2018 US researchers revealed a significant compromise of US critical national infrastructure by Russian intruders, who appeared to be within the presence phase of their operations; collecting intelligence and traversing networks to establish capabilities that could then be activated when needed. It is difficult to assess whether the very notion of intrusion into critical infrastructure serves as a prelude to a future attack, considering the balance between their value as intelligence targets and the risk involved in attacking them.

As per the five-step model presented in the first chapter, disinformation activities and aggressive leaks of classified intelligence do not meet the threshold of a warfare-level attack. Such operations often already fail to meet the *impact* criteria but are also primarily not conducted for military-strategic *goals*. As such, these are active tools of political coercion that ebb and flow even outside the context of conflict, though they may intensify as hostilities flare. However, presence-based attacks against critical infrastructure—such as the numerous OCOs against Ukraine—are worth inspecting in full. Similarly, high-impact attacks such as NotPetya are important to explore, as they embody the dangers of using an event-based capability within a presence-based operation.

Event-Based Capabilities

Russia is arguably the most publicly prolific government deployer of event-based capabilities. The sheer number of widely attributable offensive incidents between ostensibly Russian elements and adversary nations is unparalleled. Much like Russia's wider predilection towards relying on mercenary and sub-national operators to achieve its military-strategic goals,[60] many of Russia's event-based attacks rely on intermediaries. This in turn exemplifies one of the core ideas of event-based capabilities, as previously discussed—they must be robust, scalable, and intuitive to use by various threat actors and for different activity types.

In mid-2007, NATO had yet to seriously address the threat from offensive network operations. With a relatively subdued Russian threat, and focus on major counterinsurgency operations in rural Afghanistan, the risk seemed comparatively low. For Estonia—a relatively fresh inductee into NATO—Russia had consistently been the primary adversary, especially as the democratic Estonian government whittled down Soviet symbology from its streets.[61] Yet when the government sought to relocate the Bronze Soldier, a statue erected to commemorate Soviet victory in the Second World War, it resulted in an unexpected surge of unrest within the small country. Starting late April 2007, physical protests were accompanied by an in increasingly determined and coordinated offensive cyber campaign against Estonian networks.[62]

Attacks came in waves and were aimed at Estonian government websites and internet infrastructure, seeking to cripple the country's global connectivity. Initially, targeting was sporadic and uncoordinated, with the attack vectors limited to basic traffic flooding tools meant to crudely overwhelm remote servers.[63] By 9 May, the campaign had attracted international botnets which were large for this time, and capable of generating a far more significant and sustained traffic load. Numerous websites were temporarily inaccessible until the attack fizzled within a few days.[64] The attacks were treated by Estonia and its NATO allies as a serious incident and resulted in the near-immediate establishing of the NATO Cooperative Cyber Defense Centre in Tallinn, accelerating a subsequent decade-long process to formulate strategic guidelines on operating in cyberspace.[65] Due to the widespread attribution to the Russian government, the 2007 attacks on Estonia are often controversially hailed as the first observable case of "cyberwar".

The Estonia attacks do not meet a reasonable threshold of warfare. Utilising the 5-step classification model, the attacks sacrificed quality in favour of quantity, although they did indeed seek to affect government *targets*. The *impact* of the attacks was relatively marginal and certainly transient, but it did briefly dent the internet-dependent Estonia's access to online banking and commercial services. Attribution of the *attackers* is murky; while the attacks clearly started in a largely undirected fashion, efforts crystallised into a cohesive effort indicative of more meaningful resources. Yet, it is unclear whether the Russian government's reliance on fomenting public unrest and employing non-government proxies to act on its behalf qualifies this fully as warfare. In the absence of a strong, direct link to government accountability, this incident cannot effectively be responded to under the guise of warfare. The deniability factor firmly underpinning the attacks fits Russian strategy of avoiding direct confrontation and maintain a sub-conflict relationship with its adversaries.

Rather than an indication of Russian cyber-prowess, the 2007 Estonia campaign exemplifies Russia's inability to use network attacks for tangible strategic gain. Observed at a distance, the campaign—while operationally successful at marshalling offensive resources—failed to affect political coercion on the Estonian government and alter their behaviour, at least not positively for Russia. The impact of the attacks themselves was minimal. Internationally, the attacks tightened the alliance between Estonia and its Western neighbours, with the political fallout resulting in a more determined NATO emphasis on cyber defence and the establishment of a dedicated cyber-defence centre in Tallinn itself. To its neighbours, the attacks cemented the perception of Russia as an increasingly proactive belligerent. As such, it can hardly be called a success.

The Estonia campaign and the 2008 Russo-Georgian war are often viewed in tandem. This is understandable, as both events happened in rapid succession, featured a presumed Russian aggressor, and heavily incorporated low-end disruptive network attacks. At the time, even the Georgian government—in one of its post-war official reports—labelled the attacks levied against it as cyberwar.[66] As Georgia was markedly less advanced than Estonia in its internet infrastructure, a sustained barrage of disruptive attacks against it both had less and more impact, in different respects.

There is more evidence of a consolidated effort in the Georgian cyber offensive. Early on, the website "stopgeorgia.ru", among others, was leveraged to offer attack tools and targeting instructions, designed to coordinate efforts.[67] Contributions to the campaign included the alleged operational support of the Russian Business Network,[68] a then-notorious Russian criminal group with indeterminate affiliation to government elements. Attack methodology included a host of low-yield event-based capabilities that were reportedly pre-pre-existing[69] including denial of service botnets, various common injection techniques, and website defacements.[70] The attacks appeared to target government entities; temporarily crippled Georgian internet access; were at least supported by Russian government elements;

seemed geared towards supporting the warfighting efforts, and occurred within the context of broader Russo-Georgian hostilities. As such, whether successful or not, the network component of the Georgian conflict does indeed pass the threshold of warfare.

Examining the results, however, they seem underwhelming. The information component of the Georgian military campaign was arguably effective at limiting the government's capacity to communicate with its citizenry, perhaps even hampering some efforts at organising a defence.[71] Yet these attacks were peripheral at best to the overall war effort. Comparatively limited exposure to the internet and a minimal capacity for command and control enabled joint warfare meant that the potential for generating operational effects through network attacks was inherently limited. Much like Russia's wider strategy towards Georgia, the cyber component proved lacking, under-considered and mismatching to the adversary.

Arguably the most effective strategic element in both the Georgian and Estonian campaigns was the vindication that Russia could aggressively operate in the information space against its adversaries with no notable consequence. Despite determined attribution, and regardless of the disproportionality of the attacks, relations between Russia and the West normalised rapidly after the Georgian and Estonian campaigns.[72] In part, this was due to Russia's aggressive dominance of the information space, allowing it to portray a relatively unchallenged narrative that contributed to the appearance of a "fair" intervention by Russia. As such, it was more a success of their wider information operations effort than any meaningful success the use of OCOs.

The civil war crippling Syria since 2011 resulted in a vacuum of power that attracted numerous regional and global powers, including Russia. This provided an opportunity to field-test new equipment and offensive techniques developed during Russia's modernisation programs.[73] Embattled Syrian president Bashar al-Assad narrowly avoided defeat thanks to direct military aid provided by Iran, Hezbollah, and Russia.[74] Increasingly fragmented rebel groups were propped by NATO powers, including Turkey and the United States. The Islamic State and several other jihadi groups sought dominance over wide swathes of territory. The combination of uncertainty and numerous entities fighting for conflicting agendas has resulted in numerous bouts of combat. Russia's involvement included several relevant cases of electronic and perhaps network-warfare that signalled Russia's capacity to manoeuvre tactically.

Russian doctrine relies heavily on the employment of electronic warfare to counterbalance conventional deficiencies in offensive armament.[75] This stems from an accurate estimate of Russian asymmetries in respect to their Western adversaries. It is also an attempt to deny these adversaries their reliance on network-enabled joint warfare, which leans on their own technological superiority. Such capabilities exemplify just how thin the differences may be between electronic warfare and cyber warfare. They both attempt to influence transmitted information on the electromagnetic and virtual levels, respectively.

In early 2017 several British Royal Air Force (RAF) pilots reported encountering attacks against their onboard GPS, allegedly from ground-based Russian systems.[76] These exploited vulnerabilities in GPS targeting that were designed to defeat the RAF capacity to guide munitions towards Islamic State targets. Yet due to a dearth of details, it is difficult to assess whether the Russian attacks were electromagnetic interference or an actual attack against the RAF aircrafts' processing of GPS data. The former would constitute a well-established electronic-warfare attack vector, the latter is an indication of a mature event-based capability. In any other context, these capabilities could be neatly folded into combat operation use.

Event-based attacks against aircraft GPS subsystems is an intuitive way to incorporate OCOs into military operations. It allows pre-packaging of a repeatable capability into a deployable system, such as an anti-air battery or electronic warfare vehicle. It does not require any significant technical knowledge from the operators, save the requirement that they know when to employ the capability for maximum effect. If the capability in question was generically targeting the GPS protocol rather than exploiting a specific vulnerability in the RAF Typhoon, it could also be conveniently employed against numerous other enemy vehicles and weapons. Tactically, it is a sound investment that may help offset the risk to friendly forces from smart munitions.

A later case in early 2018 involved the Russian response to a coordinated attack against their Hmeimim air base in Syria. This reportedly jihadi-led attack involved thirteen drones seeking either to deto-nate kinetically against their targets or drop bombs overhead.[77] The technique itself was hardly unusual; jihadists had been employing makeshift drones in their attacks in the region for several years before the Hmeimim incident. Yet, the commitment of numerous assets and the coordination of the efforts were surprising. Russian air defence supposedly managed to successfully counter the attack; Pantsir-S1 air defence batteries reportedly attained kills against seven drones, while the rest were forced to land as a result of a "cyber-attack" against the GPS guidance modules.[78]

Commercial GPS modules for drones are widely available, as are various means to jam and misdirect them.[79] In most cases, an actual attack against the module is not necessary. Instead, exploiting the fact that satellite-transmitted GPS signals are weak due to the distance, attackers would simply transmit a stronger competing signal that would then direct the drone to land. While this technically involves the transmission of data atop the analogue electromagnetic layer, it does not constitute an attack against the GPS component itself. Yet, because the compromise occurs due to misfed digital data rather than interference with the electromagnetic signal, it could still be con-strued as a tactical event-based capability. Alternatively, if the drones were disabled simply because the GPS signal was jammed, that would still count as classic electronic warfare.

Finally, perhaps the most interesting Russian event-based capabili-ties are their destructive attack tools, with NotPetya fulfilling a par-ticularly instructive role. In NotPetya, the Russian military intelli-gence GRU operated a joint presence and event-based campaign. A presence-based compromise of the Ukrainian accounting company MeDoc led to the backdoored software being used to launch event-based destructive attacks against a multitude of Ukrainian entities.[80] Both the capabilities and the operations are worth examining in depth.

First, it is important to identify whether NotPetya qualifies as war-fare as per the five-step model. With Ukraine as the intended target and numerous global entities suffering significant collateral damage, it is useful to examine these separately. Starting within Ukraine itself, the *targets* were sufficiently varied and impactful to qualify. The *impact* is straightforward, as the destructive payload wreaked havoc across numerous networks. As the *attackers* have been publicly identified with high confidence as GRU they certainly meet the appropriate threshold of accountability.[81] *Goals* are difficult to assess, but there was a likely strategic agenda to weaken Ukrainian resolve against Russian advances. Russia has invested considerably in operating asymmetrically and indirectly against Ukraine, inflicting a severe coercive cost while committing comparably limited kinetic resources. The NotPetya campaign can thus be attributed to the wider Russian strategic agenda. Finally, the *relationship* between Russia and Ukraine includes bouts of combat and a forceful occupation of the Ukrainian territory of Crimea. As such and when reviewing all five parameters, The NotPetya offensive network operation appears to be an element of cyber-warfare.

It is less evident that Russia engaged in warfare against other nations affected collaterally by NotPetya. While the *targets* remain numerous, the *impact* highly significant, and *perpetrators* the same, neither the *goals* nor the contextual *relationship* between the affected parties and Russia merit observing these attacks as warfare. Irrespective of grievous financial harm caused, the adversarial yet non-hostile relationships between Russia and its fellow nations, alongside the lack of strategic intent in harming them, contributes to the assessment that NotPetya does not qualify as warfare against these countries and entities. Consequently, NotPetya manifests both as warfare and non-warfare depending on the affected party. The global nature of the internet and the ease in which collateral damage is affected mean that similar spillover is likely to recur in future conflict.

Arguably, NotPetya was not strategically useful. While it was immensely impactful globally, sloppy implementation and operational discipline weakened the acuity of the coercive message. NotPetya featured cannibalised protocol exploits, open-source dual-purpose tools and co-opted legitimate modules in its arsenal, creating a crude but effective capacity for rapid network propagation.[82] From a targeting perspective, technical analysis suggests that significant effort has gone into limiting the potential propagation of the malware, only to have these limitations broken by lateral movement between organisational networks instead of just within them.[83] The deceptive attempt at labelling the destructive malware as ransomware almost immediately failed, as it was made apparent that the malware writers had no effective capacity to either withdraw ransom money sent to them or to subsequently unlock encrypted files. Across all operational and strategic parameters, the malware campaign failed to achieve a strategic coercive or deterrent effect.

Russia is clearly intent on incorporating event-based capabilities into its strategy. Coinciding with major political strife, such attacks are wielded with increasing sophistication in order to increase coercive pressure and attempt to weaken public resolve. Some parameters indeed coincide well with the before-mentioned approach for utilising event-based capabilities; they were used to subvert existing asymmetries and target weaknesses seemingly endemic to liberal Western-leaning democracies. The capabilities were robust and sufficiently generic as to be effectively delegated for use by mobilised external parties, thereby increasing plausible deniability. Yet, the overwhelming majority of event-based attacks have failed to achieve their presumed goals.

This is perhaps due to Russia wielding event-based capabilities under circumstances more befitting presence-based operations. Offensive tools meant for limited scope tactical attacks were used en masse to attempt a strategic effect, falling short at doing so. Poor attempts at deception only hindered efforts at coordinating attacks and targeting, which are crucial in event-based attacks. Where event-based attacks thrive at potentially having a localised effect, they were improperly deployed as to create cascading collateral damage at an unprecedented scale. The resulting media attention, scrutiny, public attribution, and international backlash proved antithetical to the limited operational goals originally desired. Instead of diffusing the use of event-based capabilities to external parties, tight operational control integrated into military doctrine could have assisted in augmenting the coercive effect of the attacks. A misunderstanding of how these capabilities could be useful severely attenuated their utility.

Perhaps Russian decision-makers simply do not care; the cost of burning capabilities is perceived to be low and of relatively limited consequence. If the goal is simply to establish Russia as a prolific, aggressive threat actor capable of reaching its target, then it has arguably done so. However Russia internally perceives the outcomes of its attacks, they have often proved detrimental to its longer-term goals. A consistent Russian capacity to successfully deploy its capabilities against an adversary without multiple failures would surely be more alarming to its adversaries.

Presence-Based Capabilities

The combined Russian effort to penetrate networks is pervasive and diverse, spanning over two decades, hundreds of targets and numerous evolving capabilities. Several national agencies have committed extensive resources towards the compromise of adversary assets to promote its grand-strategy. As such, ample evidence exists when examining how Russia engages in presence-based operations. Its aggressiveness and willingness to employ offensive network capabilities reveal several advantages, but also key weaknesses. Succeeding in intelligence operations is one matter, successfully weaponising adversary networks to a strategic benefit is markedly another.

Early indications of Russian malicious network activities can be traced at least as far back as 1996. In those years, operators who were later traced back to Russian IP addresses were ransacking numerous US networks with abandon. The FBI-led investigation into what they called Moonlight Maze revealed that intruders were performing unabated lateral movements between universities, government institutions, and military networks.[84] The elaborate operation took years to purge and necessitated a large-scale counter-operation codenamed Buckshot Yankee by the US team that spawned it. The Agent.BTZ malware used to facilitate the elaborate intrusion campaign was fairly complex for its time, with artefacts from its code linking it to an evolutionary chain that persists even today with the Turla malware.[85] The operational practices, technical acumen, and the tools used to facilitate the breach have all grown greatly since the early days of the campaign.

Moonlight Maze was perhaps the first indication that Russia was willing and able to compromise adversary military networks. It reflected the alarming interconnectedness of US military networks at the time, lack of awareness and best practices towards ensuring their safety, and the hold Russian intelligence persistently maintained over

165

these networks. Even as network operations were in their infancy, several phases out of the operational life cycle—including the preparation, engagement, and the presence phases—were already routinely being carried out by government operatives for a military-strategic agenda. The logical leap separating Moonlight Maze from a presence-based attack was merely the employment of a destructive module that could wipe all infected endpoints, potentially crippling US joint operations capacity.

Other valuable case studies bely the Russian proven record of supporting military operations with network compromise. In 2016, US security firm CrowdStrike reported that malware had compromised a popular Android application which was used by Ukrainian military personnel to optimise firing times for the Soviet-era D-30 howitzer.[86] The application, which assisted in calculating targeting parameters for the artillery, had been bundled with a malicious tool called X-Agent since 2014. The X-Agent malware has been frequently associated with GRU. Public-domain analysis of the malware did not indicate that its operators sought to disrupt the actual calculations, instead gathering targeting intelligence to facilitate subsequent kinetic operations.[87] As before, only a lack of intent prevented the weaponisation of the artillery app compromise; a decision to use the malware solely for an intelligence-gathering objective rather than an offensive one was the sole parameter denying it a role as an instrument of network warfare.

Russia has similarly exhibited an evolved capability for operations against cyber-physical networks. Most notable of these perhaps is the sustained activity against the Ukrainian energy grid. As political strife continues and conflict ensues over disputed territory, Russian activity against critical infrastructure has, over time, become both pervasive and improved in quality. The ostensibly Russian "Dragonfly"[88] campaign exposed by security company Symantec in 2014 was a comprehensive espionage operation targeting industrial control system (ICS) networks, stopping short of operating offensively against them.[89] In 2015, a presence-based operation against the Ukrainian energy grid employing the BLACKENERGY 3 malware did not directly target the supervisory equipment itself. Instead, it achieved its effects through an in-depth understanding of the associated networks, and by leveraging destructive malware against the adjoining corporate network used to oversee the industrial equipment.[90] Three years of operations by various threat groups within the Russian intelligence community have resulted in accrued expertise, technical capability, operational maturity, and intelligence on the Ukrainian energy grid.

All these culminated in the network attacks that temporarily crippled a Ukrainian power substation in December 2016. The CRASHOVERRIDE—or Industroyer—malware proved to be a modular framework that leveraged previously gathered experience to facilitate targeting a variety of established ICS protocols, and generate visible, high-impact effects. CRASHOVERRIDE contained a specific module designed to conduct data wipes on industrial control systems, thereby rendering them unusable.[91] It demonstrated the capacity of Russian operates to successfully complete the four-step life cycle of a presence-based operation. CRASHOVERRIDE incorporated extensive preparation both in targeting and capability crafting, successful engagement with the target, an extensive presence phase with lateral movement towards the critical network, and an effects phase resulting in the desired degradation of the adversary.

CRASHOVERRIDE supports a perspective that positions the Russians as relatively mature but lacking in their ability to create strategically meaningful effects. The limited, high-visibility use of the malware revealed its existence and capacity to researchers worldwide, far before it was able to achieve any meaningful strategic outcome. Where such a presence-based operation could have been leveraged to create cascading failures throughout the Ukrainian energy grid, it instead triggered a localised event of limited operational impact.[92] As assessed, this was possibly not due to a lack of desire, but rather a failure of planning. In the preparation phase of the operation, lack of access to an environment that truly simulated the target environment may have led to incorrectly implemented offensive modules.[93] As a result, where the operation could have resulted in increased coercive pressure or deterrence due to perceived Russian potency in network operations, it instead revealed Russian over-eagerness. A powerful presence-based operation was wasted on use with little perceptible value. While particularly notable, CRASHOVERRIDE is not the only instance in which a presence-based capability was used sub-optimally.

In 2015, the French television channel TV5Monde was knocked offline for eighteen hours. In an unusually impactful network attack, an entity calling itself the Cyber Caliphate assumed responsibility. Cyber Caliphate began posting cautionary posts via TV5Monde's social media accounts, calling for French soldiers to leave territories controlled at the time by the Islamic State.[94] Operationally, it was a remarkable presence-based operation which initially succeeded in promoting the public perception that the Islamic State and its supporters could affect networks on a visible scale. It would have been—finally—a viable case study of cyber-terrorism, were the perpetrators who they claimed to be.

The TV5Monde hack was scrutinised for details in an attempt to unmask the attackers. Subsequent efforts by security companies FireEye[95] and Trend Micro[96] quickly revealed technical indicators linking the operational infrastructure used in the TV5Monde attack to the infrastructure previously associated with the Russian GRU. The initially clever deceptive operation against French critical infrastructure was designed to weaken French military-strategic resolve to operate in the Middle East, but it instead galvanised it. Similarly, the clumsy attempt at deception cemented the notion that Russian operators lack operational maturity in their attacks. While the French government and its allies have not directly responded to the hack, it undoubtedly further signalled that the Russians are an aggressive adversary worth defending against. Overall, the strategic utility of the hack proved minimal, and perhaps even orthogonal to the Russian agenda.

The strategic inclination for deception by way of network attacks is not limited to the TV5Monde hack. As the Winter Olympics in PyeongChang ramped up in early 2018, the organisers found themselves on the receiving end of a well-planned disruptive network attack against their infrastructure. As networks faltered, drones were grounded and the official Olympics website was disabled,[97] security companies worldwide scrambled to analyse the available forensic data and produce findings. Security company Intezer quickly pointed out code similarities between the "Olympic Destroyer" malware and previous campaigns conducted by groups affiliated with Chinese intelligence.[98] Other signs suggested the malware used originated in North

Korea. It soon became clear that neither indicators of attribution were reliable.

Further research into the Winter Olympics attack revealed that the forensic evidence was likely planted by the attackers as a "false flag" to misdirect investigators. Russian security company Kaspersky claimed with confidence that the attackers deliberately sought to impersonate North Korea by falsifying a technical fingerprint associated with North Korean network intrusion operators.[99] Cisco's Talos published additional information claiming similarity between the Olympic Destroyer malware and OCOs that previously targeted Ukraine.[100]

Irrespective of the degree of attribution to Russian state involvement, the deceptive campaign failed. Tell-tale signs of purposeful misdirection were discovered within days, rendering the effort inert. Rather than committing to impersonating a single attacker, the malware developers instead borrowed components from several. Those too fell apart under scrutiny, indicating a lack of capacity to fully produce malware convincingly capable of impersonating another.

It is further unclear what the underlying goal was in the Olympic Destroyer campaign. One curious hint emanates from the lack of effect rather than its presence. While disruption did take place, researchers suggested that the destructive capacity of the operation was far greater than executed, indicating that operators were perhaps interested in political messaging more than wanton destruction.[101] Like previous Russian operations, Olympic Destroyer fell short of achieving its stated goals and contributing to the Russian strategic mission. The underwhelming results of destructive OCOs should be alarming to Russian strategists. On several occasions, including NotPetya, Olympic Destroyer, and TV5Monde, operators came dangerously close to—or passed—the threshold of warfare against several global adversaries. That is a high degree of risk for a comparatively low-value potential. Considering the numerous mistakes and lacklustre operational security exhibited in these attacks, Russia should perhaps treat OCOs with far more care. Proper integration into strategy, tighter oversight, and dedication of requisite technical resources could both help reduce the risks and increase the odds of success for each one of these campaigns.

As all presence-based operations start off as intelligence campaigns, those too are instructive. Chapter 1 noted the elaborate intrusion campaign that compromised remote IT management company SolarWinds in 2020. Rather unusually for such a high-profile incident, none of the private sector intelligence companies that had analysed the malware and its associated telemetry chose to attribute it to any known threat actors. The primary publicly available source of attribution came from US Secretary of State Pompeo, who noted "We can say pretty clearly that it was the Russians that engaged in this activity".[102] In the absence of disclosed evidence, it is difficult to corroborate US attribution. Yet if the designation of the Russian government as responsible is correct, it merits analysis.

The SolarWinds breach was a long-term, thorough supply-chain operation granting the intruders access to possibly thousands of organisations. Some of SolarWinds' users were immensely valuable targets by their own right, including numerous government agencies.[103] Other targets provided fresh operational opportunities in their own right, by providing access to other service providers used by still more organisations. This was exactly the case with Microsoft, who confirmed their source code was viewed by the perpetrators of the SolarWinds breach;[104] the attackers had used the breach to pivot into Microsoft's networks.[105] Only the aggressive breaching US cybersecurity company FireEye eventually resulted in the full unravelling of the SolarWinds campaign.[106] While the operation could not be classified as an attack, the wasteful loss of a campaign due to overzealous targeting is emblematic of the Russian approach; bold, capable, and aggressive, but perhaps so aggressive as to overshoot success.

Finally, it would be irresponsible to discuss Russian OCOs without mentioning the extensive Russian campaign in the runup to the 2016 US presidential elections. The elaborate multi-pronged campaign featured numerous moving parts, including a network breach of the Democratic National Convention, a large-scale disinformation effort against the US public, and several leaks targeting politicians perceived to be hawkish and anti-Russian, such as Hillary Clinton and Victoria Nuland. While the nuances of the campaign were intricate and exceed the scope of this work, there are several key takeaways that are pertinent towards understanding Russian OCOs and their wider approach to information operations.

The first is that the elections campaign definitively demonstrated Russia's willingness to thoroughly violate another nation's sovereignty through network activities. By targeting national elections, Russian decision-makers chose to compromise the core tenet of a democracy, a feat seemingly guaranteeing some form of retaliation. Yet even as it did so, Russia committed to operating at a sub-conflict level, leveraging soft tools and influence operations to achieve its goals rather than directly attacking voting infrastructure itself and altering results. As previously assessed, the elections hack was a significant breach of US sovereignty but ultimately not a component of warfare due to the purely political goal, absence of direct offensive impact, and lack of a pre-existing conflictual relationship between the US and Russia. This is in stark contrast to the reported 2014 attacks against Ukrainian voting infrastructure, that attempted to overly cripple voting and alter the outcome of the election.[107]

The second takeaway is that while Russia arguably succeeded strategically,[108] it still failed operationally. Deconstruction of its various operations led to high-confidence attribution by both private sector researchers and the US government itself.[109] Russia's attempt at operational deception to thwart attribution efforts was poor quality, falling apart under even minor scrutiny by journalists.[110] Russian efforts almost single-handedly resulted in an immense international surge of effort towards countering information operations, hampering future efforts. In this sense, it almost seemed as if the Russians were surprised by their own success; the operations were designed to weaken US resolve en masse rather than be individually successful in impacting the course of US politics. Perhaps it was the maelstrom of existing American social issues and political grievances that created a turn of events in line with Russian desires.

Russian success was arguably overly reliant on luck and circumstance, but it did not need to be so. As before, tighter integration into decision-making cycles and a comprehensive doctrine guiding operators in their actions could have muddled investigations, hampered high-confidence attribution, and prevented galvanisation of the US Congress against perceived Russian interventionism. Russian information operations once again exhibited a significant resistance towards integrating the tactical, operational, and strategic elements.

Joint Operations

Russia is one of only few nations poised for success in realising the potential of OCOs. An aggressive pursuit of information operations, a willingness to engage in controversial behaviour, a relative lack of restraint by its intelligence agencies, and a storied history of manipulating information and perception all make Russia an incredibly prolific operator. Yet despite having all chances of success and the perception of unstoppable campaigns against the heart of Western interests, it routinely falters in its ability to achieve its goals through network activities.

Russian OCOs fail to achieve what is expected of them throughout their operational life cycles. In the initial preparation phase, they do well to develop offensive technical capabilities, but then fall short of crafting credible deceptive identities. They similarly do not dedicate enough resources to the prevention of infrastructure reuse, which could hamper subsequent adversary attribution efforts. Throughout their presence phase, operational security is often lacking, leading to premature compromise. In the final effects phase, they often activate their offensive payloads in poor form, resulting in both limited lasting impact, loss of sensitive capabilities, and even the occasional accidental reveal of operational intent. In other cases, a misapplication of force results in severe undesired collateral damage, perhaps outstripping the utility of the operation itself. These limitations are a deciding characteristic of both their presence and event-based efforts.

Russia has integrated event-based operations surprisingly well into military-political conflict, yet these operations do not contribute sufficiently to their tasks. In both the Georgian and Estonian conflicts, event-based attacks against adversaries were occurring daily alongside additional kinetic and diplomatic efforts. The Russian ability to craft, disseminate and facilitate targeting of pre-packaged, resilient event-based attack tools is a positive indication of its capacity to muster forces. Yet these forces have been applied in a manner incongruent with Russian strategic aims, contributing minimally to the overall efforts. However, there is indication of improvement as Russian event-based activities in Syria appear to be more effective and integrative.

If the Russians were to treat event-based OCOs the same as they treat electronic warfare, far more promising results could be yielded.

As researchers often suggest, Russia has one of the most elaborate, well-crafted and dangerous electronic-warfare capacities globally.[111] This is in part due to thorough integration through its order of battle, including within infantry battalions and other mobile force structures. Considering their numerous characteristic similarities, network warfare could benefit from a comprehensive integration doctrine similar to that of electronic warfare. This could result in a far better application of network warfare towards military needs.

Russian presence-based operations have demonstrated a highly evolved capacity against technically complex targets, including those operations that require assistance from subject matter experts. Presence-based malware frameworks have proven to be modular and capable of exploiting a variety of targets towards achieving high-impact events. Experience in compromising military and critical infrastructure targets spans at least two decades, suggesting a substantial maturity in pursuing such adversaries. As such, Russia is uniquely positioned to be successful.

What Russia lacks is strategic utility. Presence-based operations have often been used as a form of hazy political signalling, or at times thinly veiled strategic misdirection. Both have resulted in a decidedly underwhelming contribution to Russian interests, instead either consolidating adversary support, providing crucial insight into Russian capabilities, or compromising valuable offensive tools. Increased discipline and congruence with a broader military or even political strategy could have made far better use of these tools for a longer-term impact.

The overall issue with Russian OCOs is their failure to apply core strategic principles. While they do well to target adversary centres of gravity in the form of critical infrastructure, military targets, and even the population itself, they falter on other guiding principles. Deception is often poorly exercised, and often used when it is either subject to immense scrutiny or altogether unnecessary. The indirect approach is often avoided in favour of tangling directly and overtly with well-defended enemy assets. Russian brashness and under-calculated aggressiveness in its OCOs demonstrates low contemplation of conservation of force. The consequence of this approach is a decidedly low success ratio, and a broad failure to strategically offset the existing military asymmetries. Most importantly, Russia's approach to OCOs increases the risk of armed conflict instead of reducing it. Rather than being an effective component of reflexive control, crude attempts at misdirection in the face of aggressive attacks against critical infrastructure are dangerous missteps, which could eventually cross an undesired threshold. Successive attacks that face consequences increase the onus of response on the affected parties; NATO, or even the US on its own, would eventually be compelled to respond.

Russia would do well to learn from its own history. Soviet information operations are notorious, and while some were more effective than others, the overall blanket of disinformation effectively hid Soviet conventional deficiencies and strained global alliances. From the military angle, a thorough commitment towards integrating electronic warfare resulted in asymmetry-impacting capabilities and a deterrent that persists to this day. Learning from history, committing to the operational life cycle, and cautiously integrating OCOs into existing doctrine could provide Russia with the tools to uniquely upset an adversary's capability to resist its influence across all phases of conflict.

7

ASSERTING CHINESE DOMINANCE

The waters of the South and East China Seas teem with geopolitical friction. The seas represent a crucial nexus for maritime international trade, a concentration of valuable natural resources, and a series of land features that—when militarised—threaten a wide swathe of territory. To the nations straddling this tight space, the seas represent both an unparalleled opportunity for regional influence and an implement of sovereignty. Through centuries of conflict precipitated by both foreign and domestic forces, an uneasy status quo emerged in the latter part of the twentieth century. While many elements of this balance are increasingly tenuous, one of its most explosive aspects is between two nations that were once one. Both the People's Republic of China (PRC) and the Republic of China (ROC, or Taiwan) once viewed themselves as the "true China". Yet where the former now seeks reunification, the latter increasingly pulls towards independence.

This trajectory puts both nations at dangerous odds. As the two countries drift further apart, options for eventual resolution increasingly narrow towards a possible military scenario depicting a Chinese attempt to forcefully reclaim Taiwan as its sovereign territory. The probability of such a scenario is neither remote nor improbable. Taiwan appears to be gradually distancing itself from the notion of peaceful reunification; in 2016, voters in Taiwan gave the Democratic Progressive Party (DPP) its first ever legislature majority. In 2020,

Taiwanese voters reiterated and intensified their support for independence, in part due to tensions with mainland China and its mounting pressure against Hong Kong sovereignty.[1] At the same time, an empowered PRC shows increasing signs of regional assertiveness and a keen desire to pursue its goals militarily.[2] Should it ever take place, a cross-strait conflict would have far-reaching consequences. It would inherently involve numerous regional forces, including South Korea, Japan, and perhaps Taiwan's closest remaining ally—the United States.

A Chinese amphibious invasion—a necessity to achieve "armed unification" of Taiwan—would be among the most complex military manoeuvres to pull off successfully. Such a campaign requires elaborate logistics, effective mobilisation of landing forces, and rapid suppression of an entrenched enemy engaged in an existential battle for its own territory and identity. Time works against the attacker. Should an initial landing effort fail, defenders would have the opportunity to rally, marshal resources to subdue intruders, call upon regional allies to aid, and affect international diplomatic pressure. When US President Joe Biden was asked in October 2021 whether the United States would come to Taiwan's aid should it be attacked by China, he responded "Yes, we have a commitment to do that".[3] A Chinese invasion of Taiwan is likely to trigger a significant response from the United States, a Taiwanese supporter in all but treaty.[4] This may result in rapid deployment of a US carrier strike group to act both as a deterrent to hostilities, and a potential mobile strike force should conflict indeed commence. China would therefore have two key goals; subdue Taiwanese forces as rapidly as possible and prevent third parties from effectively deploying and contributing to the war effort.

The Chinese People's Liberation Army (PLA) is well aware that rapidly subjugating Taiwan is a monumental feat.[5] Achieving this would require Chinese mobilisation on a previously unseen scale. A crucial component of the effort would include overcoming entrenched Taiwanese defenders and either deterring or defeating US interdiction forces. Only complete strategic success can enable this, and as such the PLA is investing heavily in the facilitation of its potential victory. One notably novel element of this effort is the 2015 establishment of the Strategic Support Force (SSF), created to "maintain local advantages in the aerospace, space, cyber, and electromagnetic fields" while also providing "attack and defense in cyber and electromagnetic spaces."[6] The enigmatic foundation of the non-frontline SSF was accompanied by the understanding that network warfare must also be conducted by combat deployed troops at all levels,[7] thus splitting the responsibility between strategic and operational needs.

This chapter will argue that China has developed a doctrinally mature approach to conduct effective military OCOs, but it currently lacks crucial operational experience. The foundation of the SSF, which unifies operational, technical, and electromagnetic expertise from various areas of the PLA, is a recognition of both the high potential value of OCOs and the underlying difficulties in employing them effectively. This is followed by the understanding that even as the PLA rapidly increases its conventional capabilities, it continues to face a daunting challenge from both Taiwan itself and regional US reaction forces. The PLA must therefore lean heavily on strategic principles enabled by cyber operations; attacking via the indirect approach, minimising disadvantageous asymmetries, achieving operational surprise, and directly targeting centres of gravity.

The Taiwan contingency is a lens through which to observe Chinese capacity for OCOs. By examining the unique strengths and weaknesses of its enemies alongside the developments the PLA has pursued to counteract and exploit them respectively, we can in turn glean a better understanding of the potential utility of military OCOs within a significant Chinese-led military campaign. Subduing Taiwan is by no means the only possible use case for Chinese OCOs; it is simply a helpful exercise in the absence of meaningful, concrete examples.

The Chinese Civil War was a violent insurrection of the Communist party against the reigning nationalist Kuomintang party. Spanning more than two decades from 1927, the resulting conflict stretched into and beyond the Second World War. Fierce fighting devastated the Chinese mainland before Kuomintang forces were largely routed by their Communist counterparts in 1949. Loss of the mainland resulted in a full Kuomintang retreat to the island of Taiwan, where they re-established a breakaway government optimistically named the Republic of China. With victory now in hand, the Communist party re-carved the battered mainland in its image, establishing the People's Republic of China.

Both parties initially claimed that they were the "true China". The contest over legitimacy soon became a linchpin of the Cold War, which unsurprisingly saw a Soviet-backed PRC counterbalanced by a US-supported Taiwan. However, as years progressed and the Chinese mainland recovered, it became clear that Taiwan's desired global recognition grew distant. Eventually, while the US maintained its stalwart partnership with its island ally, it officially recognised the PRC as China, a significant move reinforcing the mainland's claim on the UN Security Council seat. China and Taiwan continued to co-exist in a state of pervasive friction. It is a zero-sum game in which political support for one country means denouncement of the other.[8] The two nations remain inexorably tied to each other's spheres of influence as both seek to define a political narrative on their own terms. Relations between the PRC and the United States similarly flare hot and cold over Taiwan, with the US seeking to retain its influence over the pro-Western strategically-located island.

Throughout the 1990s, an assertive US presence in the Pacific saw Taiwan thoroughly enmeshed in an alliance with the US, further straining relations between it and the mainland. This escalated several times to the brink of armed conflict, culminating in the 1996 Third Taiwan Straits Crisis. This resulted in two US carrier strike groups responding to the area in response to PLA missile tests, and a subsequent PRC de-escalation by way of concession.[9] It was an educational moment for the PLA; it poignantly internalised that conventional military parity was—at the time—infeasible against the highly coordinated, advanced US military. New approaches had to be developed, and new capabilities attained rapidly.

The bilateral relationship between the PRC and the US has ebbed and flowed over the last five decades. Previously seen as a Soviet-era adversary, relations gradually warmed as the Chinese economy increased its interdependency with the global market, standing out as a powerful conduit for trade and manufacturing. This view of China is by no means steady, however. The assertive Chinese push towards establishing soft power and advancing its economic interests, including through its globe-spanning Belt and Road Initiative, has been met with suspicion by many, including the US administration under President Donald Trump. At the same time, the PRC's diplomatic relationship with Taiwan has fluctuated between relative warmth and bristling friction. These trends have largely been commensurate with perceived Taiwanese advances towards a fully-fledged independent democratic government, an unacceptable outcome for the mainland's central communist government. The notion of an independent Taiwan is so offensive to China that officials have frequently and pithily claimed that such moves would result in an overwhelming response.[10]

The PRC's rapid growth has been accompanied by an expected increase in Chinese political expectations and the will to pursue them. In part, this manifests in moves to project power in the South China Sea through various contentious proceedings, while more forcefully asserting claims to the disputed territories therein. Controversies abound, including claims of sovereignty over the Spratly Islands, The Diaoyu islands—a 200-nautical-mile range defined by the PRC as its Economic Exclusion Zone (EEZ)—and of course the island of Taiwan itself. All of these claims are strongly opposed by other regional nations and the United States, who view such assertions as a threat to the freedom of trade and navigation throughout the area.[11] China now claims rights to regulate and enforce all maritime traffic within large swathes of the South and East China Seas. The resulting friction has led to several widely covered entanglements with fishing boats, foreign naval vessels and US military aircraft.[12]

Recent developments include the militarisation of reclaimed islands off China's littoral space.[13] In one such example, the PRC paved a 3,000-metre runway, clearly geared towards hosting large military fixed-wing aircraft.[14] Since 2016, military reclamation efforts of contentious islands—in particular the Spratly Islands—have accelerated, including the construction of runways, island defences, and listening posts.[15] This ratcheting up of adversarial behaviour has been poorly received by regional actors, perceived as part of a PLA grand-strategy to project power, enable rapid military response, and illegitimately enforce claims of regional sovereignty.

These longstanding strategic developments reflect the PRC's active defence strategy.[16] As the 2015 Strategic White Paper explains at length, and subsequent publications reinforce, the PRC does not actively seek conflict but rather swift superiority in the face of perceived grievance.[17] This grievance can take many forms; a military incident, Taiwanese independence, or overt assertions of dominance on disputed territories. It is therefore perceived as strategically desirable to preposition PLA forces for maximum effective deployment in their potential time of need. At the same time, the United States has not remained idle in light of these developments. Over the last decade, the Indo-Pacific narrative within the US and UK has undergone several major revisions to attempt to both contain and accommodate Chinese regional aspirations. These were all bound together in a strategy labelled during the Obama administration as "The Asia-Pacific Rebalance".[18] Since then, the focus had rebranded as safeguarding a "free and open Indo-Pacific", signalling a shifting tapestry of alliances and geopolitical circumstances. With the trade war narrative escalating during the Trump administration and increased military activities on both sides within the region continuing into the Biden administration, the US has increasingly viewed China as one of its primary strategic adversaries.[19]

Evolving For the Information Era

The PLA now fields an increasingly modern and versatile military. Previous notions of focusing on numerical quantity and land-based force mobilisations have largely been cast aside in favour of securing China's littoral borders, ensuring interests within its Exclusive Economic Zone (EEZ),[20] deterring US regional involvement, and potentially enabling coercive unification with Taiwan.[21] This has far reaching implications on the size, nature and manner of PLA force deployment. While the overall personnel capacity is gradually decreasing,[22] troop readiness and equipment quality have been rapidly increasing.[23] China has taken great strides to accommodate modern threats by modernising its doctrine, force building approaches, and strategies. Steps undertaken include reorganising PLA command structure and drastically altering resource allocation.[24] China therefore presents a more promising visage of military readiness, but one which has yet to be battle tested.

Observing the last decade of Chinese doctrine strongly suggests an evolution of priorities. Namely, it has gradually been transitioning from the "People's Army" approach towards a more quality-centric, network-aware, joint operations grand-strategy. Its new approach recognises the qualitative advantages of adversaries such as the United States or Taiwan, and appropriately attempts to alleviate any such asymmetries.[25] This observation has been accompanied by the realisation that the next conflict is far less likely to be a massive land-based battle,[26] but rather a chain of smaller engagements that take place off mainland China's shores.

Active defence is a significant component of PLA doctrine.[27] It outlines that the PRC will supposedly never be the initiator of armed conflict, while retaining the right to proactively act against perceived threats should the need to do so arise. This rather murky language is open to interpretation and leaves a measure of leeway to PRC politicians when directing the use of force. A perceived grievance in the form of a bold Taiwanese step towards independence could readily be interpreted as an opening salvo,[28] one breaching existing agreements and therefore inviting direct military countermeasures. Alongside China's increasingly aggressive manoeuvres in its littoral seas, it is acknowledged by observers that the PLA continuously strives towards enabling armed unification if need be.[29]

PLA references to weaponising information are no idle chatter. Chinese forces have been aggressively working to advance joint warfare capabilities alongside the capacity to target those in their adversaries.[30] Similar to the Russian and United States' militaries, the PLA has identified that the added value of joint operations results in tremendous benefits to the modern force; they must therefore both catch up to their adversaries[31] while seeking to deny them those same advantages. At the heart of this process is the primacy of information as the medium permeating all warfighting. It is a massive potential benefit to those who wield it effectively, but can also create crippling dependencies and deep systemic vulnerabilities. The latest version of the highly influential official Chinese publication *Science of Military Strategy* reversed the PLA's previous tendency to deny the use of offensive cyber operations.[32] Instead, it both acknowledged and embraced OCOs as a potential differentiator in the modern battlefield. More literature soon followed specifically discussing aspects of OCOs and their role within military strategy.[33]

The PLA's new Strategic Support Force represents the crystallisation of its new approach to informatisation.[34] By consolidating aspects of military network operations into a unified entity, the PLA has acknowledged both the significance and difficulties in pursuing such capabilities. The new order of battle includes, among others, elements from the First Department (operations), Second Department (intelligence), Third Department (technical reconnaissance), and Fourth Department (electronic countermeasures and radar).[35] China's approach seems to echo two key trends in their observation of cyber warfare; that information superiority permeates all operational domains,[36] and that deployed forces must be capable of pursuing network operations themselves.[37] By seemingly assigning responsibility for operations of strategic worth to the SSF, while allowing other forces to conduct tactical cyber operations to achieve local goals, they in fact assign responsibilities roughly analogous to the division between event-based operations and presence-based operations respectively. This uniquely positions the PLA as comparatively doctrinally mature force when it comes to the integration of OCOs.

China's greatest challenge in conducting effective OCOs is its lack of experience. This dearth manifests in two complementary aspects and a corollary. The first aspect is a lack of experience in waging combat operations at all, especially as a joint force relying on networked assets. While PLA military exercises and strategy increasingly integrate joint operations,[38] unlike the United States or Russia it has precious little experience in fully engaging a committed adversary. The second aspect is its lack of experience in conducting offensive network operations, even though the PLA and its civilian counterparts are notoriously prolific employers of network espionage operations.[39] There is little publicly available evidence suggesting that military network forces are routinely engaging in attacking networks to promote strategic-political goals. As a corollary to these two points, the PLA has little experience in applying OCOs alongside kinetic combat operations against an adversary. This places the PLA at a relative disadvantage, as operational maturity is accrued by experience. Practice makes perfect.

By radically altering its approach to information operations without field-testing the process, the PLA lacks visibility into its own faults. Establishing the SSF and assigning information operations to deployed forces is promising, yet it is unclear if the PLA is capable of achieving any of its stated goals or how it intends to do so. The force structure is new, the technologies they will employ are at least in part novel, the challenges they face are fresh, and the adversaries are highly capable. To put it mildly, the degree of uncertainty surrounding any Chinese military action is immense. The Taiwan contingency shines a light on the myriad challenges PLA planners and operators are likely to face when attempting to apply OCOs to an intricate scenario where they could be immensely beneficial.

It is a momentous challenge to attempt strategic surprise in a Taiwan contingency. Taiwan routinely prepares for this eventuality, with most of its military forces uniformly dedicated towards curtailing it. Project 2049's Ian Easton similarly acknowledged that "China's leaders have good reason to assume their intentions will be discovered by Taipei well in advance of the attack. Strategic deception is viewed by the Chinese military as desirable, but probably not attainable. Tactical deception, on the other hand, is seen as vital".[40] A pre-emptive Chinese "active defence" manoeuvre would likely initiate escalation of hostilities.[41] Such a scenario is likely to be predicated by a perceived grievance inflicted by Taiwan's government; either an overt policy shift towards independence, or a lesser political conflict spiralling out of control. Escalation of hostilities may indeed be limited to political exchanges or sabre-rattling, providing measures to depressurise tensions are successful. However, should the PRC assess the timing is right to attempt an overt military operation, the PLA will seek to swiftly overwhelm Taiwanese forces before regional US assets have time to mobilise and respond. Subsequently, the overall strategy sought out by the PLA would be to solicit a Taiwanese surrender as swiftly as possible. Delay would entail far greater losses, regional unrest, ally mobilisation and international pressure. Surprise, albeit difficult, would certainly help delay US intervention and disrupt Taiwanese mobilisation.

PLA planners are well aware that the US is a major roadblock to any potential military campaign in the Pacific. Capably trained, experienced, and well suited to joint operation of massive firepower at relative accuracy, The US military is a difficult adversary to face directly and conventionally. The US maintains one of its heftiest armed presences in the Indo-Pacific as part of its obligations towards world security, and safeguarding economic and geo-political interests while honouring commitments to regional alliances. This deployment includes fixed assets, mobile platforms and bolstering the operations of friendly forces in the region.[42]

The regional US presence is overseen by the United States Indo-Pacific Fleet Command (INDOPACOM). It in turn strategically directs the various regional components, including some relevant to the contingency; forces in Guam, Japan, South Korea, Taiwan, and most critically the US Seventh Fleet. The various combatants enable a host of varied fire and support roles. These include the decidedly offensive, such as missile cruisers; superiority missions such as fixed-wing aircraft operating from both carrier and bases, and the deployment of special forces. Sea-based assets are complemented by land-based fixed resources, primarily operating out of Japan and South Korea. The largest of these is the Kadena Air Base in Okinawa, hosting the US Air Force's 18th Wing alongside other personnel topping 20,000 active members. The aggregation of various US capabilities in the region translates to a sprawling defensive posture spanning multiple locations and capabilities.

US forces operating in the region benefit from an intricate mesh of interconnectedness between the various fielded elements. From tactical units comprised of infantry, warplanes, surface combatants and submarines to higher echelons such as forward operating bases, regional commands and continental agencies, all are intertwined through the various implementations of the Department of Defense Information Network (DoDIN). The network[43] serves as a compartmentalised, multi-tiered and multi-protocol communication grid tasked with facilitation of all manners of data transfer.[44]

There are dozens of military networks of varying roles in constant use by US forces. Some are globe-spanning networks such as the unclassified NIPRnet, the classified SIPRnet, or the top-secret intelligence-sharing JWICS network.[45] Others are specific to the region of operation, such as the Joint Tactical Information Distribution System (JTIDS), used by US forces and their allies for communication between military assets.[46] Others still are localised to either a single strike group or a limited operational frame, including closed radio groups and internal dedicated datalinks. The transmission mediums for such networks are as diverse as the networks themselves, ranging from military satellites, commercial satellites, and terrestrial radio to fixed fibre-optic links. The overall interconnectivity of these networks is a convoluted patchwork of connections, constructed piecemeal over the decades to support increased requirements for joint force operations.

From China's perspective, the complete destruction of US forces in the region is both unlikely and unnecessary. Rather—and this approach is sponsored by PLA doctrine[47]—the optimal solution is an indirect one that inhibits US ability to intervene in a timely manner. This would be accomplished through three primary components: (1) degrading US military joint operations, effectiveness; (2) deterring the US from mobilising due to overwhelming odds of casualties; (3) directly reducing the available capabilities and assets of the adversary. This strategy has been labelled "Anti-Access/Area-Denial" (A2AD) in some US official and academic publications.[48] As there is currently no conventional military parity between both actors, the PLA has stated it will seek asymmetrical advantages by way of accurate missile capability, anti-satellite weaponry and indeed, cyber-warfare. Each of the above objectives stands to benefit from a pre-prepared well-crafted offensive network capability, arguably a significant symmetry equaliser. Notably, the US claims to have a "suite of capabilities" allowing it to continue operating within A2AD restrictions, likely a reference to OCOs as well other means.[49]

Offensive cyber operations only partially fit an A2AD strategy. On one hand, it is viable to deny adversary access to a region, and freedom of action within it, through aggressive event-based attacks against its networks. Event capabilities are meant to be reusable and are therefore more robust, thereby implying a persistent threat of denial to would-be intruders. In contrast, presence-based operations are principally carried out afar from the battlefield; thus the geographic aspect inherent to A2AD applies less to presence-based operations. They may be used de facto to facilitate area denial of adversary forces, but such attacks do not as neatly conform to a "threat radius" as other A2AD components often do.[50]

Degrading the joint operational capacity of a thoroughly networked adversary entails striking at communication nerve centres. For a modern military force, these are embodied by "Command, Control, Communications, Computers, Intelligence, Surveillance, Reconnaissance" (C4ISR) nodes. These are information hubs tasked with coordinating the operation of local military assets, absorbing and disseminating intelligence, conducting mission tasking, locating and designating targets, and redirecting assets as required. While modern military craft are more than capable of operating alone or within their tactical frame,[51] they are surely to be outmatched by even a relatively inferior opponent whose moving parts are working in relative operational harmony. PLA strategy recognises the immense value embodied by C4ISR as a target,[52] specifically one that—when successfully compromised—would potentially reduce enemy preparedness.[53]

Offensive network operations offer an enticing promise, provided they are employed effectively. OCOs provide a wide range of options, contrary to the rather straightforward, destructive qualities of kinetic weaponry. Even the US Secretary of Defense has acknowledged the viability of OCOs in limiting the US military's ability to respond to threats, claiming in 2019 that "malicious cyber activity puts us at risk by eroding our capabilities and disrupting our ability to operate once conflict ensues".[54] There is also plenty of motivation not to pursue attacks even if useful networks are breached. Compromising C4ISR for surveillance rather than kinetically attacking it affords unparalleled enemy situational awareness. As command and control nodes concentrate combatant activity in the region, the operational intelligence value of compromising them may be crucial to both understanding and countering enemy deployments. By turning a network operation into an attack, an adversary can possibly attain a meaningful battlefield advantage. Whereas physically destroying command systems will immediately trigger countermeasures and lead to conflict escalation, OCOs may enable subtle disruption or manipulation of crucial systems.[55]

Overpowering Taiwan

Taiwan as an adversary presents numerous challenges and opportunities for OCOs. On one hand, historically limited access to military equipment acquisition and a reliance on ageing platforms means that the networked attack surface is significant. Persistent use of the same technologies, networks, and systems leaves neighbouring China with plenty of opportunities to both develop presence-based operations and conduct research and development on event capabilities. Conversely, that same reality means that Taiwan's military has physical redundancies and are likely more capable of conducting combat operations even bereft of networked command and control.

Against the Taiwanese military, the stated PLA doctrine could seek to severely degrade anti-air assets, cripple command and control capabilities, disable defensive assets such as aircraft, ships and missile sites, and decapitate the military and civilian hierarchy. The opening salvo is designed to facilitate follow-up operations and erode the Taiwanese military's capability to resist the more casualty-prone phases. A successful opening salvo has the potential to significantly shorten the duration of conflict and reduce PLA losses; both elements are defined as highly desirable by PLA leadership. To achieve this, the PLA is investing heavily in methods and capabilities that would increase the potency of the early stages of conflict, including ballistic missiles, amphibious landing hardware, air power, and cyber operations.

Some presence-based operations could yield returns before the shooting starts. Due to the limited size of its defence-industrial complex, Taiwan presents a fairly small supply chain attack surface. Many of the national security research and development programs are spearheaded by the National Chung-Shan Institute of Science and Technology (NCSIST). Considering China's proclivity for targeting adversary defence contractors,[56] it probably attempted to similarly target the NCSIST. Successful compromise could result in intimate access to Taiwan's homegrown technology, with the added potential benefit of introducing vulnerabilities that could be exploited in conflict time, even against air-gapped networks. The notion of bundling malicious software with equipment is not new to China; it has been previously implicated in a variety of supply chain compromises that either include its own hardware or tampering with existing components.[57] Thus, the operational expertise theoretically exists, yet it remains unclear whether China could wield it to impact Taiwanese defence networks.

A kinetic approach would rely on a mixture of shock operations and massive mobilisation. Extensive literature has been written on the PLA's capacity to effectively render airbases out of commission.[58] Strike vectors would include a high volume of short-range ballistic missiles (SRBMs), land-attack cruise missiles (LACMs) and air-launched cruise missiles to overwhelm defences and penetrate hardened aircraft shelters, while simultaneously blanketing runways with specialised munitions designed to disable take-off capabilities. This brute-force approach stands to be highly effective against exposed targets, but Taiwan's armed forces have expected it for decades, and have appropriately developed countermeasures of varying usefulness. These include several highly resistant sprawling military command centres built into various mountainsides. The mountainous topography of Taiwan allows for effective diffusion of command and control, which then cannot be easily knocked out by ballistic missiles or air attacks. Assessments suggest over 2 kilotons of conventional yield would be required to breach compounds such as the Hengshan Military Command Centre in the Yuan Mountain,[59] a momentous task even for the burgeoning PLA Rocket Force.

OCOs are uniquely advantageous in this instance by subverting physical barriers. In order to effectively serve as command and control centres in wartime, hardened complexes must be networked both to military platforms and other installations. As a result, they can be compromised via their datalinks from an external source, through a supply chain compromise, or even internally by a sympathetic soldier infecting internal nodes. A successful presence phase could potentially cripple Taiwanese defensive coordination. As previously explored, the preparation phase for such presence objectives can be pursued in peacetime, as it may take months or years to penetrate, manoeuvre, and weaponise adversary command and control networks. Presence-based capabilities may offer incremental benefits by way of gleaned intelligence on Taiwan's order of battle and force disposition.

Presence attacks could also offer early benefits against Taiwanese integrated air defence systems (IADS). The efforts to defend Taiwan's skies must be rapidly subverted to enable PLA air operations and protect amphibious landings. Taiwanese air defence platforms include an interconnected mixture of domestic and foreign-acquired hardware; Patriot batteries, short-range Antelope systems, and Tien Kung (Sky Bow) systems. As per a 2016 RAND report, while the PLA Air Force is increasingly capable of rapidly subduing its ROC counterpart, air defences potentially present a greater challenge with high attrition rates for attacking PLA aircraft.[60] Presence-based capabilities could temporarily interfere with air defence systems by partially disabling them or otherwise degrading the situational awareness they provide, allowing PLA aircraft and missiles to penetrate and defeat radars and missile batteries. Such an attack could yield a strategic benefit, but also requires significant skill and resources; it is unclear if China could successfully execute all phases of such an intricate presence-based operation. Penetrating secure military networks, remaining covert, and developing intricate offensive tools meant to impact specific military software and hardware is no mean feat. It requires familiarisation with the platforms affected, technical skill to develop both exploitation and effect capabilities against them, and operational finesse to avoid exposing the presence campaign prematurely.

Beyond the initial phases of the campaign, options for OCOs become somewhat more limited against Taiwan's forces. With the desire to shorten the conflict, it is likely that any sensitive military networks within the PLA's grasp would already be attacked in the opening phases of conflict. If successful, Taiwanese forces might be in disarray and scrambling to wipe clean their affected systems and networks, but generating new presence attacks might become more difficult. This does not immediately imply, however, that targeting networks cannot assist PLA efforts to attain air superiority and forge a path for expeditionary forces to successfully establish beachheads on Taiwanese soil. Furthermore, a body entrusted with metered use of offensive cyber operations such as the SSF may choose to reserve some presence-based operations for this phase of the campaign. Overall operational opportunities are still plentiful, including affecting Taiwanese systems meant to provide situational awareness. Impacting these could assist incursion forces in surprising entrenched defenders and keeping them in disarray. Event-based operations could be activated to isolate pockets of resistance as these are targeted by

PLA forces, providing cover for kinetic operations and degrading defensive capabilities.

Alternatively, event-based capabilities against Taiwanese military equipment may retain their potency throughout the course of the conflict. A few observations work in China's benefit in this regard. The first is that Taiwan relies heavily on US-made equipment such as the F-16 multirole fighter. As the F-16s that US forces actively use and their Taiwanese equivalents share technical characteristics, event attacks developed against US communication protocols may work well against their Taiwanese counterparts.[61] Second, while Taiwan's modernisation programs continue apace, they are hampered due to political sensitivities that often prevent the US from significant sales of high-quality offensive platforms,[62] even if those sensitivities have subsequently waned during the Trump administration. As a result, Taiwanese forces are mostly limited to updating increasing quantities of ageing systems such as the F-16[63] and the M60 Patton tank. This means that event-based capabilities developed against Taiwanese equipment may retain their potency for an extended time, allowing the PLA to gradually accumulate an arsenal of such options. Lastly, indigenous platforms such as the F-CK IDF-1 fighter aircraft would need to use systems and protocols compatible with their US-made counterparts in order to facilitate joint operations, thereby extending some of the event-based attack surface to them as well.

Directly attacking civilian infrastructure might also become an attractive option. While direct-fire resources are embattled with Taiwanese forces and potentially allied US assets, remaining presence capabilities are purportedly standing by. Uniquely, the possibility for disruption of civilian life is both possible and mentioned in PLA doctrine.[64] As wholly conquering Taiwan militarily is a lengthy, arduous endeavour likely to provoke deep civil unrest, coercing the government into capitulation is a far more promising scenario. As a corollary, while military forces attempt to cripple Taiwan's self-perceived chances of victory, network forces can begin to erode the public's morale, stamina and wherewithal. As previously presented, network attacks are unlikely to affect coercion on their own; yet coupled with a withering kinetic campaign, they may serve as a caustic agent corroding Taiwan's national fortitude.

Perhaps most easy to target would be Taiwan's internet infrastructure. This could be done in several ways, including straightforward bombing operations against fibre landing points, and satellite communication facilities used by internet service providers. Alternatively, China could fall back to tried and true methods. In April 2010, provider China Telecom rerouted a sizeable proportion of global internet traffic through China for 15 minutes, in a phenomenon called BGP hijacking. The impact included a full redirection of traffic associated with a broad range of networks, including some sensitive—though ostensibly encrypted—Western government networks.[65] Several additional such incidents followed throughout the decade.[66] While the occurrence was brief and—due to its technical characteristics—observers are unconvinced that it was intentional, targeted takeovers of global or regional network traffic remains a distinct possibility. The inclusion of traffic hijacking attacks against Taiwanese internet providers could help sow confusion and limit the ability of the island to communicate internally and externally.

Attacking Taiwanese civilian infrastructure through network attacks has several potential benefits. These include ambiguity, deniability and diffusion. Ambiguity is the complexity derived from accurately identifying the nature and source of the attack. Whereas a kinetic attack is immediately visible and detectable, a subtle cyber-attack against sewage treatment, electricity manufacture or civilian logistics is far harder to trace. Deniability relates to the attacker plausibly distancing himself from the attack, again a feat nearly impossible with a war-time missile strike. While one can argue that deniability isn't always necessary in a state of warfare, a desire to avoid global escalation with perceived war crimes against civilians remains strong. Controversial actions may be blamed on "patriotic hackers" seeking to aid their country in a time of conflict, or even attributed to war-time chaos.[67] Ultimately, PRC seeks reunification of Taiwan with the mainland rather than its destruction. Alienating the population of Taiwan will only serve to complicate future attempts to enforce PRC sovereignty over Taiwan. Finally, diffusion is the subtle art of conducting network operations over time. Rather than singularly striking a target, a clever operation can continuously influence systems by ebbing and flowing within a single presence-based intrusion. By mimicking natural system behaviour and periodically introducing interference, an attacker can negatively impact the target over time without being discovered.[68] In contrast, conventional attacks immediately elicit adversary attention if successful, and therefore subtlety is rarely a requirement or even a possibility.

Defeating Network Centric Warfare

Due to the unique blend of geopolitical circumstances, China cannot rely on targeting networks to persistently keep its actions below the threshold of warfare.[69] As Taiwan drifts further away from its notional grasp, peaceful options towards reclaiming the island shrink. Yet even if conflict will eventually be deemed necessary, that does not mean that cyber operations have no potential contribution. On the contrary, they may facilitate the strategic surprise otherwise denied to the PLA, or assist in whittling down defenders and their capacity to operate as a joint force. Taiwan's military relies on many of the same doctrinal principles and technologies as its Western benefactors, whose shadow it was built under. It is therefore useful to examine how the new Chinese approach to intangible warfare may be used to counter both US and Taiwanese advantages in the region.

US C4ISR capabilities include three main components that bear the brunt of such activity; aerial platforms such as AWACS[70] planes, fixed bases in Japan such as Kadena or Sasebo, and the US Navy's command ship in the Pacific. All three assets are defended by fighter craft, anti-air batteries, ground detachments and assorted naval vessels. An effective strike against the unified C4ISR capability is possible, if enough Chinese resources were dedicated to the task of overwhelming the defenders, but it is a highly costly endeavour and perhaps not the most optimal course of action. Directly targeting US defences would inherently result in far greater casualties and increased hostility between the PLA and US forces, further decreasing chances of clean de-escalation.

It is useful to look at overall vulnerability through a layered model. The first meaningful layer is the physical, which includes the actual medium through which communication is conducted. The second can be labelled the data layer, which signifies the passage of a single encoded piece of digital data from origin to destination. The third layer can be designated the stream layer, defined as continuous communication between two or more nodes. The fourth and final layer is the application layer, which entails reflecting communicated data to operators and allowing them to respond accordingly. Each of these layers can be compromised through different attack vectors, culminating in a highly varied attack surface.

For C4ISR systems, the physical medium is mostly radio-frequency electromagnetic transmissions. In modern times, these are mostly encrypted and jam-resistant[71] to prevent opportunistic listeners and standard jamming techniques. However, assuming the message has been successfully received and decrypted by the other side, there is no originator verification in some US protocols.[72] Consequently, there is potential vulnerability to spoofing attacks, as long as correctly configured, viable equipment is employed.

Several layers of data sharing and situational awareness protocols are used by various types of US military hardware, constituting the overall C4ISR image. These include the JTIDS (Joint Tactical Information Distribution System), an encrypted, jam-resistant radio frequency data transfer architecture. The system implements the widely-used Link-16 protocol[73] to facilitate both data and tactical communication between various friendly assets operated by the US and its NATO allies.[74] The specifications for Link-16 are freely available online, including frequencies, message codes, and protocol options.[75] If the PLA can subvert the locally used encryption through any means, reconstructing and altering the traffic is fairly intuitive. The lengthy JTIDS project is currently spearheaded by contractors such as British defence company BAE Systems. Significantly, BAE was previously breached in 2009 by malicious state-aligned actors.[76] The hack reportedly resulted in data exfiltration, including data pertinent to the F-35 Lightning II Joint Strike Fighter program, the costly collaborative efforts geared towards producing the next-generation multirole warplane.

The data layer is easier to influence via an event attack, in what is essentially a digitised version of electronic warfare. For better or worse, the US regularly publishes a full account of some of its command and control protocols in detailed technical specification, removing the need for an adversary to reverse engineer the protocol. In one such example, modern communication between Link-16 endpoints includes "J12.0" messages, defined in the protocol handbook as "Mission Assignment" messages.[77] Put plainly, a hostile participant in the network could potentially re-task friendly craft and assign them new targeting information. Even if the human operator identifies a targeting anomaly, tracking friendly assets is often accomplished via the standardised Link-16 and Link-22 protocols, among others. Due to the pervasive need for real-time performance, there are limited security measures in place to ensure message legitimacy. As the protocol details, each Link-16 active node is expected to continuously report back its position and sensory output[78] to assist in the overall battlefield awareness. As such, a new Link-16 node could register on the current network as a command-capable node. Tracking the origin of a network compromise is difficult in the middle of combat operations.

It is likely that this particular example of possible vulnerability will not survive the test of time. It is an instructive anecdote, suggestive of how communication protocols used for decades may be turned against their users. While the modern military technology life cycle increasingly incorporates better practice for information security, it is incredibly difficult to prevent the existence of such vulnerabilities over the entire operational life cycle of all technology used.

The stream layer holds potential for a different variety of offensive options. If a hostile entity has successfully been introduced into an enemy Link-16 or Link-22 network, it can theoretically begin to transmit conflicting messages to all participants and command units. A repeated attempt at overloading the regional Link-16 deployment is equivalent to a battlefield denial of service (DoS) attack. The resulting chaos would be difficult to distinguish in the heat of conflict, especially if a capable attacker is both generating a great deal of traffic and simultaneously spoofing its origin.

Finally, compromising the application layer—specifically the terminals used by US forces—is the most plausible scenario for a presence-based operation. As seen in architecture diagrams, C4ISR nodes are highly networked to support common services offered by DoD networks, including intelligence sharing, data communication and logistical support.[79] As a result, attacker lateral movement towards these networks—where a great deal of damage can be done—is expected. Once there, a compromised terminal means the attacker can effectively manipulate any message both received and transmitted. Targets can disappear off screen, and fake messages can be cascaded from the system and onto the network. This could complicate attempts to launch missiles, assign targets to friendly assets, and conduct battlefield assessments.

The critical hardware and software that makes up US-deployed electronics is primarily developed by its expansive military-industrial complex. These are spearheaded by several corporations such as Lockheed Martin, Northrop Grumman and Raytheon, who commonly win the lucrative contracts. Examples are plentiful, including the AN/SLQ-32 Electronic Warfare Suite used by multiple naval platforms and developed by Raytheon, and the Navy's Distributed Information Operations System, designed by Lockheed Martin to facilitate battlefield situational awareness.[80] The latest version of the AN/SLQ-32, released in 2016, was flagged as "not survivable due to cybersecurity deficiencies", and exhibiting "poor software reliability" by the Department of Defense's auditors.[81] Lockheed Martin was also notoriously breached in 2011 by a foreign entity, during which copious amounts of sensitive intellectual property were exfiltrated from their internal networks.[82] While security and awareness have increased, so has the rate and quality of the subsequent attempted attacks on the company's assets.[83]

An interesting example of the operational appeal of an OCO against a C4ISR node is the US Navy's Aegis Destroyer. The ship class is entrusted with both air defence and missile defence roles for naval strike groups. As such, these combatants are fitted with a highly capable combat management environment named the Aegis Combat System, or ACS.[84] This is a catch-all phrase encompassing the destroyer's radars, targeting, and ordnance capabilities. A key feature of the ACS is its built-in support for additional functionality through software upgrades, intended to keep it cutting-edge through its operational life cycle. Considering the centrality and interconnectedness of the Aegis platform, the opportunity and potential advantages of OCOs against it are abundant.

At the heart of the ACS is Lockheed Martin's AN/SPY-1 radar, which is managed by the Navy's standard AN/UYQ-70 computer terminals.[85] As documentation reveals, modern iterations of the project have taken strides towards Commercial-Off-The-Shelf (COTS) solutions by adopting well-used architectures that would be cheaper to maintain and upgrade.[86] Updating software versions on shipboard components requires a hefty cycle of preliminary adaptation and testing, which means that if a vulnerability is disclosed in the Solaris operating system variant used aboard the AN/UYQ-70, it could take several long months before it is subsequently patched out. A shift to standardised, commercial products gives the adversary greater access to documentation, software, and hardware samples, reducing the level of complexity required to craft a suitable event-based capability.[87] Once developed, these may retain their efficacy for a number of years, and are not likely to be mitigated during conflict short of turning off the system itself. A successful event or presence-based attack against the Aegis Combat System could disrupt the trust between operator and machine, thereby increasing confusion and decreasing overall combat effectiveness.

In another valuable example, the modern versions of the Tomahawk missile used by US forces incorporate full-duplex satellite networking.[88] The Tomahawk both transmits telemetry in flight and is fully capable of receiving remote commands, such as retargeting parameters or an abort order. Data communication is facilitated over a network aptly named "Tomahawk Strike Network" (TSN), which reportedly allows "anybody who has the authority to log-on… [and] take control of the missile".[89] The granularity of control is significant enough to allow inflight retargeting of the missile, by an operator situated in a wholly different facility than the operator who originally launched the missile. Plans are underway to even further integrate the missile with its surroundings. This will be accomplished by allowing its targeting module to receive sensory output from friendly assets such as UAVs and land radars, while also integrating the entire network to work over the aforementioned Link-16 protocol.

Multi-tiered integration of Tomahawks into a software-managed, remotely-controllable environment means the cyber-attack surface is massive. Indeed, if Chinese network forces successfully compromise

a TSN control node—a tall order but not impossible—they can effectively neutralise Tomahawks en-route to strike PLA missile bases and
limit US ability to intervene in the conflict prior to the US Navy's
arrival on the scene. Interference may be subtle; rather than having
missiles veer off course or return to their senders, simply increasing
their circular error probable (CEP) by a few metres would result in
difficult-to-detect errors. Even if US operators are eventually alerted
to a compromise, they will nonetheless be compelled to bring the
TSN down pending a forensic investigation in order to avoid possible
friendly fire incidents or any further mishandling of launched
Tomahawks. For the duration of the conflict, the damage to combat
readiness and efficacy would have already been done. Trust in the
platform would be impaired, which is possibly an even more damaging prospect than any concrete threat to the missiles themselves.

In the final phase of conflict, a more cautious outlook becomes
crucial as all parties involved scramble to adapt and return to operational capacity in the wake of conventional strikes and OCOs used
against key networked assets. Some networked platforms would have
possibly been sufficiently compromised as to be extricated from their
respective operators' trust circles, therefore rendering them ineffec-
tive while they are thoroughly scrubbed. Conversely, intensive measures to rapidly restore reliability to critical networks[90] would likely
return some operational capacity, with future attacks much harder to
execute. US forces routinely practice recovery procedures, reducing
the overall length of the operational cycle.[91] This will perhaps give the
US and Taiwan a slight edge as networked capabilities gradually
return, once again enabling the overwhelming force of the US joint
warfare apparatus to function as required.

China's strategy entails using critical capabilities in pre-emptive
decapitating strikes against US and Taiwanese forces. As a corollary,
key network operations against hardened military targets may have
been spent, leaving PRC capability arsenal rather limited, if not altogether eliminated. A different vector of attack—less easily countered—now becomes far more attractive for PLA operators. This
would be keyed towards continued degradation of still active US and
Taiwanese military forces, while physical forces continue their
engagement. Event-based denial attacks would be key, centred

around hampering the quality of communications between networked components. Several considerations contribute to the likelihood and success probability of such attacks. Firstly, copious amounts of US military traffic are routinely channelled through public communication mediums.[92] This includes the civilian internet, commercial satellites, and various other multi-use platforms. Even if the transmitted traffic itself is encrypted—which it often is—disruption of the entire channel via a concentrated traffic flooding could cause reduced service for the medium itself, effectively knocking the communicating parties offline. Secondly, denial attacks are feasible even in the absence of persistent network presence, as they can be launched against the medium itself, much like internet-based distributed denial of service (DDoS) attacks commonly wielded by low-level hacker-activists and cybercriminals against websites. Thirdly, Taiwan presents a relatively limited attack surface, as the small nation has comparatively smaller network architecture that may be overwhelmed by PLA operators.

Towards Rapid Resolution

China must exploit asymmetry in its military operations if it wants to win. Whether it is trying to reclaim Taiwan, deter US freedom of navigation operations, project power across the South and East China Seas, or forcibly claim other disputed islands in its vicinity, conventional strength matters but is often not the best approach. The region arguably presents the greatest concentration of modern military power in the world, fragmented across numerous capable stakeholders. As the Taiwan Straits and the adjoining seas represent critical economic interests for all, any conflict waged would ideally be short and decisive. The PLA is clearly pursuing this direction by developing a host of capabilities and doctrine to rapidly offset the advantages of its adversaries.

With its expeditious pursuit of modernisation, the PLA has invigorated its doctrine to reflect lessons learned from its global adversaries.[93] Conflicts waged by and against the US prove both the importance of integrated, network-laden operations, and the unique attack surface that they afford. OCOs are emerging as one of several asymmetric capabilities that may subvert the advantages of a technological force that is heavily reliant upon its qualitative superiority. Creating pockets of operational space can mean the difference between swift victory and protracted stalemates, the latter clearly detrimental to Chinese objectives. Domestic military literature including *Unrestricted Warfare*[94]—and more importantly *The Science of Military Strategy*[95]— does well to reflect the understanding that information shapes the battlefield. Dominance in the information space may permeate into other operational spheres. The establishment of the Strategic Support Force (SSF) has been opaque to external observers, but similarly indicates an understanding that offensive efforts against information networks must be consolidated if they are to be done right. Rather than claiming cyberspace as a domain, Chinese doctrine recognises that information operations and OCOs are present in all other facets of conflict across the conventional domains. It is about exacting value and efficiency, rather than codifying combat in networks as a separate entity with its own governing rules. This early maturity may assist China in rapidly integrating presence and event attacks into combat operations.

As with many PLA capabilities, it is unclear how this Chinese maturity extends beyond the theoretical. Creating organisational entities and organising strategic thought is important, but not nearly sufficient for operational success. In contrast to its prime competitor in the region, the United States, China lacks prerequisite experience in conducting OCOs, joint military operations, and amphibious campaigns. Considering the previously discussed complexity of both presence-based and event-based operations, it is unclear whether, at this point in time, they will serve the nascent SSF well if required. However, the PLA benefits from extensive experience in network espionage campaigns against applicable defence industries, government organisations, and supply-chain companies. This in turn may offer a degree of familiarity, which would reduce the overhead in weaponising these networks against their owners.

Against Taiwan, the PLA would aim to rapidly achieve victory and avoid an extended period of conflict, which would inevitably drag allies such as the US into the fray. The PLA's campaign objectives mesh well with the potential benefits of OCOs; achieving strategic and tactical surprise, bypassing concentrations of forces, and degrading both networked defences as well as the capacity to wage joint operations. For the PLA, OCOs may find a natural slot alongside ballistic missiles in limiting Taiwan's ability to marshal an effective defence before it is too late. Similarly, a measure of ambiguity allowed by OCOs and their ability to afford pin-point targeting may offer value in creating an expanded anti-air, access-denial (A2AD) envelope against interceding US forces. By creating confusion, degraded situational awareness and reducing hard power projection, the PLA could prevent US carrier groups and land-based assets from effectively assisting Taiwanese defenders until it is too late, or at least too late to prevent initial amphibious landings.

Finally, while the Taiwan Contingency presents a uniquely complex military scenario for the PRC, its principals reverberate through other potential cases. The idea of a sudden campaign to rapidly overtake territory is common in PLA military thought, and other regional powers have expressed concerns that this strategic approach may be applied to their own disputes. In an alarming scenario for Japan and its US ally, the PRC has previously threatened to overtake the disputed Senkaku Islands—or Diayou by their Chinese name—by what they called "a short, sharp war". This portends exactly the type of conflict in which OCOs could increase the fog of war in service of a limited island-hopping campaign against unprepared defenders. Whether the PLA is capable of pulling off such an elaborate joint campaign remains to be seen, but indications at least suggest that this is its strategic intent.

8

APPROXIMATING THE IRANIAN THREAT

It is not necessary to be the best, just good enough. The Iranian government has shown the world this time and time again in its confrontations with the country's capable adversaries. If your primary agenda is to win through tenaciously surviving mounting pressure, direct conflict can be prevented if it seems like it would result in catastrophe. Empowering regional proxies, conducting attacks against civilian infrastructure, threatening critical resources, and essentially wiring geopolitical explosives across an entire region are all valid methods to stave off enemies who have a great deal to lose. It is an attempt at deterrence by punishment. Whether it is successful is a whole other matter.

Iranian military doctrine is deeply ideological. It centres around regime preservation and draws heavily on decision-making that cascades down from Supreme Leader Khamenei.[1] There is a deep fundamental understanding among Iranian military planners that the country's main adversaries are all conventionally superior and technologically advanced. To ensure its long-term security and ability to both maintain and disseminate the Islamic Revolution, it must survive through a sustained, intricate campaign of deterrence.[2] Iran knows it may pre-empt conflict if it successfully convinces its enemies that a military attack against it would set the entire Middle East ablaze, and drag the attackers into a costly, uncontrollable campaign.

Iranian doctrine aspires to deterrence through asymmetry.[3] The methods to achieve this are numerous, but rely on projecting power beyond its borders through means meant to counter traditional military strengths. This includes heavy employment of regional proxies for combat and terror operations,[4] significant investment in ballistic missile capabilities, dense air defences, use and export of domestically produced offensive drones, and low-end swarming tactics. Offensive cyber operations fit into this approach as they allow Iran to reach enemies beyond its conventional military capabilities, exact a cost, and retaliate proportionately to attacks—at least by its own perception.[5]

Iran was likely unsurprised to see that its nuclear ambitions were targeted by Western sabotage operations. The Iranian government had learned from the errors of its regional counterparts; Iraq and Syria both had their nuclear facilities bombed by Israeli aircraft, and the Libyans had dismantled their program only to be on the receiving end of a coalition bombing campaign several years later. In contrast, successfully establishing yourself as a nuclear state remained one of the heftiest levers of substantive deterrence in the modern era. So Iranian planners built nuclear facilities that were both separated and embedded deep underground, ensuring that any attempt to excise them would amount to a sizeable—and costly—military campaign. The threshold of militarily thwarting the Iranian nuclear threat was set high. It was natural that industrious adversaries would seek to skirt it.

The result was revealed in 2010 when Stuxnet was inadvertently discovered. Allegedly carried out jointly by Israel and the US, Stuxnet was a multi-year attempt at physically sabotaging Iranian centrifuges through a self-propagating presence-based operation meant to tamper with the SCADA software that governed centrifuge safe use.[6] It was an unprecedented effort to skirt the threshold of war by way of a presence-based operation that would avoid ringing any war alarms. Whether Stuxnet was successful is a subject of much debate, well outside of this book's intended scope.[7] In the long term, it does not matter. Stuxnet was a clear signal to the Iranians that networks were a means of projecting offensive power, a way to target the vulnerable underside of an adversary without engaging their conventional defences. In those respects, Stuxnet was an inspiration to Iran as much as it was a threat.

This was by no means the only operation targeting Iran. Its enemies continued to target valuable networks for intelligence collection and device wiping, prompting a rapid increase in domestic investment in network offence and defence of its own. Between the United States, Israel, the United Kingdom, and others likely viewing Iran as a principal enemy, the nation was doggedly targeted by cyber operations. Consistent offensives and siphoned intelligence likely contributed to increased Iranian focus on network activities. Iran already had many of the prerequisites to accomplish this: a private sector dotted with IT companies eager for business in a sanctions-afflicted ecosystem; ideologically supportive civilians with operational experience; universities eager for funding and governmental support; highly networked adversaries ripe for targeting; and a well-established penchant for using proxies for international clandestine operations. Except for the actual experience in conducting cyber operations, Iran was primed for success.

This is a unique aspect of Iranian offensive cyber operations. We observe many of Iran's capabilities through a civilian lens simply because its strategic culture demands it. Most countries rely on the private sector to provide capabilities, expertise, and in some case operational execution. For a country that explicitly sets conducting operations by proxy as a core doctrinal principal, civilian operators, developers, and researchers have become a direct extension of the country's will. OCO proxies are essential to Iranian military planning and its ability to pursue objectives.

Saudi Arabia took the early brunt of Iran's budding offensive capabilities. In August 2012, one of the globally largest producers of oil—Saudi Aramco—rapidly began to lose access to its networks. As endpoints across the company winked out in droves, a group calling itself the "Cutting Sword of Justice" awkwardly claimed responsibility for the attack, claiming it to be an act of hacktivism.[8] Attribution of the malware—soon dubbed "Shamoon"—was initially cautious but gradually grew consensual; the aggressive move was the work of government-sanctioned Iranian operators. The operation did not attempt to target Saudi Aramco's vast industrial control networks governing oil production itself. Instead, Shamoon was a crude wiper, blazing through Saudi Aramco's corporate networks with impunity, corrupting hard drives before moving on. It was neither particularly novel nor ingenious malware, riddled with flaws and with few perceptible methods of evading analysis or detection.[9] Shamoon was an event-based capability activated through a presence-based operation. As is often the case with less capable threats, it borrowed components and code from previously publicised tools.[10] It was not impressive, but it was good enough to accomplish its tactical goals.

It is impossible to decouple the Iranian use of offensive network operations from geopolitics. The Islamic Revolution of 1979 was at least in part born of a fundamental hostility to Western values, forcibly installed by way of a coerced regime change in 1953 that led to the oppressive but US-friendly Shah regime. The internal backlash was severe, and the Iranian government that followed could not ensure its survival through overwhelming strength, soft power, or invasive militancy. Beyond securing its own prosperity, the Iranian government has the additional goal of propagating Shi'ite Islam by exporting its revolution, and combatting those who would challenge the attempt.[11] To accomplish such lofty goals, Iran adopted the path of resistance; pursuing prosperity through pushing against external forces and manipulation of the region to its benefit. The path of resistance meant adopting a deep pragmatism that permeates all major Iranian undertakings—from its choice of allies, its pursuit of national undertakings, the relationship between the public and private sectors, and how it exercises force regionally and globally.[12] These elements are all evident in the storied Iranian history of cyber operations.

This chapter offers the argument that Iran engages in offensive cyber operations as a means to project power below the threshold of warfare. In the absence of projecting a credible conventional threat, Iran seeks to incur financial, material, and reputational harm to its adversaries through presence and event-based attacks primarily aimed at civilian targets. There are numerous parallels between Iranian acts of state-sponsored terrorism and how it conducts cyber operations. Both are enshrined within its national security strategy and meant to exact a cost from otherwise capable adversaries by targeting their population directly. In further similarity to its other clandestine operations, Iran may call upon proxies, allies, or other loosely affiliated third parties to conduct OCOs on its behalf. This offers advantages in deniability and diversity, but the notable lack of cohesive operational culture meaningfully impacts the quality of operations. Iranian operations are varied and often achieve a short-term desirable effect, but they are disjointed and unlikely to meaningfully contribute to a strategic change in status in Iran's favour. Operationally, Iranian tools are a far cry from those of other top-tier threat groups. They are good enough to often succeed to some degree, but their shortcomings have denied them spectacular victories so far.

Perhaps most importantly, this chapter's case study is offered as a stark contrast to its three counterparts. Where the US codified the military domainhood of cyber, Iran views it as a tool within a broader asymmetrical arsenal. Where Russia views information as a full-spectrum commitment to long-term success, Iran conducts numerous high-visibility vanity operations. Where China consolidates its offensive forces under a unified coordination umbrella with public–private bridges, Iran co-opts voluntary civilians, commercial entities, and regional allies. Technological and operational superiority are replaced with aggressiveness and attempted deterrence. Iran is unique compared to other countries' approaches to offensive cyber operations because it employs them differently as part of its military strategy. The outcome is an instructive blend of advantages and disadvantages.

Iran clearly can manage operations against sprawling networks. In 2013, a network operation targeting the US unclassified Navy Marine Corps Intranet was publicly disclosed as an Iranian campaign dating back to August 2012. While the extent of the breach and what it accomplished remain unclear, it took defenders several months to effectively purge the intruders from the network.[13] Even in the absence of offensive action, and supposedly without pivoting to classified systems, it demonstrated decent operational capacity. Nascent use of simple, widely available tools can often compromise even high-profile networks operated by a well-resourced adversary.

Iran has demonstrated more than just the capacity for full-cycle network operations. In 2013, US Air Force counter-intelligence officer Monica Witt crossed the Rubicon and defected to Iran, bringing sensitive knowledge and experience with her. After simmering in her distaste of US foreign policy and a gradual radicalisation process, she had reconciled her dissonance by fully abandoning her former life in favour of the Islamic Republic. As her 2019 indictment later revealed, she proceeded to provide Iranian intelligence officers with ample classified information about her work.[14] This information was then weaponised by Iranian network operators who constructed targeting packages aimed at infecting Witt's former colleagues with intelligence gathering malware. While the details provided in the indictment offer glimpses of evidence that the targeting itself was poorly designed and noisily carried out,[15] it is still remarkable evidence of the ability to conduct blended operations. Iran had pursued a unique opportunity to exploit the fruits of human intelligence to facilitate network operations.

The Iranian threat has prolific aggressive potential. Clearly capable of diverse, effective targeting of vulnerable civilian targets, it can demonstrably inflict harm. By encouraging its own private sector and aggressively recruiting from academia, Iran can tap into many of its available resources. As Iran is increasingly geopolitically cornered, it lashes out through all means available to deter further enemy action. Iran consistently seeks to incur costs through offensive network operations, a notion that makes it dangerous enough to its adversaries. Over time, the associated tooling of its OCOs has visibly improved, as accrued experience tends to do. Practice has not made Iran perfect by any means, but it has improved the violent reach of an already capable regional power. It is essential to not over-correct and portray Iranian offensive cyber capabilities as top-tier, but it is similarly crucial not to underplay their capacity to do harm. Sometimes doing well enough is all it takes.

Fighting by Proxy

Iran has a storied history of employing allied third parties against its enemies. It can be easy to view this behaviour as insidious or underhanded, where in reality it is more likely fuelled by a pervasive pragmatism born of necessity. In some of these cases, support was lent to ideologically or religiously incompatible allies of convenience;[16] in others, cooperation was more empathically aligned with like-minded regional parties.[17] Iran quickly realised that, in order to shape a desirable Middle East, it must actively project power beyond its borders, in ways its conventional forces or standard diplomacy are limited.

APPROXIMATING THE IRANIAN THREAT

A predilection for using proxies offensively works well with offensive cyber operations. In Iran's case, several assessments will become evident to support this claim. The first argument is that barring a threshold-passing nuclear breakthrough, Iran is not symmetrically equipped to deter or prevent its adversaries from achieving their own policy goals.[18] As a corollary, Iran views proxies as a strategically sound way to inflict a continuous heavy cost on its enemies, both through loss of life, investment of precious resources, and diminished will. The Iranian choice of proxies and how it wields them is a clear signal of its willingness to utilise terrorism for a political end. The strategy has valuable parallels when observing Iranian targeting of civilians and critical infrastructure through cyber operations. Lastly, offensive cyber operations can be carried out by domestic and foreign proxies with coordination but minimal direct intervention, creating at least some deniability and allowing for more flexible targeting.

The ascent of Iranian offensive cyber capabilities is linked to the nascency of its internet culture. In the early 2000s, as the prevalence of internet usage skyrocketed in Iran, online communities focusing on defensive and offensive security began to emerge. Perhaps most notorious of these is Ashiyane, a domestic forum that quickly served as a nexus for the exchange of offensive tools, techniques, and other associated information. The forum was increasingly used to identify talent, create active collaborations, channel hacktivism towards unified goals, and generally pursue an Iranian government agenda. Before its demise in 2018, numerous notable figures in the domestic security landscape were associated with Ashiyane to some extent.[19] Several tools and exploits subsequently found in Iranian network operations have at some point proliferated on Ashiyane as well. It was a compelling way for Iran to harness its online communities.

After the Islamic Revolution of 1979, Iran had found itself differently situated across multiple geopolitical chasms. Perhaps the most notable at the time was the great power rivalry of the Cold War. Where secular Shah-era Iran was the staunchest of Western allies, post-1979 Iran rose principally as a passionately Shi'a religious contrary. Spreading its Islamic revolution was the order of the day. Where its military was once flush with US and Israeli military assistance,[20] Iran now grew increasingly geopolitically isolated, and faced the prime adversary of the post-Soviet monopolar world.[21] Though Iranian industries attempted to reverse engineer available technologies, refurbish existing hardware, and pursue domestic innovation, it was not likely to be enough to shield the fledgling Islamic nation from its enemies.

Other issues further crystallised the need for alternative means of power projection. As an immediate backlash to the Islamic revolution, Saddam Hussein's Iraq invaded Iran in 1980 to sever any notion of its Islamic revolution bleeding into Shi'a-majority Iraq. Increasingly, Sunni Saudi Arabia poured vast resources and influence into the region to promote its own agendas, which Iran fiercely opposed on a matter of religious principle.[22] At the same time, Israel had successfully navigated several military campaigns against bordering Muslim nations with US support,[23] further extending Western reach within the Middle East.

Through its challenges, Iran came to recognise the value of proxies. The realisation was straightforward—Iran could not defeat its enemies through direct military force. In the Iran–Iraq war, Iran attempted to weaken its enemy and break conventional parity by providing support to Kurdish insurgents operating in Northern Iraq, equipping them and encouraging their aims of sovereign autonomy. Despite a sizeable Kurdish population in Iran that would one day seek its own autonomy as part of a cross-border aspiration of sovereign recognition, Iran supported Kurdish guerrillas in their fight against Iraq even as Iranian forces were pummelling non-sympathetic Kurdish forces and civilians.[24] Support was given as an attempt to bleed Iraqi forces and draw them away from combat operations with Iran; internal chaos was viewed as a means towards diminishing Iraqi hard power.

The years of the Iran–Iraq war also marked a significant strategic shift for Iranian military thought. In order to support Islamic movements globally, a dedicated unconventional force was needed. The Quds Force was established to serve as the vanguard of the Islamic Revolution, using asymmetrical operations and proxy support to extend Iran's reach beyond its borders.[25] It cemented a crucial strategic reliance on blended operations that used internal and external forces to directly target perceived vulnerabilities.

Iran was instrumental to the creation of the Lebanese Hezbollah. In 1982, Iranian representatives were dispatched to Lebanon to encourage cohesive Shi'a efforts against the Israeli occupation of southern Lebanon. Through religious, financial, and material support, along with extensive military training, Hezbollah grew immensely in political and military influence.[26] Within several years, the organisation would grow to be one of the largest political forces in Lebanon and a primary threat to Israel as a constant source of political turmoil within the country. As an empowered proxy, Hezbollah would repay its ally on the investment tenfold, by cementing an additional Iranian foothold and providing Iran broad access to a lethally capable force with international reach. With a penchant for conducting cross-border and international acts of terrorism, Hezbollah quickly proved itself as a deterrent to prolonged military engagement by conventionally superior military forces.

In similar fashion, Iran propped up numerous Sunni Palestinian militant groups by providing deep support in the Gaza Strip and West Bank over the course of decades. While Iranian support to these organisations has ebbed and flowed with political fluctuations, the resources and training provided were a key survival channel that facilitated Palestinian militancy against Israel, including carrying out various acts of terrorism.[27] Established ties to Palestinians buoyed Iranian aims at further weakening Israeli resolve and drawing its resources away from targeting Iran itself.

The list of regional proxies employing terrorism goes on. In another notable example, Iran has been tied to elements of Al Qaeda, collaborating with them to some extent, possibly partially through Hezbollah mediation.[28] Iranian alliances with radical organisations are tools for continuously upsetting the distribution of power in the Middle East. As such they have proven largely successful, making both attacks against civilian targets and the use of proxies to do so an appealing prospect for offensive cyber operations, where the perceived costs are even lower and the opportunities to use third parties are even more prolific.

With the fall of Saddam Hussein's Iraq in 2003, Iran was relieved of one of its greatest enemies. A powerful Ba'ath-led Iraq had previously served as a hefty counterweight to Iranian influence, limiting available resources and continuously forcing the country to act defensively. In the wake of the US-led coalition invasion and the subsequent establishment of a democratically elected Iraqi government, Iran could seek to rally support within Iraq's Shi'a population while turning the bulk of its offensive resources beyond its immediate borders. Aside from US forces in varying quantities strewn across Iraq, Afghanistan, and Pakistan, Iran no longer faced substantial transborder threats. It could therefore turn its attention to its regional nemesis Israel, its most prominent Sunni adversary Saudi Arabia, and Iran's anointed "Great Satan"—the United States. This focus on Western and Sunni adversaries was not coincidental. Iran correctly viewed these actors as the greatest threat to its continued export of the Islamic Revolution, and in some respects the most likely aggressors to test Iran's defensive posture and strategic will.[29]

Iran's initial offensives were through low-quality, high-visibility event-based operations. As previously discussed, event-based operations include a wide variety of possible options; from generic denial of service tools meant to temporarily prevent access to websites, to meticulously crafted electronic-warfare implements, which broadcast a specific digital payload to interfere with specific brands of military equipment. The simpler side of the spectrum can often be accomplished with minimal investment in engineering, intelligence, and operational research. The more intricate end of the spectrum may take immense funding and years to deploy effectively. The tooling wielded by Iran when they targeted the US financial sector between 2012 and 2013 was evidently on the simpler side. One such notable event-based operation, dubbed "Operation Ababil" by its operators, had chosen to adopt the guise of a collective of hacktivists.

The attackers called themselves the "Izz ad-Din al-Qassam Cyber Fighters". Through several waves of event attacks, they flooded US financial institutions with junk data, effectively carrying out a DDoS attack that prevented legitimate use of banking websites used daily by millions.[30] This style of attack may seem no more than a temporary nuisance, but successful DDoS attacks can exact millions in revenue loss to victims, and, in some cases, prevent the use of time-sensitive essential services. With banks falling well within the definition of critical national infrastructure, Operation Ababil's event attacks were a disruptive attack against essential civilian targets. Simple as the campaign was, it affected access and—perhaps most importantly—induced a level of concern.

In the summer of 2013, the small Bowman Avenue dam situated in Rye Brook, New York, was breached by intruders.[31] The offenders had initially sought to harvest credentials required to establish a presence within the dam's control network, gathering information about its operation. An ageing but functional sluice gate was apparently administered by a simple networked SCADA system, theoretically allowing remote operation of the dam. Yet any attempt at completing the presence-based operation against the target was disrupted by sheer luck; the control system was disconnected from the gate for maintenance. Several years later, in March 2016, a detailed US indictment against seven Iranian citizens alleged they had pursued offensive actions against the US financial sector by conducting distributed denial of service attacks against US banks. One of these individuals, Hamid Firoozi, was also credited with the breach of the Bowman Avenue dam. Interestingly, Firoozi and the other indicted Iranians were not enlisted members of Iranian security forces or intelligence community. They were staff at two domestic IT companies—ITSecTeam and Mersad Company.[32] The seemingly commercial entities were supposedly engaged in numerous network intrusions on behalf of the Iranian Revolutionary Guard Corps; they were local proxies undertaking intelligence campaigns, offensive operations and attempted terrorism on behalf of Iranian strategic aims.

There is undeniable appeal to private sector proxies, not just in Iran. Where government agencies are heavy with bureaucracies, small companies can be nimble. The private sector can compensate for shortages of capable manpower; government expenditure budgets are often easier to come by than actual hiring slots. Companies would also ostensibly use their own tools and infrastructure, which theoretically reduces the risk of exposure should an operational compromise occur. As a corollary, this separation also notionally complicates state-level attribution by introducing more distance between the government and operations carried out on its behalf—if this segmentation is done well. It is no coincidence that many of these considerations may appear familiar to many other governments, including the

United States. The burden of bureaucracy-heavy operations is commonplace; it is simply particularly visible in Iran's cyber operations. Contractors can both be a short-term blessing and a long-term curse.

The risks are numerous. Commercial entities are not subject to the same legal frameworks that public entities are, and the regulatory oversight is similarly distinct. Perhaps most importantly, public sector operational units are part of the same organisational culture as their counterparts in intelligence and other force branches. That they are intrinsically linked to the defence life cycle has immense culture significance; it means shared values, cascading priority intelligence requirements, and fighting for the same objectives under a shared banner. These cannot be discounted when ensuring that the operational hierarchy can understand its strategic role and importance. Cohesive culture is crucial to success.

There are also practical risks to an over-reliance on the private sector for operations. Intuitively, there is less control. In countries where the legal ecosystem around hacking as a business model is murky, governments may end up relying on operational outfits that moonlight as criminals. This is not a theoretical phenomenon—freelance actors and private sector companies used by the Russian and Chinese governments have been caught red handed engaging in criminal operations while off the clock from their state-sponsored intelligence or offensive operations.[33] In some cases, these for-profit criminal endeavours were done using the same tools and infrastructure. That in turn presents an outsized operational security risk to any associated states. Repurposing capabilities intended for clandestine presence-based operations to conduct money grabs or even install ransomware can cause important operations to be lost if the criminal endeavours are exposed. Any irresponsible reuse of operational capabilities increases the odds of detection.

Iranian domestic proxies are perhaps more a liability than a benefit. It is true that companies such as ITSecTeam or Mersad allow the Iranian government an additional means of pursuing operational objectives. Yet, as their cases illustrate, the cost may have overshot the benefits. In their thoroughness, the US-issued indictments demonstrate that it is possible to not only attribute the individuals involved, but also to achieve high confidence that their operations were carried out at the government's behest. Shoddy tools, poor operational security, and a lacklustre decoupling from their government associations meant that these domestic proxies presented major risk and little utility.

Iranian commercial proxies are emblematic of Iran's overall operational culture. They symbolise an aggressive overextension of capability. Proxies provide invaluable reach and capacity that are otherwise sorely missing from the organic growth of Iran's operations. However, the most capable of presence and event-based operations require more than a loose collection of entities that ostensibly share a budget and a loose national agenda. As demonstrated before, truly impactful operations demand the continuous pairing of technical, operational, intelligence, and research and development expertise. In the absence of this tight coupling, Iran will continue to be prolific in its operations and manage to land blows against its adversaries, but will remain hobbled in comparison to its higher-tier adversaries that are capable of unified effort.

Operational Trajectory

Evidence documents evolution. As in other sections throughout this book, carefully piecing together analyses of operational network activities throughout the years allows the observer to remark on the overall course charted by a country. The way in which a country adopts offensive network operations evolves at varying paces; some are more responsive to changes in the dynamics of warfare and intelligence, while others are slower to adapt. Warfare is certainly not fundamentally changed by the onset of network operations; but Iran has done well to adopt them as a key driver of international influence.

Iranian operational trajectory can be viewed under two pillars; capability and utility. The former relates to the actual perceived quality of Iranian offensive network operations. Quality is a subjective term, but there is value in assessing how, over the years, the various Iranian tools and infrastructure stack up against Iran's adversaries and other influential actors. The latter pillar, utility, relates to how well Iran has actually integrated offensive network operations into its overall decision-making and intelligence life cycle. This is a determination of cohesion of effort, and indicates just how well a country is able to wield its capabilities to achieve objectives.

Iran has a well-developed integration of offensive cyber operations into its defence posture, but its capabilities have evolved slowly, and utility remains comparatively limited as a result. Where countries such as the United States have wrestled with the extent to which offensive operations should be made readily available as a tool of policy, Iran has eagerly accepted their role as an appealing way to seize attention and exact a cost without incurring significant risk to its most valuable assets or drawing meaningful countermeasures. For a heavily sanctioned nation which has teetered on the brink of war with its capable adversaries for over a decade, network operations have proven a reliable means of power projection. Yet despite this, evidence of Iranian tools continues to show them as comparatively underdeveloped and riddled with flaws, while their operators consistently make grave errors. Gaps in capability hamstring utility.

This is not to discount Iranian operational development as insignificant. Throughout the course of the 2010s, the Iranian security ecosystem rapidly adapted towards employing effective targeting against its perceived enemies and achieving meaningful effects. For a country historically enamoured with using severe acts of marketable violence against civilian populations as a means of coercive diplomacy, network offensives now appear a natural extension. The moral justification of targeting civilians by way of attacking the networks they rely on is not a departure from earlier Iranian policy. This is indicated by Iran's use of Quds Force lethal operations against civilian targets, and the overt encouragement of terrorist organisations to dissuade regional threat actors from operating freely. If anything, operations against critical national infrastructure are often considered a so-called bloodless alternative to kinetic attacks; a promising way to enjoy the benefits of terrorism without incurring the wrath of the international community.

Attention is oxygen to terrorism. The eager coverage of Iranian operations—both by international news outlets and information security companies—has cultivated the perception of Iran as an adversary worthy of respect and caution. Attacking US banks sends a signal that even the world's reigning powers are not immune to harm. Targeting

Saudi oil production suggests that continued efforts to undermine Iranian regional power will be met by immediate force against the very lifeblood of the economy. The amplification of the underlying threat through political coverage far outstrips the actual harm done through the attacks. Rather than viewing Iranian offensive operations as a means to achieve operational objectives, it is more useful to view them as a part of Iran's attempts at the erosion of their adversaries' will.

Iranian adoption of network operations is often viewed as a reaction to Stuxnet.[34] That approach holds some merit; Stuxnet's severity, duration, complexity, and in-depth systemic compromise of a key Iranian strategic initiative had likely been an earthquake to security decision-makers. It was both an alarming new threat, and a dazzling opportunity. Stuxnet led to a momentous shift in Iranian priorities towards this new type of conflict, where Iranian defences had been subverted through asymmetrical operations, but its interests could also potentially thrive.[35] The first attacks that were more easily and directly attributable to the Iranian government started around 2012, two years after the public disclosure of Stuxnet and the Olympic Games operation that deployed it. Yet Iranian society birthed cohorts of hackers and would-be operators well before Stuxnet's pivotal moment. The Ashiyane online forum was founded in 2003[36]—several years prior to Stuxnet's reveal—and quickly became a cultivated ecosystem for Iranians interested in security from all angles. It was not alone, with several other forums popping up in Iran's domestic hacking scene. From that moment on, key elements of Iran's cyber operations culture had already begun to organically grow. It was a scene led by commercial companies[37] and loosely affiliated collectives connecting in online forums. Where in other countries this interplay occurred in parallel to the growth of network operations within the military or intelligence agencies, in Iran the private sector eclipsed the public sector, at least initially.

This growth had not gone unnoticed. From scattered testimonies and fragments of evidence it became increasingly evident that Iranian security forces expressed an active interest in harnessing the potential of communities like Ashiyane and the so-called "Iran Cyber Army" to pursue national security goals. Internally, it made sense. Relying on detached entities to conduct nuisance operations against Iran's enemies folded neatly into existing arrangements for conflict by proxy. When Ashiyane members defaced Israeli websites during the 2006 Second Lebanon War and 2009's Operation Cast Lead in Gaza,[38] it was a highly visible way of communicating that aspects of Israeli society were within reach. Even if the benefits were tangibly negligible, so was the risk. At the time there was little public awareness of how simple a defacement is to execute, and coverage of the incidents overshadowed and amplified their initial impact many times over. These were simple event-based operations; robust, scalable, and quick.

Decentralised proxies worked well against low-hanging fruit. Where security awareness was low and opportunity vast, Iranian hacking collectives quickly sifted through possible targets, racking up victims. The actual effects ranged from simple defacements[39] to leaking customer or user information. Targets were not always chosen based on their individual impact, but rather loosely on their national identification. It was the momentary triumph of quantity over quality, where hundreds of small websites compromised by automated tools[40] generated the perception of an all-out attack against Israel's capacity to defend its internet. In reality, most of the co-opted nationalist hacktivism rarely succeeded in achieving lasting impact against meaningful targets. It was digital noise and chaos.

There were a handful of more noteworthy operations. In December 2009, after unprecedented civil unrest gripped Iran, Twitter briefly saw its services compromised and taken offline. The social media network's website was replaced by an image banner claimed that "If Leader [Khamenei] gives the order—we will attack. If he asks us— we are willing to sacrifice our lives. If he asks us for restraint—we will obey".[41] As indicated by the company itself, this was not an actual attack against Twitter's services, software, or users. Instead, a compromise of Twitter's DNS records meant that any request to visit twitter.com was instead redirected to a website under the control of the attackers.[42] A disruptive, transient effect, but one that did well to draw notice. The group behind that particular operation was the "Iran Cyber Army", another loose collection of nationalist-leaning civilians supposedly working on behalf of the Iranian Revolutionary Guard Corps (IRGC).[43] The IRGC is a large, well-equipped and well-trained military force broadly parallel to the Iranian military;[44] it reports directly to the Supreme Leader. As subsequent indictments would reveal, the IRGC frequently aligned itself with civilian counterparts—both ideologically motivated and otherwise—to bolster its cyber capabilities.[45]

The early years of Iranian network operations were exploratory. Having identified the value of network operations, Iran's initial forays into targeting adversaries were crude, opportunistic, unstructured, and highly dependent on the voluntary mobilisation of domestic information security communities. Considering high-impact presence and event-based operations require an intricate interplay between intelligence, research and engineering, it should not be a surprise that Iran was unable to generate capabilities equal to those of its enemies. Much like the identification of online communities as a reservoir of talent and operational capacity, Iranian authorities made do. Iran notably punched above its weight, aggressively lashing out at Western and Arab targets in a gambit to establish itself as a force to be reckoned with; as before, it worked.

The primary goal of Iranian network operations was not direct impact. They sought instead to expand the perception of Iranian reach. By instilling onlookers with a sense of dread that Iranian hackers would directly target the soft civilian underbelly in Western countries, aggressive moves to counter Iranian nuclear aspirations would theoretically be deterred. Iranian news coverage extensively covered senior foreign officials as they claimed in 2014 that "Washington itself is mindful of the Islamic Republic's military might in the arena of information technology and communication".[46] Network operations became an overt instrument of foreign policy, slotting nicely alongside empowering militaristic proxies and supporting terrorism. As terrorism thrived with subdued coverage, so had the perception of Iranian cyber-warfare capabilities. The very same day in 2014, a *Foreign Policy* article pitting Iran and China as comparable in capabilities quoted Israeli academic Gabi Siboni, who claimed "Iran should be considered a first-tier cyber power".[47] In similar fashion, Business Insider ran a 2015 article extolling the aggressiveness and skill of Iranian threat actors, highlighting a perceived US disadvantage and quoting geopolitical researcher Ian Bremmer as he claimed that "In 10 years time, Iran's cyber capabilities will be more troubling than its nuclear program".[48]

Iranian investment in offensive network operations rapidly paid itself off through the by-product of its coverage. The quality of operations and their success rate mattered less.

Targeting US banks was a significant escalation. The aforementioned Operation Ababil, occurring throughout 2012, represented a concerted Iranian event-based attack meant to disrupt financial infrastructure. By cobbling together botnets comprised of several malware variants, the attackers were able to generate network throughput topping 100Gbps, a rarity at the time. It was sufficient to temporarily bring the websites of US banking giants—including Bank of America and Wells Fargo—to a halt.[49] There was no lasting harm to the banks except the loss in revenue, but that did not matter. The operation reflected the ingenuity of the Iranian approach; co-opt existing tools, use them efficiently to create a transient effect, and bask in the glow of success as experts and coverage discuss the sophistication of the operation. Perhaps most importantly, the Iranian operation managed to skirt the thresholds of warfare by limiting impact and inhibiting attacker attribution, crucial steps in the previously shown model.

The denial of service attacks were covered at length. Analysis conducted in October 2012 had identified that a malware toolkit called itsoknoproblembro was at least partially responsible for the generated traffic.[50] The malware was labelled "a highly sophisticated toolkit"[51] and the CEO of anti-DDoS provider Prolexic even offered that "from a DDoS perspective, they are on the level of a Stuxnet type of attack".[52] This notion—that a denial of service botnet reached the intricacy of hardware attacks crafted to physically harm specific models of interconnected centrifuges—was hard to take at face value, and indeed others were far more cautious. Information security company Talos noted that the sheer volume generated by the botnet was likely attributable to "[compromised] systems with high-bandwidth links".[53] Talos also noted several indications that the malware had glaring flaws, including chunks of the toolkit source code being widely available on text-sharing sites, and using default passwords in deployed command and control sites. In similar fashion, Dmitri Alperovich, the CTO and co-founder of security company Crowdstrike, claimed that "These are not super sophisticated attacks, but we're seeing very large, almost historic, attacks from the standpoint of the volume of traffic we're seeing".[54] Sophistication was not necessary; Operation Ababil was a triumph of targeting, effective use of available offensive resources, and news cycle co-optation.

Emboldened by success, Iran pivoted to presence-based operations. When pro-Israeli American billionaire Sheldon Adelson mused in October 2013 that nuking an Iranian desert was a plausible means of deterring Iranian nuclear aspirations,[55] the Iranian government was incensed. Adelson was widely known to have deep associations with Israeli prime minister Benjamin Netanyahu, and therefore served as another potent political force in the relationship between the United States and Israel. In February 2014, a lengthy presence-based operation against Adelson's Las Vegas Sands Casino managed to successfully wipe clean swathes of the network and its servers.[56] The operators orchestrating the attacks had initially infected Sands Bethlehem before pivoting within the network and locating the credentials of a senior network engineer. That in turn allowed the attackers to seek out Sands' headquarters in Las Vegas, where they promptly made off with numerous sensitive documents before attacking the servers themselves. As the investigation would later reveal, the Iranian operators had used a combination of well-known attack methods, repurposed widely available tools and crude destructive scripts to achieve their goals. They had also missed an opportunity to do further harm, as they had failed to capitalise on access to key infrastructure which would have allowed the operators to potentially wipe the networks of all global Sands assets. As before, the operation fell short of warfare as it targeted a single private entity. The Iranians once again demonstrated effective use of limited capabilities to safely generate a highly visible effect, in service of a political agenda. Loud, sub-optimal, but effective.

Destructive network operations were a success story. The efforts were comparatively low cost, generated visible effects, fed the perception of a powerful Iran capable of bloodying its enemies, and— perhaps most importantly—skirted a lethal military response. Considering Iran's list of enemies included several nations famous for their cyber capabilities, it was a particular triumph to exact a cost from any of them. The actual degree of complexity for each operation or how it measured against Western counterparts mattered very little. The network attacks were a sequence of operational successes, and a key part of Iran's military strategy in the absence of a military showdown with its enemies. It was therefore no surprise that Iran sought to build on these streams of successes as it replicated its approach with Shamoon, lashing out anew.

Four years after the Shamoon attacks against Saudi Aramco and Qatar's RasGas, Iran struck again. Multiple organisations throughout the Middle East were targeted by a destructive attack identified by researches as an evolution of previous Iranian efforts; they soon named it Shamoon 2.[57] The attacks similarly espoused faux-activism by employing the imagery of the body of three-year-old Syrian refugee Alan Kurdi. The attempt was likely meant to confound attribution and prevent a Saudi portrayal of the attack as meriting retaliation. Highlighting once more the Iranian emphasis on the complete life cycle of presence-based operations, the destructive malware contained credentials likely harvested by preceding intelligence operations against the chosen targets. The Disttrack malware itself was similar to the previous version employed in 2012, and had not represented an immense leap in capability. Instead, the operators used cobbled together scripts and legitimate network administration tools to infect networks at scale, delivering the destructive payload.[58]

In 2018 Shamoon was detected again. The quickly-named Shamoon 3 campaign abandoned evocative imagery as the veneer of simulated legitimacy for its attacks. Assessment of the involved malware revealed that it had indeed continued to improve on its predecessor, but at a thoughtful pace. There were no significant innovations, and no adoption of novel mechanisms for infection, dissemination, or destructive impact. Instead, any ingenuity in the campaign might have actually stemmed from its targeting. As reported, one of the targets of Shamoon 3 was Italian oil company Sapiem, known to provide services to Saudi Aramco.[59] Targeting the supply chain of their Saudi adversaries demonstrated offensive capabilities guided by intelligence collection efforts. Irrespective of the technological pace of innovation, Iran was increasingly capable of choosing its targets with nuance.

It is valuable to observe the pace of progress for Iranian operations. Tools and capabilities have certainly improved to a degree, as confirmed by a clear evidentiary trail. In the intervening years between

Stuxnet and Shamoon 3, Iranian operators proliferated greatly both in the private and public sectors. Where once the offensive security landscape was sparsely dotted with a handful of patriotic citizens eager to deface targets, by 2018 it was clear that Iran had invested heavily in homegrown capabilities, operations by proxy, and an overall strategy to collect intelligence and achieve effects through network operations. It is enough to support the notion that Iran can punch above its weight, but not that it is a first-tier power. Successes notwithstanding, Iranian tools, a decade after first being spotted, continue to trail the major players with numerous flaws and minimal novelty.

Under scrutiny, most Iranian operations skirt the threshold of warfare by varying degrees. Incidents that targeted critical infrastructure providers within Iran's enemies might have qualified, had the operational networks been compromised beyond corporate networks. Attempts at muddling government-level attribution were poorly made. Had Iran's targeted enemies chosen to respond, they likely could have established sufficient attribution to do so based on available evidence, as was seen with US indictments.

From a military perspective, Iran has successfully used OCOs to extend its strategic depth somewhat.[60] Taking the fight directly to enemy critical infrastructure and other civilian assets reduced the need for direct military friction between its conventional forces and its enemies. Cyber operations were used to signal discontent, strike at enemies, and demonstrate a capacity to exact a toll, even if perception exceeded effects significantly. A blended asymmetrical approach positioned OCOs not as a domain or even a warfighting capability, but as an inter-conflict means of power projection alongside regional proxies, investment in drones, and of course, Iranian nuclear ambitions.

Several notable incidents highlight existing gaps in the Iranian operational life cycle. In March 2019, an anonymous party began leaking malicious tooling and disclosed information allegedly pertaining to network operations perpetrated by the Iranian Ministry of Intelligence (MOIS).[61] The leakers had chosen Telegram, Twitter, and hacking forums as their mediums of choice, and it remains unclear who systematically targeted Iranian threat groups with such detailed access.[62] Commercial threat intelligence companies researching the supposed group confirmed that the leaked information appeared linked to the targets of their monitoring, codenamed APT34, or OilRig. Suffering a leak does not inherently diminish the professionalism of a given threat actor; the NSA, CIA, and Russian threat groups have all been the targets of such compromise. Instead, the scale of the leak is significant. The information published included source code for the malicious tools, targeting information, stolen credentials, and even documents unmasking specific operators.[63] Such thorough coverage constitutes an immediate unravelling of the strategic efforts of an entire threat actor. More so than any set of tools being leaked, it was a deep crisis.

A separate incident embarrassed a different Iranian group. In July 2020, IBM's X-Force reported having identified several instructional training videos hosted openly online.[64] The videos demonstrated handling of compromised accounts and exfiltration techniques, and included examples of real targets pursued by the operators. As explained by IBM researchers, their uncovering was made possible through a misconfiguration of the server used by the operators to host the videos, allowing them to be accessed by anyone who knew where and how to look. Yet the lapse in judgement ran far deeper than this; hosting several hours of training materials that included real targeting on internet-facing cloud-hosted infrastructure was an unnecessary risk to begin with. Beyond the operational security errors, the actions documented in the videos corroborated suspected notions on Iranian operators; they are effective at manual work but lack bespoke, automated tooling that would alleviate the need for much of the actions taken in the videos.

A Clear But Measured Danger

By 2022, a decade after Stuxnet, Iran has not publicly crafted anything of similar quality. This observation is not an indictment of operator capacity to cause harm through presence or event-based operations. Instead, noting the quality of Iranian tools is important to understand their relative strengths and weaknesses as a frequently analysed threat actor. Over a notably short time span, Iran had successfully built up a culture of network operations spanning multiple groups, campaigns, tools, and targets. Despite repeated unmasking of their tools and campaigns, operators have continued to compromise dozens of targets and achieve headlines by wreaking havoc on critical infrastructure providers instrumental to the resilience of their enemies. At least visibly, Iran is one of the most prolific adopters of offensive network operations as a tool of coercive diplomacy.

It is not easy to build out the complex ecosystem for offensive network operations. As covered in the second chapter, such operations draw from a rich tapestry of experience, know-how, culture, and technical infrastructure drawing heavily from electronic warfare and intelligence collection. Compared to the multi-decade deep technological expertise of the United States, Russia, United Kingdom, and even Israel, Iran had little such history to lean on. However, Iran could utilise its mastery of reverse engineering technologies, as a result of sanctions and carry-over equipment from the Shah era. Iran rapidly succeeded in cultivating an ecosystem for information security, where those interested in offensive security could get their hands dirty on behalf of their country. By emulating best practices, repurposing public tools, and adapting well known techniques, some meaningful effects were achieved that would elsewhere be carefully pursued by military units or offensive divisions within intelligence agencies. While chaotic and undisciplined, Iran's approach vastly accelerated the rate of experience accrual by operators, and afforded the government some tactical success in its endeavours.

It is not a matter of whether Iran is capable of top-tier operations. The significant observation is that the Iranians are both willing and able to achieve objectives by targeting adversary networks. In less than two decades, the Iranian government has wielded a growing array of offensive options in a manner that inflicted visible harm on enemies and contributed to its reputation as an aggressive and capable threat. Iran is not the best operator, but certainly not the worst. It is rapidly learning, willing to experiment, and eager enough to use whatever works—and thus, most importantly, it is dangerous.

The integration of network operations was born of need. With most of Iran's adversaries being well beyond its conventional reach, the Shi'ite nation instead turned to more indirect solutions to project power, destabilise its adversaries, and hopefully deter military action against its repeatedly targeted domestic nuclear programme. It would be challenging to argue that Iran has successfully deterred its enemies from moving against it by way of its relatively low-complexity network attacks. Yet through operations against Saudi Arabia, Qatar, Israel, the United States, and others, Iran has repeatedly signalled that it is willing to target critical civilian infrastructure and has the means to do so. This is an unignorable risk.

Iranian historic dependency on proxies as a means of attaining global reach contributed to its cyber posture. By encouraging commercial, academic, and crowdsourced entities to operate on its behalf with relative impunity and tacit encouragement, Iran generated an image of capacity that it does not have. In stark contrast to Western nations, predominantly reliant on their intelligence agencies and militaries for offensive network operations, Iran outsourced its force building to external third parties. This in turn afforded several notable advantages, including segregation of duties, a variety of available talent, diversity in approach and capabilities, and limited deniability. Yet outsourcing comes with a cost; Iran does not enjoy the level of control over its operators as other nations do, resulting in more operational security faults, visible gaps in professionalism, a lower rate of operational novelty, and diminished cultural cohesion.

Offensive network operations are ideal for nations keen on pursuing terrorism as a political course of action. The near-zero count of cyber operations carried out effectively by terrorist organisations has occupied many academics and analysts, who rightfully viewed such aspiration as a promising evolutionary step.[65] Real-world terrorism is hard to carry out, cyber has a lowered barrier of entry, and civilian infrastructure networks are frequently undermaintained and misconfigured. Therefore, it stands to reason that terrorism would find itself right at home in the information space. Through Iran, we are provided the missing piece—carrying out offensive operations effectively requires a strategic investment and access to technical resources that terrorist organisations simply do not often have. It is conventional wisdom in terrorism studies that territory is instrumental for an organisation's long-term success. The same is true for cyber operations; sovereignty and the resources it affords are an immense boon to those seeking out offensive capabilities.

To get the clearest view possible of Iranian intent in network operations, one simply need turn to its greatest regional adversary—

Israel. Over the course of fifteen years, Iran has encouraged, facilitated, and in some cases directly carried out a variety of network operations against the nation, with a range of objectives in mind. As previously mentioned, Israel was one of the primary targets of nationalistic Iranian faux-hacktivism in the mid-2000s, mostly materialising as a high volume of defacement attacks against scores of Israeli websites. Iranian operators were later found to have compromised Israeli military networks, allegedly on the hunt for access to missile defence networks[66]—a valuable target both for Iran and its missile-laden regional allies. Israeli authorities would confirm later in April 2020 that Iranian operators had breached a SCADA system associated with water treatment, in an alleged attempt to alter the chlorine levels in the water and poison thousands of citizens.[67] The attempted attack was followed by two more in June 2020 targeting civilian water pumps.[68] Though all of the attacks reportedly failed, they were described as sophisticated, even if there were relatively minimal details on what this level of sophistication entailed.

It is important to avoid under-estimating the threat of Iranian offensive operations. On one hand, Iranian operational outfits have proven capable enough to wreak havoc on large networks, causing widespread non-physical damage with a hefty recovery cost. Iranian presence and event-based operations have also demonstrated a meaningful level of agility, with successive targeting cycles capable of breaching adversary networks and in some cases achieving harmful objectives. As operations have often occurred in proximity to regional or global events of note to Iran, the country has been able to capably weave its offensive capabilities into an arsenal of coercive tools available to it on demand. Even more importantly, these tools have achieved at least part of what they have intended to do. They are relatively low-cost and have provided Iran with reach, increased perceptions of Iran's power, and rarely exacted meaningful consequences upon delivery. Such accomplishments are impressive for a country with barely two decades of accrued information security experience.

It is equally important not to over-estimate Iranian operations. They overwhelmingly lack the technical intricacy, innovation, and potency of their top-tier counterparts in the US and elsewhere. Their methods of choice often rely on well-known software exploits, freely available repurposed research tools, haphazardly constructed scripts, and bespoke malicious software of variable quality. Iran has frequently missed opportunities to cause far greater harm to its enemies, even when it had intended to do so. It has also suffered multiple embarrassing reveals of its operators, tools, and targeting campaigns. Overextension through proxies means that there is little cohesion in the overall effort. While proxies enable compartmentalisation, they also hinder cooperation and system growth that could have accelerated the overall increase in capability and skill. Iran certainly has the available resources to carry out offensive operations, but its roster still falls short of its enemies'.

9

A REVOLUTION IN CYBER AFFAIRS?

Technology advances at an accelerating pace. Where scientific discoveries were once few and far between, the twentieth century has reduced the turnaround rate for innovation. This trend has accelerated over the twenty-first century, in which products and devices once considered revolutionary may seem dated within five years of their inception. It is therefore understandable to contend that any model offered for operations in and against networks would not stand the test of time; these too would lose their relevance within a few years of their conception. The very agility that is required in those that operate against technology is required by those who write about it.

The characteristics of modern digitisation are accelerated by a few key trends that are expected to become prevalent over the coming years. The first is the development of machine learning (ML) and artificial intelligence (AI), software capable of independent problem-solving in a capacity exceeding existing deterministic methods. The second is tightly linked to the first and entails the meteoric rise in autonomous platforms across their many uses. Self-governing systems are becoming increasingly adept at solving complex tasks previously only handled by humans, which in turn results in more responsibilities and tasks being offset to them. The increase in autonomy provides opportunities but creates new risks and systemic vulnerabilities. Lastly, an adoption of simulated environments—augmented reality (AR), virtual reality (VR), and mixed reality (MR)—are likely to create new forms of communication, congregation, and operation that may deeply impact the interface between humanity and machines. Each one of these three trends is already seeping into military research and development, and could adversely affect the manner in which networks may be targetable in the coming future. Some of what we know will change, but much likely will not, at least not immediately.

This chapter contends that the underlying characteristics of OCOs will remain viable at least in the short and medium terms. The nature of intangible warfare will not inherently change with the next iteration of technology, but rather will be intensified. Incorporating autonomy into decision-making at all levels of warfare would further distance the ability of people to grasp its complexities, turning many essential systems into "black boxes" and thereby increasing the value of targeting them.[1] An increased reliance on autonomous platforms, rich sensory input overlaid on reality, and opaque algorithms assisting in all aspects of warfare ensures that these would become crucial targets. The notion of weaponising an adversary against itself becomes even more prevalent.

These trends in technology do not represent a break from existing themes; they represent the latest cycle of counter-innovation accompanying intangible warfare. The rise of networking has resulted in OCOs, which in turn may eventually breed counter-OCOs in the form of autonomous AI-based defences.[2] Sparks of this already exist; such so-called solutions are now purportedly included in various offerings by private sector information security companies. These claim to employ limited implementations of AI in various instances, such as network anomaly detection, malware analysis, and social network analysis. These solutions rely on the flexibility of modern learning algorithms to enrich existing approaches rather than supplant them; they are not autonomous network defence solutions. Rather than targeting sensory inputs which would in turn deceive the operators that observe them, it becomes useful to target technology at more abstract levels. At its core, automation prominently relies on data to shape its subsequent decisions. Target the data flows that shape perception for AI, and it will then incorrectly inform its operators and misalign their situational awareness. The idea of shaping human behaviour remains the same, the attack is simply distanced further from the humans being targeted.

It is important to continuously evaluate the viability of OCOs. As information security practices arguably improve over time, we theoretically expect less software vulnerabilities to be generated. As a result, networks should gradually become increasingly resilient and less susceptible to OCOs. This is, unfortunately, theoretical at this point. While a reduction in overall vulnerability may eventually be the norm, currently even decades-old techniques continue to be effective against ostensibly secure targets, and organisations large and small are repeatedly breached. Trends in increased security are countered by others, such as the proliferation of low-cost, low-quality, low-security devices that impact daily life on a greater scale.[3] Similarly, the rate of public zero-day vulnerability disclosure has skyrocketed in 2021, both due to defensive and offensive research investments.[4] Even as the rate of by-default network encryption increases,[5] individuals increasingly introduce always-on microphones, GPS tracking, and wearable cameras[6] into their lives, thereby opting in to monitoring that was previously difficult to achieve. Existing approaches to exploiting software may eventually diminish, only to be replaced by logical flaws or the "poisoning" of the algorithms[7] to which we have delegated so much responsibility.

Even as some claim software vulnerabilities to be a transient feature of computing, others remain unconvinced.[8] Certain categories of software bugs may reduce in prevalence as issues are systematically addressed, newer architectures are developed, and development practices improve. Others would remain and are inherently more difficult to stamp out. New types of vulnerabilities would likely be introduced, corresponding to new technologies and opportunities. Some may be logical in nature rather than directed at breaking code, and would instead seek to subvert the algorithms themselves by manipulating their inputs in significant ways.

The trajectory of software vulnerabilities is extrinsic to the success of OCOs. It is the increased human dependency on technology which ensures continued overall vulnerability. As humanity delegates more functions to increasingly complex autonomous systems, the very ability to understand that something is amiss decreases.[9] Already, the specific decision-making processes that guide algorithms such as neural networks are opaque to all but a handful of people. The sophistication of software logic is rapidly deepening, rendering its rationale hazier to users and even many developers. It therefore becomes plausible that, at some point, errors maliciously introduced into these processes may become unnoticeable by users and administrators. One way or another, OCOs would likely remain possible for the foreseeable future.

Despite rapid progress, humanity is only at the early stages of autonomous software and is hence limited in its capacity to assess its impact. This is no less true for the viability of OCOs, though some key indicators are already emerging. Militaries are already introducing growing quantities of autonomous platforms across all levels of operations.[10] These platforms will be more capable of performing complex tasks. Use of artificial intelligence may enable battlefield superiority in a networked world,[11] but at a cost; human ability to comprehend and directly control the elements of warfare may shrink, increasing the threat and potential value of OCOs as an operational method of choice. If more aspects of combat operations are governed by software at some level, the software itself becomes a key target worth relentlessly pursuing, either through event-based or longer presence-based capabilities.

This final chapter is understandably more limited than the preceding ones. For one, access to details on the bleeding edge military developments mentioned throughout is understandably restricted. Strategic publications indicate overall directions and developmental priorities, but lack specificity. Similarly, developments within the military sector are often correlated to advancements in the private sector, those being less opaque. At the same time and with the accelerating pace of technological advancements, trends may shift or even be rendered obsolete within a relatively limited time span.[12] Advancements form counter-innovation cycles that may be difficult to predict ahead of time. Each cycle is spun from its predecessor, which makes predicting beyond a single cycle unwieldy. Yet even modern advancements in networking, robotics, multi-domain warfighting, and artificial intelligence have all been discussed extensively since the twentieth century. As previously indicated in the chapter on intangible warfare, technological developments and their integration into warfare are firmly rooted in history. While the pace of advancement may accelerate, its products will have a traceable lineage to existing developments. The challenges of tomorrow may be different, but they will undoubtedly evolve from those of today.

Becoming Less Vulnerable

It is a point of contention whether software is becoming more secure overall. If exploitable software vulnerabilities are on the decline, network operations against secure environments—at least their engagement stages—become harder to accomplish without a greater investment of resources. If this were true, the economy of OCOs would be altered until resource costs appeared less inviting, deterring decision-makers from risking remaining capabilities within their arsenals. Opinions on the direction of the overall vulnerability of software run the gamut and include perspectives both grim and optimistic. In reality, both sides offer merit in their arguments. It is reasonable to simultaneously claim that security practices have dramatically improved over the last two decades, while acknowledging that the attack surface has remained massive, and in some cases expanded significantly. We are safer in some respects and more exposed in others. This is true both for the military use-case and the civilian.

On the positive side, major operating systems now enjoy significant security features rendering compromise increasingly difficult. This includes up-to-date versions of Microsoft Windows,[13] but also improvements to the various Linux/Unix implementations frequently used in servers and hardware deployed in military networks.[14] These improvements are accompanied by more streamlined patching cycles that mitigate risks faster than was previously possible, further degrading the long-term viability of a given attack vector. Additional security-centric operating systems and deployment solutions have been publicly available for several years, offering tighter controls and compartmentalisation of sensitive environments, limiting the overall impact of compromising specific software in some circumstances.[15] Cloud service providers may or may not increase overall security, depending on how they are used. Even as mobile devices have proliferated, they have arguably become increasingly resistant to persistent compromise. Technology company Apple has enshrined security as one of its key tenets[16] with variable success, and Android architecture has markedly improved since its market introduction.

Security mitigations applied across all levels of computing have mostly made it trickier to achieve full system compromise. Success may now require complex exploit-chaining,[17] raising the cost of success against capable defenders. Perhaps the biggest evidence of this may be seen in the booming exploit market, in which private companies seeking to purchase exploits offer rapidly increasing pay-outs.[18] Exploits against high-value targets are more expensive because of increased demand, but also due to the complexity of success. The entry price has therefore increased to levels that may deter some under-resourced militaries from engaging effectively against hardened targets. In other cases, government agencies and militaries may turn to third parties to provide them with capabilities.

In sharp contrast, even ancient techniques have retained their operational utility. Despite best efforts by network defenders to reduce its effectiveness, the underlying model of network operations has not changed: seek initial access to the network, move laterally to points of interest, exfiltrate data along the way, and pursue objectives as needed. Perhaps most pertinently to the military domain, frequently updating vulnerable hardware and software remains a key challenge in maintaining an operational environment that is relatively resistant to OCOs. The extensive testing required, associated costs, risks of decommissioning equipment for upgrades, and continuous need for interoperability with legacy equipment means that upgrades may be few and far in between. In some cases, recently detected vulnerabilities were embedded so deep in the hardware stack that only radical patching of core functionality could mitigate the vulnerability.[19] In other cases, military equipment developed opaquely by contracted providers is not subjected to the same levels of open scrutiny and may therefore exhibit flawed secure development processes. Considering the increased intricacy of a single element of military hardware, the vulnerable attack surface has possibly grown faster than corresponding mitigation efforts.[20]

Software vulnerabilities are not becoming niche. Major operating systems and service providers invest heavily in securing their products, but remain within reach of adversaries. In China, hacking contests such as the Tianfu Cup have resulted in numerous exploits against ubiquitous software, including Microsoft Windows, the latest version of Apple's iOS, Ubuntu, and more.[21] The Chinese exploit renaissance is particularly noteworthy considering recent Chinese legislation providing greater control—and potential first use—of disclosed vulnerabilities to the government.[22] The proliferation of platforms, products, operating systems, and smart devices further increases the potential attack surface for those who are able to invest. Vulnerabilities are not inherently about software. As is often repeated, the most vulnerable element in many situations is people. While the argument on the prevalence of exploitable software bugs is certainly relevant, humans retain their innate vulnerability to compromise. The evidence for this is damning; even low-quality phishing still succeeds at scale, including against high-value targets.[23] Many successful network operations were in some part facilitated either knowingly or unwittingly by a compromised individual. This reality is not likely to change in the foreseeable future.

Dependence on data is now a societal vulnerability. One anecdote that demonstrates this is the story of Strava. A benign private sector company developing fitness devices that track user performance found itself engulfed in an inferno of global attention from security researchers after it unveiled an interactive map in late 2017.[24] The map—intended to be an attractive visualisation of Strava's market penetration—aggregated anonymised user activity into heat maps, showing where its fitness trackers were used. As many observers quickly realised, the map inadvertently revealed patrol routes within military bases, the pinpoint location of sensitive facilities, and international deployments of forces.[25] It was also possible to deanonymise users with relatively minor effort. While there were core issues in how Strava implemented their data tracking, the map itself was a feature rather than a bug. It had inadvertently shown how even military forces could be threatened by misuse of civilian technologies by its members. It demonstrated a radical expansion of the threat model.

Strava was not an isolated incident; individuals and groups blindly give of themselves to technology. People willingly relinquish privacy, sensitive data, or control over aspects of their lives for perceived benefits. This is not inherently wrong or right, but the associated costs are immensely difficult to assess, especially for the average user. As people increasingly intertwine their daily routine with more connected devices, they increase their inherent vulnerability vulnerability. These devices soon become significant in shaping perceptions of reality itself; tampering with them and the data they rely on may therefore alter that perception in weaponisable ways. Military forces are no different in this respect. A growing reliance on data feeds and technology makes their users inherently more dependent, and therefore vulnerable.

One potential future example of this is the increasing use of augmented reality (AR). AR includes technology that overlays what the eyes normally see with contextual information and additional visual features. Versions of this have existed in militaries for many years—heads-up displays for aircraft that overlay targeting information and telemetry have long been operational. As the technology matures and becomes more prevalent in military use, it becomes a vulnerability in its own right. Relying on augmented displays rather than physical perception can create a dependency which could be exploited through OCOs that seek to shape what the operator sees, causing undesired behaviour. With information operations already seeking to softly impact the overall perception of reality, this trend may gradually increase in prominence.

On Intelligence

As with countless other concepts, artificial intelligence (AI) lacks a consensual definition. Even as it explodes in popularity and is brandished in corporate marketing campaigns, the boundaries of where an algorithm ends and artificial intelligence begins are blurry. Per Horowitz, one largely agreeable component of AI is that it can achieve its goals in a broader range of circumstances and environments than traditional algorithms.[26] Others designate AI as being able to solve tasks of sheer complexity normally handled only by humans, such as speech analysis and contextual decision-making.[27] The underlying approach to solving tasks more organically mimics human behaviour by gradually adapting and learning from experiences both failed and successful.[28] AI approximates facets of human intelligence using software patterns adapted to improve with further exposure to inputs.

It is important to distinguish between two primary forms of AI. Limited applications are already at play in various capacities as narrow or modular artificial intelligence, capable of solving domain-specific tasks. These systems may exhibit above-human levels of success in environments with limited variables and rules, but cannot be broadly applied against any domain without significant adaptation. Examples of this include the "Deep Blue" system that defeated Garry Kasparov at chess in 1997,[29] or the systems competing in DARPA's Grand Cyber Challenge.[30] Such systems may achieve previously unattainable performance when set to specific tasks but would not be immediately useful in solving others. The other type refers to artificial general intelligence (AGI), capable of assessing and resolving challenges across any number of domains by intaking environmental data and generating a favourable behaviour. Such systems are infinitely harder to create and would represent a momentous leap in the field if successfully developed.[31] AGI is viewed with apprehension by many who are concerned with the unpredictability of an entity capable of rapid self-evolution beyond human control or understanding.[32]

The overall appeal of AI—even narrow AI—is understandable and extends far beyond military applications. Autonomous vehicles capable of reacting responsibly to their surroundings promise a revolution in transportation. AI may assist individuals in making responsible choices daily by more objectively assessing data and achieving optimal decisions for their circumstance. At higher levels, AI may contribute to running increasingly complex networks such as so-called smart cities. This includes automated allocation of resources as they are needed, rapid detection and mitigation of faults, and aggregated feedback to operators who can then action as necessary. The motivation to adopt AI for numerous uses is mounting, and its integration into the security domain will likely be correlated with its adoption by society at large.[33]

Artificial intelligence is already being applied for numerous military uses. The People's Republic of China is investing heavily in a spectrum of AI developments in an attempt to ensure long-term technological superiority over its global competitors,[34] and plans to achieve global AI supremacy by 2030.[35] The United States has similarly identified AI as significant to its technology-led "Third Offset" strategy first penned in 2014, where such technologies serve alongside autonomous platforms in cementing US power projection in an increasingly contested geopolitical environment. In a 2016 speech by then Deputy Secretary of Defense Robert Work, he openly stated that "we believe quite strongly that the technological sauce of the Third Offset is going to be advances in Artificial Intelligence and autonomy."[36] Applications range from the tactical to the strategic, from more capable missile guidance and steering to battlefield planning. Considering their limitations to domain-specific problem solving, however, AI is not the panacea it seems to be. Much like OCOs, it will only be useful if used correctly. It requires the identification of opportunities, generation of appropriate datasets, and responsible incorporation of such mechanisms into both new and existing platforms. Perhaps even more so than in other areas, artificial intelligence offers unique opportunities in enabling both event and presence-based offensive network operations. At least in the US, this is being already attempted at some scale, where AI is folded into rapid decision-making and navigating numerous concurrent intrusion operations.[37] Since OCOs represent the repeat compromise of computers and networks, they enjoy a fairly predictable set of rules and characteristics that would fit narrow AIs. As Horowitz claimed, rather than being the weapon itself "AI is actually the ultimate enabler."[38]

The use of artificial intelligence in strategic decision-making is particularly notable. The allure certainly exists. The modern battlefield is complex and difficult to effectively assess by the human mind, which is both inherently limited in its capacity to process sensory inputs, and heavily prone to decisional bias. As such, autonomous platforms capable of objective analysis represent the potential to aspire to an objective strategic optimum.[39] Use of AI in tactical decision-making for target allocation is nascent but emerging;[40] escalating decision-making to the campaign level is enticing but exponentially more complicated. Yet handing matters of strategic consequence to AI risks turning them into centres of gravity by their own right; these platforms may eventually become pivotal to command and control, and thereby become worthwhile targets. With artificial intelligence accelerating the pace of in-conflict decision-making, it risks reducing response times until they impede the ability of the human in the loop to fully understand what they are seeing or authorising. Turning biased strategic platforms opaque is a meaningful threat that could be exploited by a capable enemy.[41] Adversarial algorithms meant to specifically prey on weaknesses manifest in artificial intelligence and machine learning are proliferating, with the field potentially facing a surge as adoption of AI and ML increases.[42]

For presence-based operations, relying on AI may assist decision-making and manoeuvring within adversary networks. By automatically assessing sensory input received from malware infections in adversary networks, viable options may be rapidly generated and actioned upon by operators.[43] The use of such platforms may shorten the presence phase by reducing the need for operator-led lateral movement within networks. Autonomous presence-based operations may also fare better at detecting fault lines within networks and vulnerable endpoints, increasing the chance of success.[44] Alongside shortened presence, reducing the chance of human error due to AI-led command and control would reduce the risk of detection. As previously explored,[45] the presence phase of an operation consists of numerous micro-cycles in which operators manoeuvre implants within compromised networks, retrieve additional information, perform assessments, and decide on follow-up actions. While creativity may be useful, these processes are often repetitive and conform to the same set of circumstances. The laundry list of activities includes maintaining operational security, identifying vulnerabilities and credentials used for further lateral movement, locating key sources of intelligence, and obtaining privileged access to the objective systems. The methods to accomplish these activities are often also repetitive and directly in response to certain telemetry; use a certain vulnerability against a certain type of endpoint, run a certain module to obtain additional data, avoid endpoints where a certain brand of anti-malware solution may be deployed. Consequently, training AI to mimic operator behaviour may not only be possible but a relatively cost-effective way to scale OCOs. By relegating all but the most complex, hard-to-breach target networks to autonomous network intrusion platforms it may be possible to generate higher-quality effects against a broader set of targets.

For event-based operations, incorporating limited-scope artificial intelligence into weapon systems may increase their robustness. A networked attack platform may be able to; (1) intelligently curate viable targets, (2) enumerate vulnerabilities per target, (3) choose applicable exploits, and (4) choose mission-relevant offensive payloads.[46] This reduces both the overhead and expertise required by deployed operators, which may reduce hesitation by commanders in employing such capabilities. If such a system can calculate the probability of success for attempting to compromise an adversary system or network, commanders can more realistically assess whether such an option is viable.

In the preparation phase, AIs may eventually contribute to both presence and event-based OCOs. As vulnerability research is often a significant component in facilitating offensive operations, an AI capable of more rapidly and thoroughly dissecting adversary protocols, software, and hardware may prove invaluable. While some automated capabilities already pervade vulnerability research,[47] these are more limited in scope and primarily serve to identify potential bugs meriting additional investigations. As previously mentioned, the vulnerability research process is expensive in both time and resources; even capable entities such as the NSA have only so much top-tier talent to assign to the task. Delegating aspects of this research to artificial intelligence approximating their human counterparts could broaden the scope of available targets, while reserving key talent to unique and novel challenges where human creativity is not easily replaceable.

AI's contribution may not be limited to assisting human operators; it may eventually supplant them. At least initially, some capable nations are seeking to augment network defence with autonomous capabilities that have reaction times and agility outstripping that of a person or group of people. To parse and assess massive quantities of data requires significant computational resources. These quantities grow as sensors improve and proliferate, a distinct characteristic of both the modern battlefield and modern networking in general. DARPA invests in many such projects, ranging from anomaly detection, improving the resilience of networks to attack, and automatic patching of vulnerabilities.[48] The innovation cycle that had bred network operations will eventually result in AI-enabled network defence as a counter-innovation.

Innovative network defence solutions would require adapted OCOs. If operators become insufficiently dynamic to overcome automated network defences, the business case for autonomous offensive capabilities may evolve organically. Beyond just facilitating rapid lateral movement, offensive platforms would need to make intelligent decisions and respond to high-tempo changes made in networks by deployed defences. This eventual maturation of OCOs into autonomy offers advantages but also considerable risk; it becomes difficult to detect the point at which operators and system developers lose effective control. OCOs may become opaque black boxes, which must be trusted to perform as desired behind enemy lines even in highly variable situations. The amount of uncertainty introduced and the necessity to cut the human operator out of the decision-making loop may prove dangerous, as the risk of cascading impact that already characterises OCOs may exponentially increase. Even after two decades of network operations, and within manually controlled operations, capable nations still frequently fail to safeguard their tools and avoid collateral damage.[49] The significance of autonomy in OCOs is therefore an unknown quantity.

Unmanned platforms have been in military service for over two decades. Their classic roles vary but they are often either precision weapons or concentrate on reconnaissance. As the underlying technologies matured and proliferated, drones have increasingly occupied additional battlefield roles, including communication, electronic warfare, logistical support, transport, and even search and rescue. They come in every size and are deployed across all physical domains; air, land, sea and space.[50] Today's unmanned platforms are more versatile and capable of enabling accurate operations at scale while minimising physical risk. With improvements in robotics and the aforementioned artificial intelligence, this trend is unlikely to reverse in the foreseeable future.

Many military platforms are transitioning from automation to autonomy.[51] In addressing the viability of such systems, the question has gradually transformed over the last two decades from "can this be done?" to "how should this be done?" The tide of autonomy cannot be prevented, only channelled in directions where it may have higher utility and lower risk; such considerations exceed the scope of this book but may become instrumental to modern warfare. For those who assess the potential impact of robotics, concern mounts to alarming levels. The US Army Training and Doctrine command views them as "potential game changers" that "can provide a decisive edge over an adversary unable to match the capability or equal the capacity."[52] Russian military scholars have repeatedly expressed that robots are due to perform myriad tasks in modern warfare, making them a key characteristic of new-generation warfare.[53] Official Chinese publications view unmanned intelligent platforms as a key reason that the traditional centres of gravity in warfare have been displaced.[54] The consensus seems to be threefold; (1) there will be an exponential increase in the use of autonomous platforms; (2) these platforms will become increasingly capable of performing complex roles, and (3) they are both a threat and an opportunity to upset existing symmetries.

Increased adoption of unmanned platforms increases a military's exposure to OCOs. Capable as they may be of feats unattainable by human beings, they also incur unique vulnerabilities. Subversion of software used by a human operator may cause undesired results and—in extreme circumstances—even physical harm, but it does not directly target the individual. Governed by software by nature, unmanned systems are fundamentally exposed to compromise by OCOs, either directly or indirectly. As aptly put by Hartmann and Steup in 2013, "[Unmanned Aerial Vehicles] must be classified as highly exposed, multiply linked, complex pieces of hardware."[55] The possibility to inflict complete combatant shutdown as a result of a successful network attack against the system is both real and significant.

The United States was caught by surprise in 2011 when a RQ-170 Sentinel UAV suddenly became unresponsive while conducting a mission within Iran's borders. Iranian media soon announced that they had not only intercepted the drone, but effectively interdicted its control channel to force it to safely land on an Iranian airstrip. While the United States did not officially acknowledge the details of the incident, the Obama administration petitioned the Iranian government for the return of the aircraft. Theories of how this came to pass were plentiful, but one likely explanation persisted; spoofed GPS data combined with jamming of US GPS signals allowed the Iranians to provide overriding coordinates.[56] This form of protocol compromise was effectively an event-based attack against the Sentinel, conducted by a second-tier regional power. The potential for more capable parties to carry out more intricate attacks against autonomous platforms is significant.

With a rich array of sensory and external inputs, unmanned systems are intrinsically vulnerable to external compromise. Sensors can be tricked into producing erroneous analysis, while external data sources such as command and control, guidance, and telemetry can be overtaken to fool a platform into behaving unexpectedly. Consequently, an over-reliance on drones could eventually create new centres of gravity in their command and control centres. If military power is increasingly facilitated by remote operation centres guiding fleets of mixed autonomous platforms, targeting these centres becomes a viable method to reduce adversary fighting strength.

The issue of targeting unmanned command and control is further exasperated by the introduction of so-called "swarming" tactics. Rather than relying on a limited number of potent high-cost systems, militaries may opt to instead deploy cheap formations in high quantity, overwhelming defenders.[57] As a thorough RAND publication from as early as 2000 explained, swarming is not uniquely related to unmanned systems and has been applied by human forces for centuries with variable success.[58] The tactic allows remediating asymmetries against well-resourced or massed adversaries with comparatively lesser effort. Such an approach is already a significant component of Iranian doctrine for dealing with the United States and Israel;[59] it is a component of modern Chinese doctrine;[60] it was used by jihadi forces in Syria to overwhelm deployed Russian forces;[61] and is actively pursued by the United States via DARPA[62] and elsewhere.

Unmanned swarms require an increased degree of autonomy as agility becomes essential. Individual craft would need to respond rapidly to changes in their surroundings and adversaries, and coordinate with adjacent and remote systems.[63] This will be made possible by extensive mesh networks governed by software. Ostensibly, operational swarm-wide oversight would in the near future remain in the hands of human operators, but those operators would not be able to directly pilot the dozens or hundreds of drones participating in any given swarm. Autonomous piloting authority would be transferred to the system itself, further extending its vulnerabilities.

Event-based attacks against deployed swarms may be devastatingly effective, and could include disrupting sensory flow, interfering with telemetry and command channels, or even sending contradictory data to the participating drones. Through this it may be possible to achieve a wide range of previously infeasible effects against deployed manned forces that rely on both machine-fed data and their own deduction and contextual situational awareness. The potential for presence-based attack is similarly broad; an adversary may gain significant advantages through targeted operations against theatre command centres used to govern drones, or against the communication infrastructure used to facilitate control of swarms.

It is possible that as unmanned systems proliferate in quantity and significance to modern societies, the spectrum of network warfare will commensurately shift. Where today even impactful network intrusions are often discarded as unworthy of countermeasures by the victim, the future may alter this calculus. If unmanned platforms become a new form of critical infrastructure, even non-offensive intrusions against the networks that house them may spark grave alarm in victims. The dangers of the cybersecurity dilemma as posited by Buchanan[64] may be aggravated—it is impossible to determine whether breaches against networks encompassing autonomous systems and artificial intelligence are meant for intelligence collection or a corruptive attack.

Cyber as a Domain

The story of "cyber" as a domain of war stretches back less than a century. While it can be traced back to Norbert Wiener's notion of cybernetics as the persistent relationship between humanity and machines,[65] it is already becoming tenuous to envision networks as a separate man-made cyberspace, detached from other aspects of humanity and subject to its own rules. The inverse is the reality; the more networks became integrated into other aspects of warfare, the more they became intricately bound to war's innately human circumstance.

A REVOLUTION IN CYBER AFFAIRS?

It is unlikely that information and the medium that bears it will fade into irrelevance. A retreat into connectionless conflict devoid of the ever-present hunger for data seems a remote possibility. However, this does not imply that "cyber" will necessarily remain as a distinct domain or warfare, or even as a pervasive term. As networks fully permeate warfare, they may simply become implied. Cyber as a prefix will become unnecessary if everything is cyber. Military planners already struggle when accounting for the interplay of networks and the other domains; information seems to seep into all aspects of warfare, muddling the forced attempts at separation. As this intensifies, attempting to distil a perception of warfare in the so-called fifth domain will become more difficult. OCOs of varying types and intensities may eventually find their way into an integrative doctrine, but they do not inherently require the benefits that domainhood provides, and may indeed suffer from it. Irrespective of the assumed increased susceptibility to network attacks, the coercive ability of network operations would arguably remain limited. In such a scenario, OCOs will maintain their dependence on the physical domains for pursuing objectives, as a network attack against a defender would predominantly be followed by a kinetic force meant to subdue it.

The idea that cyber will not retain its domainhood is not revolutionary. Many other nations—primarily non-Western—already do not envision computers, the networks they form, or the logical entities they create as a separate domain of war. While nations prolific in network operations such as Russia, China, Iran and Israel have an aggressively proactive view of OCOs as a key enabler of modern military success, they primarily view the operational space as information-centric, and tasked with enabling broader strategic goals or kinetic operational forces. In this sense, the strategic perspective comparison between east and west differs from the Cold War. As the US focused on a potent air force and overall technological supremacy, the Soviet Union focused on tight air defence grids, potent rocketry, and proliferation of affordable equipment to global allies. Today, global contenders largely agree that information is pivotal for success in modern warfare, and that qualitative technological superiority is essential. They differ in how they treat the interplay of this technology with the existing tenets of warfare. Where the US and its NATO allies invest in the uniqueness of cyberspace, others espouse holistic warfare through which information is a constant, unrelenting, and crucial undercurrent.

Regardless, OCOs as presented within this book are independent of the status of cyber as a domain. Whether carried out by a dedicated military command, intelligence agency, a theatre command centre, or an individual aircraft, the characteristics remain much the same. The operational cycles that enable OCOs require a grand commitment of staff, resources, intelligence, and research. These do not necessarily need to be delivered by a dedicated domain structure, as indeed is the case for many militaries globally. As such, the models presented within this work on the integration of OCOs to military doctrine and strategy perform equally whether observing cyberspace as a domain or grasping information holistically. As long as their requirements and parameters are considered and woven into the decision-making process, they may prove their utility today or in the foreseeable future. The approach to networks may differ as maturity increases, but their characteristics remain largely similar.

CONCLUSIONS

Cyber-security is naturally a multi-disciplinary field. At its core, the field often reflects the intersection of networks with all other domains of study and practice: criminality, health, safety, finance, commerce, sociology and psychology, and others. As a corollary, cyber-warfare is the intersection of military thought, intelligence, and network security. In cyber, technical and operational aspects are uniquely inseparable and equal in importance; we cannot and should not separate them, or favour one over the other in analysis. Offensive and defensive capabilities developed without being understood by their designated users will remain unused, or perhaps misused. Similarly, military strategy that does not account for the radical changes in the operational environment over the last two decades will find itself lacking, even against seemingly disadvantaged adversaries. A reliance on warfare spearheaded by cutting-edge technology must compensate for the vulnerabilities that it creates, as these will increasingly be preyed upon by others.

It is easy to label network incidents incorrectly. Some media coverage of high-profile incidents involving networks remains laden with hyperbole, despite a generally positive trend towards nuance. Security researchers worldwide routinely lament the lacklustre progress in adopting even basic protections, and companies continue to be breached due to basic lapses in security. Password reuse, unpatched systems, misconfigured networks, and susceptibility to phishing remain potent ways of compromising even seemingly secure environments. When so many malicious activities occur within the researched space, it may become difficult to identify the subtler interplay of network operations and military activities.

Not all network intrusions are attacks, and not all attacks are warfare. While humanity has millennia of experience in distinguishing between malicious activities in the physical domains, the struggle still continues for distinctions in activity targeting networks. Recognising the differences is significant not only for policy or law-centric analysis, but also for the military domain. Determining which activities could reasonably be assigned to military agencies and units and which should remain civilian is key. Determining which activities are included when discussing national approaches to military approaches is crucial.

Even the perception of cyber itself varies wildly. Some—such as the US and its allies—codify cyber as a domain of warfighting and establish distinct commands to tackle it. Russia adopts a more pragmatic approach to OCOs, viewing them as a fragment of activity within a far broader scope of information operations occurring in both peacetime and conflict. It is a way of achieving or approaching objectives with less friction and reduced chances of kinetic escalation. For China, network operations of varying kinds are key facilitators of its modern doctrine, which envisions quickly reducing Western technological advantages in any conceivable way and enabling rapid operations. Iran views OCOs as a means of further extending its reach and employing proxies to attempt deterrence. There is no distinctly right or wrong approach; rightness is a measure determined by context. Part of the advantage of network operations lies in their flexibility; they may be applied in many ways against an adversary depending on the circumstance.

The core argument has been presented that viewing offensive network capabilities as a monolithic stretch of operational space is risky and counterproductive. Though they differ, Russian, US, Chinese, and Iranian doctrines primarily treat OCOs as a single spectrum of possibilities that share most characteristics. Yet, network operations were not all created equal. Some may be instantly deployed on the battlefield by an infantry detachment, while others are carefully managed, multi-year operations against sensitive adversary command centres. On the other hand, most offered taxonomies splice network operations across numerous variables and parameters, making them useful for post-hoc academic analysis but more limited in utility for military planners. The intention was therefore not to generate models that cover every manifestation of operating against networks—though that certainly has its value.

Cyber did not come into existence overnight. Operations carried out today are a result of counter-innovation cycles that accelerated in the twentieth century. The need to deceive and manipulate the machines on which we have become reliant intensified when militaries turned to radar for guidance and the radio for communication. As these technologies proliferated and became complex, the desire to target them and the possibility to gain value from doing so commensurately increased. This intangible warfare became doubly important as computer networks evolved into a virtual crutch within modern warfighting reliant on ever-active datalinks providing sensory data, command and control, and telemetry. Any models pertaining to network warfare must therefore account for their roots in electronic warfare and signals intelligence; cyber is essentially their technological love child.

The goal was therefore to craft concepts which categorise cyber operations in useful ways. By dividing OCOs into event-based and presence-based operations, immediate fault lines begin to surface. When subjecting each of these high-level categories to the operational processes underpinning all OCOs, it becomes clear just how much they differ, and how risky it is therefore to bundle them together. Event-based operations are primarily robust, multi-use capabilities requiring high reliability and intensive research and development cycles. Conversely, presence-based operations require permanent intelligence support, covert operational nuance, and would likely eventually compromise themselves upon use. The former befits use with deployed forces, while the latter may be best retained by intelligence agencies with an operational mandate.

This book has sought to create a robust perspective on OCOs in accumulative layers. Each layer was meant to focus discussion, exclude possibilities that muddle analysis, and offer straightforward classification criteria that are easy to implement, both in subsequent research but also in strategy for employing offensive network operations. Each of the first four chapters offers self-sufficient analysis from a different perspective. When combined, they indicate that cyber operations can tremendously benefit from implementing the lessons of other forms of technology-focused warfare. A clear typology for offensive network capabilities can then assist in further development of the overall field, by facilitating examination of how each category can be incorporated into strategy, operations, and tactics.

Chapter 1 examined the spectrum of network operations, setting boundaries around the conceptual perimeter of cyber-warfare. The chapter was intended to narrowly focus the debate itself in a two-pronged approach. The first exercise included identifying which incidents do not merit examination within the prism of offensive network operations undertaken by parties involved in warfare. This excluded a broad range of activities frequently encountered within the debate, such as intelligence collection, criminal activities, information operations, or loosely affiliated ideological hacking. The second exercise sought to offer five escalating criteria which in turn help assess which malicious incidents are worthy of inclusion within the observed dataset. This has the significant side benefit of illustrating how most malicious incidents assessed today should likely not be labelled as cyber-warfare.

Chapter 2 surveyed how each of the main characteristics of OCOs are thoroughly rooted in historic intangible warfare. The twentieth century demonstrated how a rapidly growing reliance on the electromagnetic spectrum initiated a process that culminated in what is now labelled cyber operations. Each iteration of intangible warfare—from jamming, to electronic warfare, to command and control and network-centric warfare—contributed characteristics that now accompany modern cyber capabilities. By presenting how OCOs draw from the existing strong foundation of intangible warfare, further research can be made on integrating electronic warfare and OCOs along their similarities.

Chapter 3 introduced the distinction between OCO archetypes—event-based and presence-based operations. The argument was that these two categories are simple enough to be easily usable while disparate enough to be useful. Event and presence-based capabilities were shown to have distinct characteristics across the entire operational life cycle. The resource requirements are often unique, the development process differs, targeting is undertaken with different goals in mind, and even the operational staff itself may be altogether distinct. As such, lumping these two operation types together may result in overly broad results.

Chapter 4 dissected the utility of event and presence-based operations. The goal was to interweave the characteristics of the two OCOs archetypes with strategic considerations to determine when, where, how and why they should be used. By examining how pre-existing strategic wisdom remains wholly applicable to cyber-operations, it became clear that event-based capabilities would usually fit tactical or operational needs, while presence-based operations are often best suited for strategic support and pursuit of loftier objectives.

Difference of Approach

All of the surveyed countries rely on network operations to a degree, and they all do so imperfectly. Assessing the various ways key countries fall short of optimally utilising OCOs was the goal of the subsequent four chapters. The United States exhibits top-tier technical capabilities but a limited capacity to put them into use across all operational spaces, due to bureaucratic difficulties, extreme compartmentalisation, and the isolation of OCOs within a separate military command. Capabilities therefore exist, but are often not applied where they are needed most. Russia is the most visibly aggressive user of OCOs globally, but does so with reckless abandon and insufficient consideration to their overall impact on underlying military-political objectives. Success is often incidental, limited, or a result of committing to operations in low-quality bulk. Russia has integrated OCOs so thoroughly that they are treated as simply another instrument in the information-operations toolset, with limited regard to how they could be more nuanced in order to pursue more intricate goals. China appears relatively doctrinally mature, with a cohesive structure amalgamating cyber capabilities from across different branches, but a desire to view the information space holistically is not yet backed up with any profound operational experience. Iran has aggressively layered offensive operations into its grand-strategy, making effective use of the limited resources at its disposal. While it cannot compete with some of the top performers, it is demonstrably capable of generating visible effects in service of its objectives. Iran is clearly not as well-resourced as its counterparts and has exhibited numerous operational flaws, but has also consistently punched above its weight, achieving impact against its regional and global adversaries as it sought to respond to perceived slights and alter the calculus of conflict. Each nation could benefit thoroughly from embedding the characteristics of OCOs into their doctrine, and thoughtfully constructing strategy that befits their unique situation.

Despite common concepts, implementation may still vary. Much like other aspects of military strategy, there is no one optimal approach. A smaller military such as the Israeli Defense Forces may focus heavily on event-based capabilities meant to support its strategically crucial air force, while at the same time engaging in presence-based operations to soften adversaries and impact their readiness. Taiwan may wish to disproportionately focus on presence-based operations against the PLA Rocket Force or regional command and control, in efforts to sufficiently delay any PLA advancement so that US forces or the international community may interdict any attempt to subdue it. Iran may increase its reliance on strategic event-based operations against critical targets within asymmetrically stronger adversaries to weaken political resolve and act as a de facto deterrent. OCOs can act as either a force multiplier, an operational enabler, or even as limited means of pursuing objectives; it all rests on context.

All the case studies raise one obvious question—who is better? As this book shows, the only responsible answer for that question is with another: better at what? At the time of writing, the US clearly has the most significant pairing of technical acumen, operational discipline, intelligence capacity, and subject matter expertise needed to successfully carry out operations against a highly defended adversary using bespoke technologies. If the relevant scenario was looking for the best threat actor capable of executing an intricate presence-based operation, achieving all tactical objectives, and getting away with it, the US tops that small list. Yet this is not the only viable scenario.

The Russian approach, loud as it may be, cannot be discounted. While their operations frequently get exposed and often fail to achieve their full objectives, they are a capable adversary. A willing-

CONCLUSIONS

ness to experiment, effective usage of domestic expertise, lack of concern for being exposed, multiple potent intelligence agencies, and a powerful, proven record of capable electronic warfare all add up to a dangerous reality. As seen in Ukraine, the Russians learn rapidly from their mistakes. The unshackled nature of their operational approach means that they are more likely to allow OCOs to be used as intended, which provides numerous opportunities. Were the Russians to pursue a military campaign against any of their neighbours, it is highly likely that a full spectrum of information operations would be brought to bear, with significant harm—both transient and otherwise—incurred by their victims.

Figure 3: Summary OCO assessment for case studies

	STRENGTHS	WEAKNESSES
UNITED STATES	– Proven SIGINT/OCO maturity – Tooling quality – Doctrinal integration – Resourcing investment – Dedicated R&D	– Sprawling operational bureaucracy – Rigidity of domain framework
RUSSIA	– EW capabilities – Aggressiveness, proven experience – Dedicated R&D – Maturity of combat & intel operations	– Variable operational quality – Internal political power dynamics – Limited strategic success
CHINA	– Doctrinal maturity – Force restructure – Intelligence experience – Vulnerability research	– Unproven ability to apply to military campaign – Variable operational quality
IRAN	– Co-optation of civilian resources – Successful proxy use – Operational experience – Aggressiveness	– Lower-quality tooling – Frequent operation quality issues – Limited evidence of dedicated R&D for impactful operations

The Chinese remain a relatively unknown quantity. Lack of combat experience is meaningful, but does not tell the whole story. Mature doctrine, evolved strategy, a burgeoning capacity for vulnerability research, a proven track record of intelligence operations, and a unified Strategic Support Force are all strong signals of intent. The Taiwan contingency is instructive because it shows the world what unity of purpose affords; the sheer investment in preparing PLA forces for a daunting military scenario are truly impressive. Considering the deep understanding that PLA planners have of their adversaries' respective strengths and weakness, they are likely investing heavily in capitalising on the many vulnerabilities of Western technological reliance, with the aim of carving out just enough of an effect to achieve their strategic goals. So if the question is who is most likely to effectively use OCOs in service of strategic success, the answer may possibly be China.

Iran is the least capable of the case studies observed. Doctrinal literature is limited, tooling quality is questionable, and there are multiple failings of both operational security and the ability to achieve lasting objectives. Be that as it may, the Iranians have succeeded in two key aspects. The first and perhaps most simple is that they are able to exact a cost from their adversaries. By targeting civilians and other vulnerable, exposed targets, Iran is able to create effects where it seeks to. Secondly, Iran has succeeded in signalling that it will gladly employ OCOs to counter perceived slights, political challenges, or to navigate confrontations. In this sense, it has successfully integrated OCOs into its grand-strategy like few other nations have.

Beyond the observed case studies, the degree to which nations may rely on OCOs may also understandably vary. An attempt to overly emphasise military OCOs in Israeli campaigns against the Palestinian Hamas may be fruitless.[1] While they do rely on information infrastructure, Hamas doctrine broadly assumes distrust in its own equipment and even a total command disconnect when conflict ensues.[2] While the PLA may enjoy some success in subduing Taiwanese defensive efforts by degrading their command and control, the island campaign would still incorporate multiple layers of entrenched defenders fighting bitterly and autonomously regardless of available networks.[3]

OCOs are not one size fits all. The models developed in this book are intentionally broadly scoped. This allows for flexibility in implementation, drawing from the particularities that characterise situational parameters. It also means that extrapolating from one scenario to another is risky and must be undertaken with care. What has been seen is not necessarily indicative of what will come next. The prime example for this is Stuxnet; there has been no further public disclosure of similar incidents incorporating long-term deceptive presence-based operations against energy facilities to deter nuclear aspirations of rogue nations.[4] What followed Stuxnet was an array of high-profile destructive attacks against critical infrastructure, such as in Ukraine. These attacks were often used as a tool for political signalling rather than a true ploy to covertly influence adversary grand-strategy. Any further analysis of cyber operations must therefore be tightly paired to the specific scenario being assessed. Context matters.

Further Opportunities

The work presented here has focused on introducing and testing core concepts for the military use of network operations. As such, it leaves many areas relatively undeveloped. As the use of OCOs evolves and more public evidence is provided, the role of these capabilities must be consistently re-evaluated. Each of the conceptual aspects introduced in the book may be further developed with additional works, including those that explore the implementation of event and presence-based capabilities, OCOs in low-intensity conflict, the use of OCOs by sub-national fighting groups, and the use of network operations by other actors not considered within this work. It is also advisable to pursue research into the next "swing" of the innovation cycle, which would bring autonomous platforms and artificial intelligence into the set of considerations.

The United States has been embroiled in decades of low-intensity combat operations against disparate enemies. Priorities have begun to shift back towards preparing for conflict with near-peer adversaries only in the twenty-first century, in light of a resurgent China and an increasingly belligerent Russia. Israel, Russia, Iran, the United States, Saudi Arabia, and others have been combating irregular or sub-national groups in the Middle East and elsewhere. While briefly touched on within this book, the specific considerations for using both event and presence-based OCOs against irregular forces could be explored in depth. These groups are often reliant on civilian infrastructure and a mismatched set of appropriated tactical equipment. As such, they are both more and less vulnerable to different types of operations. An analysis of effective OCOs use against sub-national groups could contribute greatly towards the robust understanding of offensive network capabilities as a whole.

Flipping that, it is also imperative to assess how sub-national groups may employ network operations. These are already occurring in limited forms, with Hamas and Hezbollah known to undertake limited offensive network operations against Israel in times of conflict.[5] While both groups enjoy a measure of autonomy and access to resources, they are inherently disadvantaged when it comes to acquiring manpower and facilitating the high-cost operational cycles that OCOs commonly entail. It thus becomes useful to explore how low-cost, high-yield OCOs may be attained by such groups, what targets they are likely to pursue, and how they may incorporate such capabilities to enable and augment their accomplishments in the kinetic space.

Examining the integration of OCOs into the strategy of other nations would also contribute to a broader understanding of the toolset. North Korea is a prolific user of network operations in pursuit of its grand-strategy, making it a prime candidate for case study analysis. With a penchant for network operations and an integrative approach to technological warfare, Israel remains a prime case study for any use of OCOs against adversaries. The United Kingdom has storied signals intelligence, potent tools, and a declared intention to strategically conduct OCOs. Implementing the assessment models offered throughout the book against these countries may both strengthen their validity and also offer additional insights.

Trends must consistently be accounted for and incorporated into any assessment. While some aspects of network operations have remained static over the last two decades, technological adoption and development have radically changed. With technologies set to enmesh even further into modern life and gradually increase in autonomy, they may adversely impact both the significance of OCOs and how they are carried out. Time spans are notoriously difficult in technology, so it remains challenging to tell when artificial intelligence would

CONCLUSIONS

become crucial to this field. The trajectory, however, is clear. Humanity has already peaked in its ability to process the reams of data that networks provide, and relies heavily on software to assist. With more data, sensors, and requirements, the need to delegate a greater portion of analytical processes to increasingly intelligent software is becoming clearer.

Autonomous platforms may both target and be targeted through networks. The more computerised these platforms become, the more our overall network attack surface increases. Similarly, such platforms may deliver offensive payloads of their own, potentially assisting in acquiring access to air-gapped networks within well-defended territories. The implication of autonomous platforms should therefore be gradually explored as their adoption increases and data emerges; they may increasingly become a significant component in OCOs, both as target and offensive platforms.

The future role of artificial intelligence in military OCOs is indeterminate. As with autonomous platforms, they may both be targets of or a vehicle for offensive operations. Much like the adoption of networks and computers, a gradual dependency on artificial intelligence may emerge as they become more adept at solving complex tasks and achieving optimal battlefield results. This may extend to all tiers of military thought; the tactical, operational, and strategic. Where tactical, limited-scope AI may assist in situational-responsive weapons guidance, strategic AI may help direct resources and operational planning at the theatre level. At the same time, other forms of narrow AI may also be used to scale the use of OCOs, effectively shrinking the resource constraints detailed in the operational life cycle. In adherence to the cyclical mentality, narrow AI may also be used for network defence, thereby drastically reducing the success rate of deterministic approaches to OCOs which are currently so often relied upon. All such applications—and their potentially unique impact—could be thoroughly explored in further research.

Finally, defensive operations are crucial to operating in and against networks. While they were largely excluded from this work, assessing the offence–defence balance would make for a valuable supplement. The various strategic approaches to OCOs can and should be countered with risk mitigation approaches. These include a frank reconsideration of the Western approach to technology-laden superiority, in which equipment is often superfluously networked. As a response to the utility of military OCOs presented here, perhaps it would be wise to consider the vulnerability of certain platforms and systems prior to networking them. This type of analysis is particularly relevant in light of some ongoing debates, such as current discussions about the US nuclear arsenal.[6]

This book generates some answers and numerous additional questions. If strategic planners and military analysts more thoroughly understand the utility of OCOs, they may do better at steering the analytical conversation and doctrinal construction towards the right questions. One certainty is that networks will both retain and increase their prevalence in all military affairs. Recognising this, it is essential to continuously challenge how OCOs are used, and offer additional questions meant to advance their study to currently uncharted areas.

NOTES

INTRODUCTION

1. Geoff Dyer, "US Launches Online Assault Against Isis," *Financial Times*, 6 April 2016, https://www.ft.com/content/4d98edd0-fba5-11e5-b3f6-11d5706b613b
2. Dyer.
3. David E. Sanger, "U.S. Cyberattacks Target ISIS in a New Line of Combat," *The New York Times*, 24 April 2016, http://www.nytimes.com/2016/04/25/us/politics/us-directs-cyberweapons-at-isis-for-first-time.html
4. Ash Carter, "A Lasting Defeat: The Campaign to Destroy ISIS," Cambridge, Mass: The Belfer Center, October 2017, 33.
5. Carter, 33.
6. Carter, 33.
7. Dina Temple-Raston, "How The U.S. Hacked ISIS," NPR, 26 September 2019, https://www.npr.org/2019/09/26/763545811/how-the-u-s-hacked-isis
8. Jeremy Fleming, "GCHQ Director's Speech at CYBERUK 2018," CYBERUK, 12 April 2018.
9. United Kingdom Government, "National Cyber Force Transforms Country's Cyber Capabilities to Protect UK," GOV.UK, 19 November 2020, https://www.gov.uk/government/news/national-cyber-force-transforms-countrys-cyber-capabilities-to-protect-uk
10. The State Council Information Office, "China's Military Strategy" 2015, https://web.archive.org/web/20191219232501/http://eng.mod.gov.cn:80/Press/2015-05/26/content_4586805.htm.
11. Robert M. Lee, "Potential Sample of Malware from the Ukrainian Cyber Attack Uncovered," SANS Industrial Control Systems Security Blog, 1 January 2016, https://ics.sans.org/blog/2016/01/01/potential-sample-of-malware-from-the-ukrainian-cyber-attack-uncovered
12. Cybereason, "Paying the Price of Destructive Cyber Attacks," 2017, 2, https://www.cybereason.com/paying-the-price-of-destructive-cyber-attacks.

13. David Maynor et al., "The MeDoc Connection," *Cisco Talos* (blog), 5 July 2017, http://blog.talosintelligence.com/2017/07/the-medoc-connection.html

14. US Press Secretary, "Statement from the Press Secretary," The White House, 15 February 2018.

15. NCSC, "Russian Military 'Almost Certainly' Responsible for Destructive 2017 Cyber Attack," UK National Cyber Security Centre, 15 February 2018, https://www.ncsc.gov.uk/news/russian-military-almost-certainly-responsible-destructive-2017-cyber-attack

16. The 5-step cyber-warfare assessment model will be detailed in the first chapter.

17. US Joint Chiefs of Staff, "Joint Publication 3–12: Cyberspace Operations," 8 June 2018, 41.

18. Gregory J. Rattray and Jason Healey, "Categorizing and Understanding Offensive Cyber Capabilities and Their Use," in *Proceedings of a Workshop on Deterring Cyberattacks: Informing Strategies and Developing Options for U.S. Policy* (Washington D.C.: National Academic Press, 2010), 82–83.

19. DHS Press Office, "Joint Statement from the Department Of Homeland Security and Office of the Director of National Intelligence on Election Security," Department of Homeland Security, 7 October 2016, https://www.dhs.gov/node/23199

20. Ben Buchanan, *The Cybersecurity Dilemma: Hacking, Trust and Fear Between Nations* (Oxford: Oxford University Press, 2017).

21. Cyber-espionage and cyber-warfare routinely get fused together by media outlets and researchers, muddling the observable space.

22. See for example two reports from 2017 and 2020: US Department of Defense, "Fiscal Year 2016 DoD Programs—F-35 Joint Strike Fighter (JSF)," January 2017; US Department of Defense, "Fiscal Year 2019 DoD Programs—F-35 Joint Strike Fighter (JSF)," January 2020, https://www.documentcloud.org/documents/6780413-FY2019-DOT-E-F-35-Annual-Report.html

23. Sean Gallagher, "F-35 Radar System Has Bug That Requires Hard Reboot in Flight," *Ars Technica*, 10 March 2016, https://arstechnica.com/information-technology/2016/03/f-35-radar-system-has-bug-that-requires-hard-reboot-in-flight/

24. Lockheed Martin, "Autonomic Logistics Information System (ALIS)," November 2009, https://www.lockheedmartin.com/content/dam/lockheed-martin/rms/documents/alis/CS00086-55%20(ALIS%20Product%20Card).pdf

25. US Department of Defense, "Chinese Exfiltrate Sensitive Military Technology," 2011.

26. See for example the evolution from Duqu to Duqu 2.0, malware families from 2011 and 2015 respectively, both ostensibly attributed to Israel. GReAT, "The Duqu 2.0: Technical Details," Kaspersky Lab, 11 June 2015, https://securelist.com/the-mystery-of-duqu-2-0-a-sophisticated-cyberespionage-actor-returns/70504

27. See for example the intrusion toolset known as APT, a highly modular platform used to compromise high value targets. GReAT, "The ProjectSauron APT,"

Kaspersky Lab, 9 August 2016, https://securelist.com/faq-the-projectsauron-apt/75533

28. The actual publication of materials was carried out by a group calling itself "The Shadow Brokers", suspected to be but never confirmed as a Russian false-flag information operation.

29. At the time of writing, The Vault leaks are hosted on WikiLeaks and can be found here—Wikileaks, "Vault 7: CIA Hacking Tools Revealed," Wikileaks, 7 March 2017, https://wikileaks.org/ciav7p1

30. Danny Bradbury, "FSB Hackers Drop Files Online," *Naked Security* (blog), 23 July 2019, https://nakedsecurity.sophos.com/2019/07/23/fsb-hackers-drop-files-online

31. Catalin Cimpanu, "Source Code of Iranian Cyber-Espionage Tools Leaked on Telegram," ZDNet, 19 April 2019, https://www.zdnet.com/article/source-code-of-iranian-cyber-espionage-tools-leaked-on-telegram

32. US Joint Chiefs of Staff, "Joint Publication 3–12: Cyberspace Operations."

33. US Air Force, "Air Force Doctrine Document 3–12," 15 July 2010.

34. US Army, "Army Field Manual 3–38—Cyber Electromagnetic Activities," 12 February 2014.

35. US Joint Chiefs of Staff, "Joint Publication 5–0: Joint Planning," 16 June 2017.

36. Russian Federation, "The Military Doctrine of the Russian Federation," 25 December 2014, http://rusemb.org.uk/press/2029.

37. Anthony H Cordesman, "An Open Strategic Challenge to the United States, But One Which Does Not Have to Lead to Conflict," Center for Strategic and International Studies, 24 July 2019; The State Council Information Office of the People's Republic of China, China's Military Strategy.

38. Note the relatively limited evidence of US, UK, or Chinese offensive operations.

39. See for example the ARPANET and the rise of packet-switch networking, the tenets of which remain applicable to date.

1. PRINCIPLES OF CYBER-WARFARE

1. Ilan Berman, "The Iranian Cyber Threat, Revisited," US House of Representatives Committee on Homeland Security Subcommittee on Cybersecurity, Infrastructure Protection, and Security Technologies (2013), 2.

2. Gregory J. Rattray, *Strategic Warfare in Cyberspace* (Cambridge, Mass: MIT Press, 2001), 2.

3. Sanger, "U.S. Cyberattacks Target ISIS in a New Line of Combat."

4. Jason Crow, "The Situation Is Developing, but the More I Learn This Could Be Our Modern Day, Cyber Equivalent of Pearl Harbor." Tweet, *@RepJasonCrow*, 18 December 2020, https://twitter.com/RepJasonCrow/status/1339950863801061377

5. Keir Giles, "Russia's 'New' Tools for Confronting the West," London, UK: Chatham House, March 2016, 9.

6. Peter Elkind, "Sony Pictures: Inside the Hack of the Century," *Fortune* (blog), 1 July 2015, http://fortune.com/sony-hack-part-1

7. James B. Comey, "Addressing the Cyber Security Threat," Speech, Federal Bureau of Investigation, 7 January 2015, https://www.fbi.gov/news/speeches/addressing-the-cyber-security-threat

8. HP Security Research, "Profiling an Enigma: The Mystery of North Korea's Cyber Threat Landscape," Hewlett Packard, 16 August 2014.

9. Jenny Jun, Scott LaFoy, and Ethan Sohn, "North Korea's Cyber Operations," Center for Strategic and International Studies, December 2015, 41.

10. Nathan P. Shields, United States of America V. Park Jin Hyok, No. MJ 18–1479 (United States District Court 8 June 2018).

11. Successfully affecting military assets would still require a complete operational cycle and devotion of resources that are far beyond the capabilities of most non-state actors. See follow-up chapters for extensive elaboration on the unique calculus of OCOs.

12. US Department of Defense, "The Department of Defense Cyber Strategy," 2015, 4; US Department of Defense, "Cyber Strategy Summary," 2018, 2, https://media. defense.gov/2018/Sep/18/2002041658/-1/-1/1/CYBER_STRATEGY_ SUMMARY_FINAL.PDF

13. An important caveat here is that the threshold of warfare is not discussed from a legal perspective but from a war studies perspective. International law has distinct frameworks for concepts such as "armed attack" and its significance for allowing states to respond in force, which are not discussed herein.

14. The State Council Information Office of the People's Republic of China, China's Military Strategy.

15. Russian Federation, "The Military Doctrine of the Russian Federation."

16. UK Cabinet Office, "The UK Cyber Security Strategy: Report on Progress and Forward Plans," 2014, 13.

17. White House, "International Strategy for Cyberspace," 6 May 2011, 2.

18. CPNI, "Critical National Infrastructure," UK Centre for the Protection of National Infrastructure, 2020, https://www.cpni.gov.uk/critical-national-infrastructure-0

19. Ralph Langner, "Stuxnet—Dissecting a Cyberwarfare Weapon," *IEEE Security and Privacy* 9, no. 3 (June 2011): 49.

20. Nicolas Falliere, Liam O Murchu, and Eric Chien, "W32.Stuxnet Dossier," Symantec, February 2011, 37.

21. Michael N. Schmitt, *Tallinn Manual on the International Law Applicable to Cyber Warfare* (New York: Cambridge University Press, 2013).

22. Michael N. Schmitt and NATO Cooperative Cyber Defence Centre of Excellence, eds., *Tallinn Manual 2.0 on the International Law Applicable to Cyber Operations*, 2nd ed. (Cambridge, UK; New York: Cambridge University Press, 2017), 415.

23. Buchanan, *The Cybersecurity Dilemma Hacking, Trust and Fear Between Nations*.

24. Terry Crowdy, *The Enemy Within: A History of Spies, Spymasters and Espionage* (London: Bloomsbury Publishing, 2011), 15.

25. Roger D. Scott, "Territorially Intrusive Intelligence Collection and International Law," *The Air Force Law Review* 46 (1999): 223.
26. Scott, 218.
27. Craig Forcese, "Spies Without Borders: International Law and Intelligence Collection," *Journal of National Security Law and Policy* 5 (2011): 185.
28. FireEye, "Unauthorized Access of FireEye Red Team Tools," FireEye, 8 December 2020, https://www.fireeye.com/blog/threat-research/2020/12/unauthorized-access-of-fireeye-red-team-tools.html
29. FireEye, "Highly Evasive Attacker Leverages SolarWinds Supply Chain to Compromise Multiple Global Victims With SUNBURST Backdoor," FireEye, December 13, 2020, https://www.fireeye.com/blog/threat-research/2020/12/evasive-attacker-leverages-solarwinds-supply-chain-compromises-with-sunburst-backdoor.html
30. Catalin Cimpanu, "CISA Updates SolarWinds Guidance, Tells US Govt Agencies to Update Right Away," ZDNet, 30 December 2020, https://www.zdnet.com/article/cisa-updates-solarwinds-guidance-tells-us-govt-agencies-to-update-right-away/
31. Crow, "The Situation Is Developing, but the More I Learn This Could Be Our Modern Day, Cyber Equivalent of Pearl Harbor."
32. Brendan Pierson and Jan Wolfe, "Explainer-U.S. Government Hack: Espionage or Act of War?," *Reuters*, 21 December 2020, https://www.reuters.com/article/global-cyber-legal-idUSKBN28T0HH
33. Forcese, "Spies Without Borders," 198–99.
34. Though nations may indeed conduct specific OCOs against criminal networks as part of a broader, usually international law enforcement effort.
35. William Lynn, "Defending a New Domain—The Pentagon's Cyberstrategy," *Foreign Affairs* 89, no. 5 (October 2010): 97–108.
36. Joe McReynolds, "China's Evolving Perspectives on Network Warfare: Lessons from the Science of Military Strategy," *China Brief* 15, no. 8 (17 April 2015): 3–7.
37. R. J. Deibert, R. Rohozinski, and M. Crete-Nishihata, "Cyclones in Cyberspace: Information Shaping and Denial in the 2008 Russia–Georgia War," *Security Dialogue* 43, no. 1 (1 February 2012): 3–24.
38. US Department of Justice, "Iranian Hackers Indicted for Stealing Data from Aerospace and Satellite Tracking Companies," 17 September 2020, https://www.justice.gov/usao-edva/pr/iranian-hackers-indicted-stealing-data-aerospace-and-satellite-tracking-companies
39. Giles, "Russia's 'New' Tools for Confronting the West," 36.
40. Thomas Rid and Ben Buchanan, "Attributing Cyber Attacks," *Journal of Strategic Studies* 38, no. 1–2 (2 January 2015): 7.
41. Duston Volz and Jim Finkle, "U.S. Indicts Iranians for Hacking Dozens of Banks, New York Dam," *Reuters*, 25 March 2016, http://www.reuters.com/article/us-usa-iran-cyber-idUSKCN0WQ1JF

42. US Department of Justice, "Two Chinese Hackers Working with the Ministry of State Security Charged with Global Computer Intrusion Campaign Targeting Intellectual Property and Confidential Business Information, Including COVID-19 Research," 21 July 2020, https://www.justice.gov/opa/pr/two-chinese-hackers-working-ministry-state-security-charged-global-computer-intrusion

43. US Department of Justice, "Six Russian GRU Officers Charged in Connection with Worldwide Deployment of Destructive Malware and Other Disruptive Actions in Cyberspace,"19 October 2020, https://www.justice.gov/opa/pr/six-russian-gru-officers-charged-connection-worldwide-deployment-destructive-malware-and.

44. US Joint Chiefs of Staff, "Joint Publication 3–12: Cyberspace Operations," 8 June 2018, 9.

45. Roland Heickero, "Emerging Cyber Threats and Russian Views on Information Warfare and Information Operations," Swedish Defence Research Agency, March 2010, 17.

46. Russian Federation, "The Military Doctrine of the Russian Federation."

47. Russian Federation.

48. Tal Kopan, "DNC Hack: What You Need to Know," *CNN*, 21 June 2016, http://www.cnn.com/2016/06/21/politics/dnc-hack-russians-guccifer-claims/index.html

49. ThreatConnect Research, "Shiny Object? Guccifer 2.0 and the DNC Breach," ThreatConnect, 29 June 2016, https://www.threatconnect.com/blog/guccifer-2-0-dnc-breach

50. Thomas Rid, "All Signs Point to Russia Being Behind the DNC Hack," *Motherboard*, 25 July 2016, http://motherboard.vice.com/read/all-signs-point-to-russia-being-behind-the-dnc-hack.

51. Lorenzo Franceschi-Bicchierai, "We Spoke to DNC Hacker 'Guccifer 2.0,'" *Motherboard*, 21 June 2016, https://motherboard.vice.com/en_us/article/aek7ea/dnc-hacker-guccifer-20-interview

52. Nicole Perlroth and Savage, Charlie, "Is D.N.C. Email Hacker a Person or a Russian Front? Experts Aren't Sure," *The New York Times*, 27 July 2016, http://www.nytimes.com/2016/07/28/us/politics/is-dnc-email-hacker-a-person-or-a-russian-front-experts-arent-sure.html

53. DHS Press Office, "Joint Statement from the Department Of Homeland Security and Office of the Director of National Intelligence on Election Security."

54. Margaret Coker and Paul Sonne, "Ukraine: Cyberwar's Hottest Front," *Wall Street Journal*, 10 November 2015, sec. World, http://www.wsj.com/articles/ukraine-cyberwars-hottest-front-1447121671.

55. Nicole Perlroth and Quentin Hardy, "Online Banking Attacks Were Work of Iran, U.S. Officials Say," *The New York Times*, 8 January 2013, http://www.nytimes.com/2013/01/09/technology/online-banking-attacks-were-work-of-iran-us-officials-say.html

56. Perlroth and Hardy.

57. Volz and Finkle, "U.S. Indicts Iranians for Hacking Dozens of Banks, New York

Dam." Reuters, 24 March 2016, https://www.reuters.com/article/us-usa-iran-cyber-idUSKCN0WQ1JF

58. Kenneth Katzman and Paul K. Kerr, "Iran Nuclear Agreement," *Congressional Research Service*, 2015.

59. Fawaz A. Gerges, "The Obama Approach to the Middle East: The End of America's Moment?," *International Affairs* 89, no. 2 (2013): 313.

60. James S. Brady, "Remarks by the President to the White House Press Corps," The White House, 20 August 2012, https://www.whitehouse.gov/the-press-office/2012/08/20/remarks-president-white-house-press-corps

61. Langner, "Stuxnet—Dissecting a Cyberwarfare Weapon."

62. James P. Farwell and Rafal Rohozinski, "Stuxnet and the Future of Cyber War," *Survival* 53, no. 1 (February 2011): 23–40.

63. David E. Sanger, "Obama Ordered Wave of Cyberattacks Against Iran," *The New York Times*, 1 June 2012, http://www.nytimes.com/2012/06/01/world/middleeast/obama-ordered-wave-of-cyberattacks-against-iran.html

64. Falliere, Murchu, and Chien, "W32.Stuxnet Dossier." https://www.wired.com/images_blogs/threatlevel/2011/02/Symantec-Stuxnet-Update-Feb-2011.pdf

65. It was the alleged failure in self-restriction that caused the malware to infect many unrelated devices, eventually attracting the attention of malware investigators.

66. Thomas Rid and Peter McBurney, "Cyber-Weapons," *The RUSI Journal* 157, no. 1 (February 2012): 9.

67. Sanger, "Obama Ordered Wave of Cyberattacks Against Iran."

68. Sanger.

69. Julian E. Barnes and Thomas Gibbons-Neff, "U.S. Carried Out Cyberattacks on Iran," The New York Times, 22 June 2019, https://www.nytimes.com/2019/06/22/us/politics/us-iran-cyber-attacks.html

70. United States Industrial Control Systems Cyber Emergency Response Team, "Cyber-Attack Against Ukrainian Critical Infrastructure | ICS-CERT," ICS-CERT, 25 February 2016, https://www.cisa.gov/uscert/ics/alerts/IR-ALERT-H-16-056-01

71. Joe Slowik, "CRASHOVERRIDE: Reassessing the 2016 Ukraine Electric Power Event as a Protection-Focused Attack," Dragos, 15 August 2019, 4.

72. Russian military activity in Eastern Ukraine has not been publicly acknowledged as active warfare by the Russian government. Numerous indications of Russian troops have been rigorously denied and countered by information operations, discussed in a later chapter.

73. Waltz, Kenneth, *Man, the State, and War: A Theoretical Analysis* (Columbia University Press, 2013), 9.

74. In the Second Lebanon War of 2006, Israel favoured its air force in the early days of conflict, with limited strategic success.

75. Kenneth N. Waltz, "The Origins of War in Neorealist Theory," *Journal of Interdisciplinary History* 18, no. 4 (1988): 615.

76. Richard A. Clarke and Robert K. Knake, *Cyber War: The next Threat to National Security and What to Do about It*, 1st ed. (New York: Ecco, 2010), 9.

77. The final part of the chapter revisits Operation Orchard in order to examine how it is indeed one of the only instances that fully align with an acceptably standard definition of cyber-warfare.

78. Alfred Thayer Mahan, *The Influence of Sea Power upon History, 1660–1783* (Redditch: Read Books Ltd, 2013).

79. James A. Lewis, "Thresholds for Cyberwar," Center for Strategic and International Studies, 2010, 1.

80. See for example; Sam Jones, "Ministry of Defence Fends Off 'Thousands' of Daily Cyber Attacks," *Financial Times*, 25 June 2015, https://www.ft.com/content/2f6de47e-1a9a-11e5-8201-cbdb03d71480

81. Raymond C. Parks and Duggan, David P., "Principles of Cyber-Warfare," in *Proceedings from the Second Annual IEEE SMC Information Assurance Workshop* (New York: West Point, 2001), 122.

82. Rattray, *Strategic Warfare in Cyberspace*.

83. Ulrik Franke, "War by Non-Military Means," Stockholm, Sweden: FOI, 2015, 34.

84. Keir Giles, "'Information Troops'–A Russian Cyber Command?," in *3rd International Conference on Cyber Conflict*, 2011, 46.

85. John Arquilla, "Ethics and Information Warfare," in *Strategic Appraisal: The Changing Role of Information in Warfare*, ed. Zalmay Khalilzad, John P. White, and Andrew Marshall (Santa Monica, CA: RAND, 1999), 380.

86. Martin C. Libicki, *What Is Information Warfare?*, 3rd ed. (Washington DC: National Defense University, 1995).

87. US Department of Defense, "Information Operations Roadmap," 30 October 2003, 9.

88. The unique characteristics of offensive network operations are covered in a subsequent chapter, alongside their associated risks and opportunities.

89. Anthony H Cordesman, Ashley Hess, and Nicholas S Yarosh, *Chinese Military Modernization and Force Development: A Western Perspective* (Washington D.C.: Center for Strategic and International Studies, 2013), 54.

90. David S. Alberts, John Garstka, and Frederick P. Stein, *Network Centric Warfare: Developing and Leveraging Information Superiority*, CCRP Publication Series (Washington D.C.: National Defense University Press, 1999), 88.

91. Cordesman, Hess, and Yarosh, *Chinese Military Modernization and Force Development*, 1–2.

92. M. Taylor Fravel, *Active Defense: China's Military Strategy since 1949*, Princeton Studies in International History and Politics (Princeton, New Jersey: Princeton University Press, 2019), 183.

93. Cordesman, Hess, and Yarosh, *Chinese Military Modernization and Force Development*, 57.

94. Larry M. Wortzel, "PLA Command, Control and Targeting Architectures: Theory, Doctrine, and Warfighting Applications," in *Right-Sizing the People's Liberation Army: Exploring the Contours of China's Military*, ed. Roy Kamphausen and Andrew Scobell (Carlisle, PA: Strategic Studies Institute, 2007): 191, https://www.files.ethz.ch/isn/48426/Right%20Sizing%20the%20People's_full.pdf

95. The State Council Information Office of the People's Republic of China, *China's Military Strategy*; Fravel, *Active Defense*, 216.

96. Fravel, *Active Defense*, 230.

97. The State Council Information Office of the People's Republic of China, *China's Military Strategy*.

98. The State Council Information Office of the People's Republic of China.

99. US Joint Chiefs of Staff, "Joint Publication 3–12: Cyberspace Operations," 2 May 2013, 6.

100. Cordesman, Hess, and Yarosh, *Chinese Military Modernization and Force Development*, 58.

101. David E. Sanger and Mark Mazzetti, "Analysts Find Israel Struck a Syrian Nuclear Project," *The New York Times*, 14 October 2007.

102. Peter Crail, "IAEA Sends Syria Nuclear Case to UN," Arms Control Association, 7 July 2011, https://www.armscontrol.org/act/2011_%2007–08/%20IAEA_Sends_Syria_Nuclear_Case_to_UN.

103. Sean O'Conner, "Access Denial—Syria's Air Defence Network," Jane's International Defence Review, 2014, 1.

104. David Makovsky, "The Silent Strike," *The New Yorker*, 17 September 2012, http://www.newyorker.com/magazine/2012/09/17/the-silent-strike

105. "Syria: Foreign Intervention Still Debated, but Distant," *Strategic Comments* 18, no. 6 (August 2012): 1–5.

106. Makovsky, "The Silent Strike."

107. Amos Harel and Aluf Benn, "No Longer a Secret: How Israel Destroyed Syria's Nuclear Reactor," *Haaretz*, March 23, 2018.

108. Clay Wilson, "Information Operations, Electronic Warfare, and Cyberwar: Capabilities and Related Policy Issues," DTIC Document, 2007; Clarke and Knake, *Cyber War*, 10.

109. Arthur K. Cebrowski and John J. Garstka, "Network-Centric Warfare: Its Origin and Future," in *US Naval Institute Proceedings*, vol. 124, 1998, 25.

110. Cebrowski and Garstka, 25.

111. US Joint Chiefs of Staff, "Joint Publication 3–12: Cyberspace Operations," 2 May 2013.

112. Mitch Gettle, "Air Force Releases New Mission Statement," United States Air Force, 8 December 2005, http://www.af.mil/News/ArticleDisplay/tabid/223/Article/132526/air-force-releases-new-mission-statement.aspx

113. Schmitt, *Tallinn Manual on the International Law Applicable to Cyber Warfare*.

114. Steve Ranger, "NATO Updates Cyber Defence Policy as Digital Attacks Become a Standard Part of Conflict," *ZDNet*, 30 June 2014.

115. Eimi Harris, "NATO Adds Cyber to Operational Domain," NATO Association of Canada, 4 July 2016, https://natoassociation.ca/nato-adds-cyber-to-operational-domain/

116. James Schneider and Lawrence L. Izzo, "Clausewitz's Elusive Center of Gravity," *Parameters*, September 1987, 56.

2. CHARTING INTANGIBLE WARFARE

1. This is not always the case. Some modern capabilities target ubiquitous "generic" technologies, such as Microsoft Windows or popular communication protocols.
2. Rattray, *Strategic Warfare in Cyberspace*, 17.
3. Rattray, 66.
4. Rattray, 165.
5. Carl Von Clausewitz, *On War*, 3rd ed., vol. 1 (London: N. Trubner & Co, 1873), 73.
6. Hew Strachan, "The Battle of the Somme and British Strategy," *Journal of Strategic Studies* 21, no. 1 (March 1998): 81.
7. Jonathan Shimshoni, "Technology, Military Advantage, and World War I: A Case for Military Entrepreneurship," *International Security* 15, no. 3 (1990): 190.
8. James W. Rainey, "Ambivalent Warfare: Tactical Doctrine of the AEF in World War I," *Parameters* 13, no. 3 (1983): 34–46.
9. Shimshoni, "Technology, Military Advantage, and World War I," 211.
10. Shimshoni, 207.
11. Reginald V. Jones, "Scientific Intelligence," *Journal of the Royal United Service Institution* 92 (1956): 55–56.
12. Jones, 61.
13. Burton et al., *The Strategy of Electronic Warfare*, 4.
14. Jones, "Scientific Intelligence," 61.
15. Alfred Price, *Instruments of Darkness* (Barnsley, UK: Frontline Books, 2017), 82–83.
16. Also called Wotan in German encoded communications
17. Robert Cockburn, "The Radio War," *IEE Proceedings A Physical Science, Measurement and Instrumentation, Management and Education Reviews* 132, no. 6 (1985): 426.
18. Jones, "Scientific Intelligence," 63–64; Cockburn, "The Radio War," 428.
19. Robert Watson-Watt, "Battle Scars of Military Electronics—The Scharnhorst Break-Through," *IRE Transactions on Military Electronics* 1, no. 1 (March 1957): 19.
20. Watson-Watt, 22.
21. Cockburn, "The Radio War," 428.
22. Cockburn, 430.
23. Watson-Watt, "Battle Scars of Military Electronics—The Scharnhorst Break-Through," 24.
24. Cockburn, "The Radio War," 427–28.
25. Cockburn, 429.
26. Cockburn, 426.
27. Cockburn, 429.
28. John Boyd, "The Essence of Winning and Losing," ed. Chet Richards and Chuck Spinney, Pogo Archives.
29. Robert S. Bolia, "Overreliance on Technology in Warfare: The Yom Kippur War as a Case Study," DTIC Document, 2004, 53, http://oai.dtic.mil/oai/oai?verb=getRecord&metadataPrefix=html&identifier=ADA485884

30. Lawrence Whetten and Michael Johnson, "Military Lessons of the Yom Kippur War," *The World Today* 30, no. 3 (March 1974): 109.

31. Dov S. Zakheim, "The United States Navy and Israeli Navy: Background, Current Issues, Scenarios, and Prospects," Center for Naval Assessment, 2012, 4, https://www.academia.edu/63285786/The_United_States_Navy_and_Israeli_Navy_Background_Current_Issues_Scenarios_and_Prospects

32. Israeli Navy, "אתר חיל הים", מלחמת יום הכיפורים–1973, accessed January 23, 2017, http://www.navy.idf.il/1274-he/Navy.aspx

33. Bolia, "Overreliance on Technology in Warfare," 51.

34. Nadav Safran, "Trial by Ordeal: The Yom Kippur War, October 1973," *International Security* 2, no. 2 (1977): 151.

35. In information security, the overall susceptibility of a network to malicious activities is called its "attack surface".

36. Burton et al., *The Strategy of Electronic Warfare*, 1.

37. Burton et al., 1–2.

38. USSR Exercise Control Staff, "Task for the Operational Command Staff Exercise Soyuz-75 for the 4th Army," Cold War International History Project, Polish Institute of National Remembrance, March 1975, 9, http://digitalarchive.wilsoncenter.org/document/113511

39. USSR Exercise Control Staff, "The Operational-Tactical Exercise of Allied Fleets in the Baltic Sea, Codenamed VAL-77," Cold War International History Project, Polish Institute of National Remembrance, 1977, 13, http://digitalarchive.wilsoncenter.org/document/114599

40. US Army, "U.S. Army Field Manual 100–2–1: Soviet Forces," Headquarters of the Department of the US Army, 16 July 1984, 178.

41. US Army, 178.

42. Eliot A Cohen, "A Revolution in Warfare," *Foreign Affairs* 75, no. 2 (1996): 39.

43. Cordesman, Hess, and Yarosh, *Chinese Military Modernization and Force Development*, 34.

44. Norman B. Hutcherson, "Command & Control Warfare: Putting Another Tool in the War-Fighter's Data Base," Alabama, United States: Air University Press, 1994, 1.

45. At the time of the cited report, Clapper was the director of the US Defense Intelligence Agency (DIA).

46. James R. Clapper Jr, "Challenging Joint Military Intelligence," DTIC Document, 1994, 94.

47. US Army, "Army Regulation 525–20: Command & Control Countermeasures (C2CM)," US Army Headquarters, 31 July 1992.

48. Chairman of the Joint Chiefs of Staff, "Memorandum of Policy No. 30: Command and Control Warfare," March 8, 1993.

49. Chairman of the Joint Chiefs of Staff, 2.

50. US Army, "U.S. Army Field Manual 100–6: Information Operations," Headquarters of the Department of the US Army, August 1996, 37.

51. Ron C. Plucker, "Command and Control Warfare—A New Concept for the Joint Operational Commander," DTIC Document, 1993, 6.

52. Plucker, 16.
53. Plucker, 21.
54. US Joint Chiefs of Staff, "Joint Publication 3–13: Command and Control Warfare (C2W)," February 7, 1996, 5.
55. Plucker, "Command and Control Warfare—A New Concept for the Joint Operational Commander," 2.
56. Alberts, Garstka, and Stein, *Network Centric Warfare*, 5.
57. Alberts, Garstka, and Stein, 7.
58. Alberts, Garstka, and Stein, 12.
59. Anthony H Cordesman, "Chinese Strategy and Military Power in 2014," Center for Strategic and International Studies, November 2014, 122.
60. Cordesman, 126.
61. US Department of Defense, "U.S. Department of Defense Directive 3600.01—Information Warfare," November 1992, 1.
62. Specifically, the revised Directive 3600.1 from 1996 defines Information Warfare as "IO conducted during time of crisis or conflict…".
63. US Department of Defense, "U.S. Department of Defense Directive 3600.01—Information Operations," December 1996, 1–2.
64. US Department of Defense, "Information Operations Roadmap," 5–8.
65. Thomas Rid, *Rise of the Machines: The Lost History of Cybernetics* (London: Scribe Publications, 2016).
66. John Arquilla and David Ronfeldt, "Cyberwar Is Coming!," *Comparative Strategy* 12, no. 2 (1993): 141–65.
67. Arquilla and Ronfeldt, 28.
68. Arquilla and Ronfeldt, 30.
69. US Department of Defense, "Information Operations Roadmap."
70. Jeffrey Caton L., "Army Support of Military Cyberspace Operations," Strategic Studies Institute, January 2015, 5–7.
71. Caton, 8.
72. US Strategic Command, "U.S. Cyber Command (USCYBERCOM)," 30 September 2016, http://www.stratcom.mil/Media/Factsheets/Factsheet-View/Article/960492/us-cyber-command-uscybercom
73. Dan Kuehl and Leigh Armistead, "Information Operations: The Policy and Organizational Evaluation," in *Information Operations* (Washington D.C.: Potomac Books, 2007), 18–20.
74. Dorothy E. Denning, *Information Warfare and Security*, 4th ed. (Reading: Addison-Wesley, 1999).

3. TARGETING NETWORKS

1. A previous version of this chapter was published and presented at the 2018 CyCon X conference. For the original publication, see Daniel Moore, "Targeting Technology: Mapping Military Offensive Network Operations," in *2018 10th*

International Conference on Cyber Conflict (CyCon) (Tallinn, Estonia: IEEE, 2018), 89–108.

2. This is a descriptive example rather than an exhaustive accounting of all presence-based operations. Some elements are constant, such as the need for pivoting within networks and the use of malicious software. Others, such as the modularity of the malware and the need for bespoke offensive components, depends on the nature of the operation and the capabilities of the offender.

3. Mandiant, "APT1—Exposing One of China's Cyber Espionage Units," 2013, 1.

4. Mandiant, 3.

5. Mandiant, 6.

6. Not all companies have shied away from granular attribution. As recently as August 2018, US company CrowdStrike has attributed an intrusion campaign they labelled "Stone Panda" to specific individuals and buildings in China. See Adam Kozy, "Two Birds, One STONE PANDA," *CrowdStrike Blog* (blog), August 30, 2018, https://www.crowdstrike.com/blog/two-birds-one-stone-panda/. In the government space, the United States now favours individual-level attributions against adversary military service members as part of criminal indictments.

7. Kaspersky Lab, "Taste of Topinambour: Turla Hacking Group Hides Malware in Anti-Internet Censorship Software," July 15, 2019, https://www.kaspersky.com/about/press-releases/2019_taste-of-topinambour.

8. Sean Kimmons, "Cyber Teams Throw Virtual Effects, Defend Networks against ISIS," US Army, 15 February 2017 http://www.army.mil/article/182400/cyber_teams_throw_virtual_effects_defend_networks_against_isis

9. US Joint Chiefs of Staff, "Joint Publication 3–12: Cyberspace Operations," 2 May 2013

10. Chairman of the Joint Chiefs of Staff, "National Military Strategy for Cyberspace Operations," December 2006.

11. US Department of Defense, "Fiscal Year 2019 DoD Programs—F-35 Joint Strike Fighter (JSF)."

12. US Department of Defense, "Information Operations Roadmap."

13. James R. Clapper Jr, Michael S. Rogers, and Marcel Lettre, "Statement on Foreign Cyber Threats to the United States," US Senate Armed Services Committee (2017), 5.

14. Eric M. Hutchins, Michael J. Cloppert, and Rohan M. Amin, "Intelligence-Driven Computer Network Defense Informed by Analysis of Adversary Campaigns and Intrusion Kill Chains," *Leading Issues in Information Warfare & Security Research* 1 (2011): 4–5.

15. Kim Zetter, "How Digital Detectives Deciphered Stuxnet, the Most Menacing Malware in History," *Wired*, 11 July 2011, https://www.wired.com/2011/07/how-digital-detectives-deciphered-stuxnet

16. Zetter.

17. Langner, "Stuxnet—Dissecting a Cyberwarfare Weapon."

18. Falliere, Murchu, and Chien, "W32.Stuxnet Dossier," 3.

19. Farwell and Rohozinski, "Stuxnet and the Future of Cyber War."

20. Sharon Weinberger, "Is This the Start of Cyberwarfare?," *Nature* 474, no. 7350 (2011): 142.

21. A thorough account of the unique circumstance of Stuxnet is provided in Kim Zetter, *Countdown to Zero Day: Stuxnet and the Launch of the World's First Digital Weapon* (Broadway Books, 2014).

22. DNI US, "Cyber Threat Framework Lexicon," Office of the Director of National Intelligence, 2013, 2.

23. Matthew Monte, *Network Attacks & Exploitation: A Framework* (Indianapolis, IN: John Wiley & Sons, Inc, 2015), 20.

24. Jack L. Burbank et al., "Key Challenges of Military Tactical Networking and the Elusive Promise of MANET Technology," *IEEE Communications Magazine* 44, no. 11 (2006): 39–42.

25. The idea of separating a network from all other networks is called "air-gapping", and is a widely accepted methodology of reducing a network's potential attack surface.

26. US Joint Chiefs of Staff, "Joint Publication 1–02: DoD Dictionary," May 2017, 234.

27. US CYBERCOM JIOC, "Improving Targeting Support to Cyber Operations," US Cyber Command, 30 November 2016.

28. In the United States, the NSA is a civilian agency. In the Israeli example, it is the military unit 8200. Irrespective of this designation, some countries are increasingly migrating offensive cyber capabilities to a dedicated military force, such as in the case of US Cyber Command.

29. US Joint Chiefs of Staff, "Joint Publication 2–0: Joint Intelligence," 22 October 2013, 24–25.

30. NSA, "SID and DIA Collaborate Virtually on Russian Targets," 18 May 2004.

31. Michael R. Gordon, "Despite Cold War's End, Russia Keeps Building a Secret Complex," *The New York Times*, 16 April 1996, https://www.nytimes.com/1996/04/16/world/despite-cold-war-s-end-russia-keeps-building-a-secret-complex.html

32. NSA, "SID and DIA Collaborate Virtually on Russian Targets."

33. Gregory Conti and David Raymond, *On Cyber: Towards an Operational Art for Cyber Conflict* (New York: Kopidion Press, 2017), 181–82.

34. Gregory Coile, "WIN-T SATCOM Overview Briefing," Program Executive Office Command Control Communications-Tactical, 2009, 5.

35. Non-Secure Internet Protocol Router Network (NIPRNET) and Secure Internet Protocol Router Network (SIPRNET), US Department of Defense networks used for unclassified and classified communications between and within partner organisations.

36. Lynn Epperson, "Satellite Communications Within the Army's WIN-T Architecture," Program Executive Office Command Control Communications-Tactical, 6 February 2014.

37. Bruce T. Crawford, James J. Mingus, and Gary P. Martin, "The United States Army

Network Modernization Strategy," Committee on the Armed Services (2017), 6–8, http://docs.house.gov/meetings/AS/AS25/20170927/106451/HHRG-115-AS25-Wstate-CrawfordB-20170927.pdf

38. Rattray, *Strategic Warfare in Cyberspace*, 171.
39. Joseph S. Nye, "Cyber Power," DTIC Document, 2010, 5.
40. Monte, *Network Attacks & Exploitation*, 124.
41. Martin C. Libicki, *Cyberdeterrence and Cyberwar* (Santa Monica, CA: RAND, 2009), 83.
42. There are many ways to accomplish this, including identifying unique patterns that appear in the malicious files themselves, analysis of the command-and-control infrastructure used by the malware to communicate with its operators, and mapping distinct behavioural signals both on the device and network levels.
43. The degree of operational compromise on detection could also be mitigated through operational security practices; limiting artefacts in the malware and debugging symbols, obfuscation techniques, limiting superfluous presence in adversary networks, and compartmentalising infrastructure between targets.
44. David R. King and Joseph D. Massey, "History of the F-15 Program: A Silver Anniversary First Flight Remembrance," *Air Force Journal of Logistics*, 1997, 11, https://epublications.marquette.edu/cgi/viewcontent.cgi?referer=https://www.google.com/&httpsredir=1&article=1048&context=mgmt_fac
45. John A. Tirpak, "Making the Best of the Fighter Force," *Air Force Magazine* 90, no. 3 (2007): 44.
46. Kyle Mizokami, "Why the Air Force Is Buying a Bunch of F-15s Even Though the F-35 Is Coming," Popular Mechanics, 19 February 2019, https://www.popularmechanics.com/military/aviation/a26413900/air-force-buying-new-f-15/
47. Kaspersky Lab, "The Regin Platform: Nation State Ownage of GSM Networks," 24 November 2014, 3.
48. Ryan Gallagher, "The Inside Story of How British Spies Hacked Belgium's Largest Telco," *The Intercept* (blog), December 13, 2014, https://theintercept.com/2014/12/13/belgacom-hack-gchq-inside-story/
49. Marcel Rosenbach, Hilmar Schmundt, and Christian Stöcker, "Source Code Similarities: Experts Unmask 'Regin' Trojan as NSA Tool," *Spiegel Online*, 27 January 2015, sec. International, http://www.spiegel.de/international/world/regin-malware-unmasked-as-nsa-tool-after-spiegel-publishes-source-code-a-1015255.html
50. Kaspersky Lab, "The Regin Platform: Nation State Ownage of GSM Networks," 23.
51. A generic destructive payload is one that works on common hardware/software, such as the ubiquitous wipers and file encryption routines used in publicly disclosed attack operations and criminal ransomware campaigns.
52. Dan Goodin, "Group Claims to Hack NSA-Tied Hackers, Posts Exploits as Proof," *Ars Technica*, 16 August 2016, https://arstechnica.com/information-technology/2016/08/group-claims-to-hack-nsa-tied-hackers-posts-exploits-as-proof/

53. Ellen Nakashima and Craig Timberg, "NSA Officials Worried about the Day Its Potent Hacking Tool Would Get Loose. Then It Did.," *Washington Post*, May 16, 2017, sec. Technology, https://www.washingtonpost.com/business/technology/nsa-officials-worried-about-the-day-its-potent-hacking-tool-would-get-loose-then-it-did/2017/05/16/50670b16-3978-11e7-a058-ddbb23c75d82_story.html

54. Microsoft, "Microsoft Security Bulletin MS17–010—Critical,"14 July 2017, https://docs.microsoft.com/en-us/security-updates/securitybulletinsummaries /2017/2017bulletinsummaries

55. Symantec, "What You Need to Know about the WannaCry Ransomware," 23 October 2017, https://www.symantec.com/blogs/threat-intelligence/wannacry-ransomware-attack

56. National Audit Office, "Investigation: WannaCry Cyber Attack and the NHS," accessed 1 January 2018, https://www.nao.org.uk/report/investigation-wannacry-cyber-attack-and-the-nhs

57. Symantec, "Petya Ransomware Outbreak: Here's What You Need to Know," 24 October 2017, https://www.symantec.com/blogs/threat-intelligence/petya-ransomware-wiper

58. Symantec, "Internet Security Threat Report," February 2019.

59. NIST, "NVD—CVSS Severity Distribution Over Time," 2020.

60. Lillian Ablon, *Zero Days, Thousands of Nights: The Life and Times of Zero-Day Vulnerabilities and Their Exploits.* (Santa Monica, CA: RAND, 2017), 11.

61. Kimberly Underwood, "The Army Evolves Its Formations for Cyber and Electronic Warfare," *SIGNAL Magazine*, 21 October 2020, https://www.afcea.org/content/army-evolves-its-formations-cyber-and-electronic-warfare

62. Marco Giannangeli, "Russians forcing RAF to abort missions in Syria by 'hacking into' their systems," *The Daily Express*, 15 January 2017, http://www.express.co.uk/news/world/754236/russia-raf-bombers-syria-hacking-missions-military-army

63. DNI US, "Cyber Threat Framework Lexicon," Office of the Director of National Intelligence, 4.

64. Buchanan, *The Cybersecurity Dilemma: Hacking, Trust and Fear Between Nations*, 76–84.

65. Neil Barrett, "Penetration Testing and Social Engineering: Hacking the Weakest Link," *Information Security Technical Report* 8, no. 4 (2003): 56–64.

66. John Markoff, "SecurID Company Suffers Security Breach," *The New York Times*, 17 March 2011, sec. Technology, https://www.nytimes.com/2011/03/18/technology/18secure.html

67. Uri Rivner, "Anatomy of an Attack," *Speaking of Security—The RSA Blog* (blog), 1 April 2011, https://blogs.rsa.com/anatomy-of-an-attack/

68. Christopher Drew, "Lockheed Says Hacker Used Stolen SecurID Data," *The New York Times*, 3 June 2011, sec. Technology, https://www.nytimes.com/2011/06/04/technology/04security.html

69. This aligns nicely with US military doctrine that situates Cyber and Electromagnetic Activities (CEMA) as a unified operational function, see US Army, "Army Field Manual 3–38—Cyber Electromagnetic Activities".

70. US Army, 30–32.

71. NSA, "Computer Network Operations—GENIE," 2013, https://www.eff.org/files/2015/02/03/20150117-spiegel-excerpt_from_the_secret_nsa_budget_on_computer_network_operations_-_code_word_genie.pdf

72. Zachary Fryer-Biggs, "Secretive Pentagon Research Program Looks to Replace Human Hackers with AI," Yahoo! News, 13 September 2020, https://news.yahoo.com/secretive-pentagon-research-program-looks-to-replace-human-hackers-with-ai-090032920.html

73. DNI US, "Cyber Threat Framework Lexicon," 5.

74. DNI US, 5.

75. Jeff Malone, "Intelligence Support Requirements for Offensive CNO," in *Cyber Warfare and Nation States Conference* (Canberra, Australia, 23 August 2010).

76. Herbert S. Lin, "Offensive Cyber Operations and the Use of Force," *Journal of National Security Law and Policy* 4 (2010): 64.

77. Malone, "Intelligence Support Requirements for Offensive CNO," 16.

78. GCHQ, "Full-Spectrum Cyber Effects," 2012.

79. Dragos, "CRASHOVERRIDE: Analysis of the Threat to the Electric Grid Operations,"12 June 2017, 9.

80. Dragos, 10.

81. Dragos, 4.

82. CRASHOVERRIDE is the cryptonym provided by private security company Dragos. It is also commonly known as Industroyer, as labelled by ESET.

83. Monte, *Network Attacks & Exploitation*, 63.

84. Nicole Perlroth, Mark Scott, and Sheera Frenkel, "Cyberattack Hits Ukraine Then Spreads Internationally," *The New York Times*, 27 June 2017, sec. Technology, https://www.nytimes.com/2017/06/27/technology/ransomware-hackers.html

85. See for example US Press Secretary, "Statement from the Press Secretary."

86. Dale Peterson, "Offensive Cyber Weapons: Construction, Development, and Employment," *Journal of Strategic Studies* 36, no. 1 (February 2013), 123.

87. Rattray and Healey, "Categorizing and Understanding Offensive Cyber Capabilities and Their Use," 79.

88. Richard A. Clarke, "War From Cyberspace," *The National Interest*, 2009, 32.

89. Adapted from the US Military's taxonomy of "deceive, degrade, deny, destroy, or manipulate", see US Army, "Army Field Manual 3–38—Cyber Electromagnetic Activities," 17. Libicki similarly speaks of attacks meant at eruption (target illumination), disruption, and corruption. See Libicki, *Cyberdeterrence and Cyberwar*, 145.

90. US Joint Chiefs of Staff, "Joint Publication 1–02: DoD Dictionary," 229.

91. Army Headquarters, "US Army Field Manual 3–13—Information Operations," November 2003, 7.

92. US Army, "Army Field Manual 3–38—Cyber Electromagnetic Activities," 9.

93. Sarkar Samit, "Massive DDoS Attack Affecting PSN, Some Xbox Live Apps," *Polygon*, October 21, 2016, https://www.polygon.com/2016/10/21/13361014/psn-xbox-live-down-ddos-attack-dyn

94. Jasper Hamill, "Bank-Busting Jihadi Botnet Comes Back To Life. But Who Is Controlling It This Time?," Forbes, June 30, 2014, https://www.forbes.com/sites/jasperhamill/2014/06/30/bank-busting-jihadi-botnet-comes-back-to-life-but-who-is-controlling-it-this-time/#3df4bb0f6f07

95. Falliere, Murchu, and Chien, "W32.Stuxnet Dossier"; Farwell and Rohozinski, "Stuxnet and the Future of Cyber War."

96. The classic approach to warfare—most commonly codified by Prussian strategist Carl von Clausewitz—favours destruction as the sole means of achieving military coercion.

97. See for example the 2012 Shamoon attack, in which a presumably Iranian attacker wiped thousands of computers at Saudi's national gas company, Aramco; Christopher Bronk and Eneken Tikk-Ringas, "The Cyber Attack on Saudi Aramco," Survival 55, no. 2 (May 2013): 81–96.

98. Langner, "Stuxnet—Dissecting a Cyberwarfare Weapon."

99. Falliere, Murchu, and Chien, "W32.Stuxnet Dossier."

100. The "Aurora Experiment" was a 2007 showcase by the Idaho National Laboratory in which they demonstrated how a software vulnerability in generators used by critical infrastructure providers can rapidly cause physical damage to the point of complete equipment destruction. The experiment video and associated documentation is available online as "Operation Aurora," MuckRock, July 3, 2014, https://www.muckrock.com/foi/united-states-of-america-10/operation-aurora-11765/#files

101. The German steel mill attack is a 2013 incident in which attackers supposedly caused significant physical damage to a furnace by way of a presence-based operation, see Robert M. Lee, Michael J. Assante, and Tim Conway, "German Steel Mill Cyber Attack," SANS ICS, 2014.

102. OODA loop—A process in which combatants Observe, Orient, Decide, and Act. Military vernacular for conceptualising decision-making process in combat. See Boyd, "The Essence of Winning and Losing."

103. These vulnerabilities indeed exist, see for example US Department of Defense, "Aegis Modernization Report Program—Fiscal Year 2016," 2017, 3.

104. Buchanan, The Cybersecurity Dilemma Hacking, Trust and Fear Between Nations.

105. Gallagher, "F-35 Radar System Has Bug That Requires Hard Reboot in Flight."

106. US Department of Defense, "Fiscal Year 2015 DoD Programs—F-35 Joint Strike Fighter (JSF)," January 2016, 35.

107. US Department of Defense, "Fiscal Year 2016 DoD Programs—F-35 Joint Strike Fighter (JSF)," 48.

108. US Department of Defense, "Fiscal Year 2019 DoD Programs—F-35 Joint Strike Fighter (JSF)."

109. Lin, "Offensive Cyber Operations and the Use of Force," 66.

110. Myron Hura et al., "Chapter 9—Tactical Data Links," in Interoperability: A Continuing Challenge (Santa Monica, CA: RAND, 2000), 107–21.

111. US Department of Defense, "Fiscal Year 2016 DoD Programs—F-35 Joint Strike Fighter (JSF)," 70.

112. Currently for the US, the F-22 and the F-35.

113. US Department of Defense, "Fiscal Year 2016 DoD Programs—F-35 Joint Strike Fighter (JSF)," 71.

114. Lockheed Martin, "Autonomic Logistics Information System (ALIS)."

115. Diana Maurer, "F-35 Sustainment: DOD Needs to Address Key Uncertainties as It Re-Designs the Aircraft's Logistics System," Committee on Oversight and Reform (2020), 1.

116. US Department of Defense, "Fiscal Year 2019 DoD Programs—F-35 Joint Strike Fighter (JSF)."

117. US Department of Defense, "Fiscal Year 2016 DoD Programs—F-35 Joint Strike Fighter (JSF)," 96.

118. US Department of Defense, 96.

119. Sydney J. Freedberg Jr., "ALIS Glitch Grounds Marine F-35Bs," *Breaking Defense* (blog), 22 June 2017, http://breakingdefense.com/2017/06/breaking-alis-glitch-grounds-marine-f-35bs

120. US Department of Defense, "Fiscal Year 2019 DoD Programs—F-35 Joint Strike Fighter (JSF)," 13.

121. Government Accountability Office, "DOD Needs a Strategy for Re-Designing the F-35's Central Logistics System," March 2020, 29, https://www.gao.gov/assets/710/705154.pdf

122. US Department of Defense, "Fiscal Year 2019 DoD Programs—F-35 Joint Strike Fighter (JSF)," 38.

123. US Department of Defense, "Fiscal Year 2020 DoD Programs—F-35 Joint Strike Fighter (JSF)," December 2020, https://www.dote.osd.mil/Portals/97/pub/reports/FY2020/other/2020DOTEAnnualReport.pdf

4. VIRTUAL VICTORY: APPLIED CYBER-STRATEGY

1. Arquilla and Ronfeldt, "Cyberwar Is Coming!," 41.

2. Buchanan, *The Cybersecurity Dilemma Hacking, Trust and Fear Between Nations*, 78–81.

3. Buchanan, 81–82.

4. Martin C. Libicki, "Why Cyber Will Not and Should Not Have Its Grand Strategist," *Strategic Studies Quarterly* (Spring 2014): 23–39, https://www.airuniversity.af.edu/Portals/10/SSQ/documents/Volume-08_Issue-1/Spring_2014.pdf

5. Lawrence Freedman, *Strategy: A History*, (Oxford: Oxford University Press, 2015), 228–29.

6. US Joint Chiefs of Staff, "Joint Publication 3–0: Operations," 11 August 2011, 126.

7. B.H. Liddell Hart, *Strategy*, 2nd rev. ed (New York: Meridian, 1991), 323.

8. J.F.C. Fuller, *The Foundations of the Science of War* (Hutchinson & Company, 1926), 204.

9. This is in part the reason special forces are viewed as favourable enablers of the economy of force. They require minimal preparation, are dynamic and flexible assets, and can assist conventional forces by reducing the danger to them. See US Joint Chiefs of Staff, "Joint Publication 3–0: Operations," 126.

10. Max Smeets, "The Strategic Promise of Offensive Cyber Operations," *Strategic Studies Quarterly* (Fall 2018): 90–113.

11. See analysis of the operational effects phase in the previous chapter.

12. The volatile value and sensitivity of presence-based operations is reflected by the level of approvals they often require. See for example US Army, "Army Field Manual 3–38—Cyber Electromagnetic Activities," 38.

13. Langner, "Stuxnet—Dissecting a Cyberwarfare Weapon."

14. Study of the targeted protocol would be necessary to effectively spoof its control messages.

15. Edward A. Smith, *Effects Based Operations: Applying Network Centric Warfare in Peace, Crisis, and War* (CCRP, 2005), 43.

16. David Bisson, "Hacker Admits to Stealing Military Data from U.S. Department of Defense," Tripwire, 16 June 2017, https://www.tripwire.com/state-of-security/latest-security-news/hacker-admits-to-stealing-military-data-from-u-s-department-of-defense/

17. Nye, "Cyber Power," 9.

18. US Joint Chiefs of Staff, "Joint Publication 5–0: Joint Planning," 114.

19. Some nations have more international communication links to the global internet than others.

20. John Borland, "Analyzing the Internet Collapse," MIT Technology Review, 5 February 2008, accessed 27 January 2018, https://www.technologyreview.com/s/409491/analyzing-the-internet-collapse/

21. David E. Sanger and Eric Schmitt, "Russian Ships Near Data Cables Are Too Close for U.S. Comfort," *The New York Times*, 25 October 2015, sec. Europe, https://www.nytimes.com/2015/10/26/world/europe/russian-presence-near-undersea-cables-concerns-us.html

22. The most common way of crippling internet access by affecting global routing is by maliciously targeting the Border Gateway Protocol (BGP). This protocol is used to govern the way internet service providers and other major networks communicate between one another. As this protocol relies on trust, an abuse of this trust may cause temporary large-scale diversions of internet traffic away from its intended course. This has happened several times in the past, though it is somewhat unclear whether previous incidents were malicious or accidental.

23. This is especially true considering the rise of "cloud computing", distributed networks that dynamically allocate resources to users. Amazon is one of the globally largest providers of such services, to clients that include the US government. On this risk, see for example David Midson, "Geography, Territory and Sovereignty in Cyber Warfare," in *New Technologies and the Law of Armed Conflict*, ed. Hitoshi Nasu and Robert McLaughlin (The Hague: T.M.C. Asser Press, 2014), 78.

24. Ashley Deeks, "The Geography of Cyber Conflict: Through a Glass Darkly," *International Law Studies* 89 (2013): 5.

25. Deeks, 13–14.

26. The gradual development of Russian offensive network operations is assessed at length in the dedicated subsequent chapter.

27. Stephen W. Korns and Joshua E. Kastenberg, "Georgia's Cyber Left Hook," *Parameters* 38, no. 4 (2008): 66–67.

28. Liddell Hart, *Strategy*, 34.

29. Clausewitz, *On War*, 1:168.

30. Sun Tzu, *The Art of War*, (Brooklyn, NY: Sheba Blake Publishing, 2017), 32.

31. Clausewitz, *On War*, 1:284.

32. Conti and Raymond, *On Cyber: Towards an Operational Art for Cyber Conflict*, 34.

33. Libicki, "Why Cyber Will Not and Should Not Have Its Grand Strategist," 34.

34. Rattray, *Strategic Warfare in Cyberspace*, 21.

35. H. R. McMaster, "The Human Element: When Gadgetry Becomes Strategy," *World Affairs* 171, no. 3 (2009): 35.

36. Rattray, *Strategic Warfare in Cyberspace*, 192.

37. US Army, "Army Field Manual 3–38—Cyber Electromagnetic Activities," 40.

38. Conti and Raymond, *On Cyber: Towards an Operational Art for Cyber Conflict*, 46.

39. US Army, "Army Field Manual 3–38—Cyber Electromagnetic Activities," 13.

40. Shipping giant Maersk was one such victim, with allegedly 4,000 servers and 45,000 personal computers requiring full reinstallation. See Richard Chirgwin, "IT 'heroes' Saved Maersk from NotPetya with Ten-Day Reinstallation Blitz," *The Register*, 25 January 2018, https://www.theregister.co.uk/2018/01/25/after_notpetya_maersk_replaced_everything/

41. Rattray and Healey, "Categorizing and Understanding Offensive Cyber Capabilities and Their Use," 79.

42. Max Smeets, "A Matter of Time: On the Transitory Nature of Cyberweapons," *Journal of Strategic Studies*, February 16, 2017, 1–28.

43. David S. Alberts, "Agility, Focus, and Convergence: The Future of Command and Control," Office of the Assistant Secretary of Defense for Networks and Information Integration, 2007, 23.

44. Anthony H. Dekker, "Measuring the Agility of Networked Military Forces," *Journal of Battlefield Technology* 9, no. 1 (2006): 3.

45. US Joint Chiefs of Staff, "Joint Publication 5–0: Joint Planning."

46. Smith, "Effects Based Operations," 54–60.

47. US Army Headquarters, "Army Doctrine Publication 3–0: Operations," October 2011, 2.

48. Monte, *Network Attacks & Exploitation*, 124–25.

5. AMERICAN CYBER SUPERIORITY

1. Bradley Graham, "Military Grappling With Rules For Cyber Warfare," *Washington Post*, 8 November 1999.

2. Defense Science Board, "Task Force on Cyber Deterrence," Department of Defense, February 2017, 14.

3. Defense Science Board, 14.

4. Space and Missile Systems Center Public Affairs, "Counter Communications System

Block 10.2 Achieves IOC, Ready for the Warfighter," United States Space Force, March 13, 2020, https://www.spaceforce.mil/News/Article/2113447/counter-communications-system-block-102-achieves-ioc-ready-for-the-warfighter

5. US Air Force, "Department of Defense Fiscal Year 2021 Budget Estimates: Air Force, Research, Development, Test & Evaluation, Space Force," February 2020.

6. US Department of Defense, "Cybercom to Elevate to Combatant Command," 3 May 2018, accessed 10 June 2018, https://www.defense.gov/News/Article/Article/1511959/cybercom-to-elevate-to-combatant-command/

7. US Joint Chiefs of Staff, "Joint Publication 3–12: Cyberspace Operations," 8 June 2018.

8. Stephen Townsend, "Accelerating Multi-Domain Operations: Evolution of an Idea," US Army Training and Doctrine Command, 2018.

9. TRADOC, "Multi-Domain Battle: Evolution of Combined Arms for the 21st Century 2025–2040," US Army Training and Doctrine Command, October 2017, 4–5.

10. TRADOC, 3.

11. US Department of Defense, The Department of Defense Cyber Strategy, 3.

12. US Cyber Command, "Achieve and Maintain Cyberspace Superiority: Command Vision for US Cyber Command," 23 March 2018, 8.

13. PPD-20 has supposedly been revised in August 2018 by the Trump administration, in a bid to loosen restrictions on OCOs. See Eric Geller, "Trump Scraps Obama Rules on Cyberattacks, Giving Military Freer Hand," POLITICO, 16 August 2018, https://politi.co/2MSWCnS

14. United States Government, "Presidential Policy Directive 20—U.S. Cyber Operations Policy," October 2012, 6.

15. United States Government, 9.

16. United States Government, 9.

17. US Joint Chiefs of Staff, "Joint Publication 3–12: Cyberspace Operations," 8 June 2018, 27.

18. US Joint Chiefs of Staff, "Joint Publication 3–13: Information Operations," 20 November 2014.

19. US Joint Chiefs of Staff, "Joint Publication 3–13.1: Electronic Warfare," 8 February 2012.

20. US Joint Chiefs of Staff, 11.

21. NATO, "Allied Joint Publication 3.20: Allied Joint Doctrine for Cyberspace Operations (Edition A)," NATO Standardization Office, January 2020, 23.

22. US Joint Chiefs of Staff, "Joint Publication 3–12: Cyberspace Operations," 8 June 2018, 22–24.

23. While the US Army is the primary organ within the US armed forces to use CEMA as a viable term, it now features prominently in doctrinal publications in other Western forces. See for example UK Ministry of Defence, "Joint Doctrine Note 1/18: Cyber and Electromagnetic Activities," February 2018.

24. US Army, "Army Field Manual 3–12: Cyberspace Operations and Electromagnetic Warfare," August 2021.

25. Kyle Rempfer, "Army Cyber Lobbies for Name Change This Year, as Information Warfare Grows in Importance," *Army Times*, 16 October 2019, https://www.army-times.com/news/your-army/2019/10/16/ausa-army-cyber-lobbies-for-name-change-this-year-as-information-warfare-grows-in-importance/

26. Air Combat Command Public Affairs, "ACC Discusses 16th Air Force as New Information Warfare NAF," US Air Force, 18 September 2019, https://www.af.mil/News/Article-Display/Article/1964795/acc-discusses-16th-air-force-as-new-information-warfare-naf/

27. US Cyber Command, "JFHQ-C Certification: Framework to Operationalize the JFHQ-C," October 2013.

28. Ellen Nakashima, "U.S. Cybercom Contemplates Information Warfare to Counter Russian Interference in 2020 Election," *Washington Post*, 25 December 2019, https://www.washingtonpost.com/national-security/us-cybercom-contemplates-information-warfare-to-counter-russian-interference-in-the-2020-election/2019/12/25/21bb246e-20e8-11ea-bed5-880264cc91a9_story.html

29. Paul M Nakasone, "A Cyber Force for Persistent Operations," *Joint Forces Quarterly* 92 (2019): 10–14.

30. US Department of Defense, "Cyber Strategy Summary," 1.

31. US Department of Defense, 1.

32. Jim Garamone, "Esper Describes DOD's Increased Cyber Offensive Strategy," US Department of Defense, 20 September 2019, https://www.defense.gov/Explore/News/Article/Article/1966758/esper-describes-dods-increased-cyber-offensive-strategy/

33. US Cyber Command, "USCYBERCOM 120-Day Assessment of Operation GLOWING SYMPHONY," 15 June 2016.

34. Mark Pomerleau, "Cyber Command Shifts Counterterrorism Task Force to Focus on Higher-Priority Threats," *Army Times*, 4 May 2021, https://www.armytimes.com/cyber/2021/05/04/cyber-command-shifts-counterterrorism-task-force-to-focus-on-higher-priority-threats/

35. Temple-Raston, "How The U.S. Hacked ISIS."

36. Mark Pomerleau, "US Is 'Outgunned' in Electronic Warfare, Says Cyber Commander," C4ISRNET, 10 August 2017, https://www.c4isrnet.com/show-reporter/technet-augusta/2017/08/10/us-is-outgunned-in-electronic-warfare-says-cyber-commander/

37. NSA, "Getting Close to the Adversary: Forward-Based Defense with QFIRE," 3 June 2011, 7.

38. NSA, 2.

39. NSA and USSTRATCOM, "National Initiative Protection Program—Sentry Eagle," 23 November 2004, 8.

40. NSA, "Case Studies of Integrated Cyber Operation Techniques," 13.

41. The Shadow Brokers are suspected by some as at least partially affiliated with the Russian government, see James Risen, "U.S. Secretly Negotiated With Russians to Buy Stolen NSA Documents—and the Russians Offered Trump-Related Material,

Too," *The Intercept* (blog), 9 February 2018, https://theintercept.com/2018/02/09/donald-trump-russia-election-nsa/. For coverage of their public communications and content, see Comae, "The Shadow Brokers: Cyber Fear Game-Changers," Comae Technologies, July 2017. There is currently no official or public high-confidence attribution for the Shadow Brokers.

42. The files can be found at the Wikileaks site, see Wikileaks, "Vault 7: CIA Hacking Tools Revealed." The individual behind the Vault 7 leak was eventually revealed to be former CIA software engineer Joshua Schulte. There is no indication that he was knowingly working in the service of a foreign government at the time. See Adam Goldman, "New Charges in Huge C.I.A. Breach Known as Vault 7," *The New York Times*, 19 June 2018, sec. US, https://www.nytimes.com/2018/06/18/us/politics/charges-cia-breach-vault-7.html

43. Martin reportedly hoarded over fifty terabytes of classified data. Despite a chronological correlation, he was never publicly linked to the Shadow Brokers. See Scott Shane, "Ex-N.S.A. Worker Accused of Stealing Trove of Secrets Offers to Plead Guilty," *The New York Times*, 1 January 2018, sec. US, https://www.nytimes.com/2018/01/03/us/politics/harold-martin-nsa-guilty-plea-offer.html

44. Microsoft, "Microsoft Security Bulletin MS17-010—Critical."

45. In the NotPetya malware, EternalBlue played a secondary role in infecting endpoints. It principally relied on other mechanisms for its success.

46. UK Foreign Ministry, "Foreign Office Minister Condemns North Korean Actor for WannaCry Attacks," GOV.UK, 19 December 2017, https://www.gov.uk/government/news/foreign-office-minister-condemns-north-korean-actor-for-wannacry-attacks

47. US Press Secretary, "Statement from the Press Secretary."

48. Freedberg Jr., Sydney J., "Wireless Hacking In Flight: Air Force Demos Cyber EC-130," *Breaking Defense* (blog), 15 September 2015, https://breakingdefense.com/2015/09/wireless-hacking-in-flight-air-force-demos-cyber-ec-130/

49. David Cenciotti, "The Amazing Growler Ball 2020 Video Teases An EA-18G Cyber Attack Capability That Is Yet To Come," *The Aviationist* (blog), 2 September 2020, https://theaviationist.com/2020/09/02/the-amazing-growler-ball-2020-video-teases-an-ea-18g-cyber-attack-capability-that-is-yet-to-come/

50. Matthew Keegan and Stephen Leonard Engelson Wyatt, Method and system for a small unmanned aerial system for delivering electronic warfare and cyber effects, United States US20180009525A1, filed 15 March 2016, and issued 11 January 2018.

51. Keegan and Wyatt.

52. John Keller, "Navy and Air Force Choose DRFM Jammers from Mercury Systems to Help Spoof Enemy Radar," Military & Aerospace Electronics, 18 June 2014, https://www.militaryaerospace.com/articles/2014/06/mercury-drfm-jammer.html

53. Paul C. Hershey, Robert E. Dehnert JR, and John J. Williams, Digital weapons factory and digital operations center for producing, deploying, assessing, and managing digital defects, United States US9544326B2, filed 20 January 2015, and issued 10 January 2017.

54. The patent documentation specifically notes a missile or a tank as viable scenarios of use.

55. Paul C. Hershey, Joseph O. Chapa, and Elizabeth Umberger, Methods and apparatuses for eliminating a missile threat, United States US20160070674A1, filed 9 September 2014, and issued 10 March 2016.

56. Paul Christian Hershey et al., System and method for integrated and synchronised planning and response to defeat disparate threats over the threat kill chain with combined cyber, electronic warfare and kinetic effects, United States US201 80038669A1, filed 28 February 2017, and issued 8 February 2018.

57. Seth L. Jahne et al., Techniques Deployment System, United States US2015 0369569A1, filed 24 June 2014, and issued 24 December 2015.

58. Jonathon P. Leibunguth, Command and Control Systems for Cyber Warfare, United States US20090249483A1, filed 30 March 2009, and issued 1 October 2009.

59. The NSA is by no means the sole proprietor of offensive network capabilities in the US Other intelligence agencies such as the CIA have their own, alongside increasing indigenous capabilities within US Cyber Command.

60. Carter, "A Lasting Defeat: The Campaign to Destroy ISIS."

61. Sanger, "Obama Ordered Wave of Cyberattacks Against Iran."

62. See Chapter Two, where Stuxnet is shown to not meet the threshold of TIAGR as it was not conducted within the spectrum of military operations.

63. Farwell and Rohozinski, "Stuxnet and the Future of Cyber War," 25.

64. Ron Rosenbaum, "Richard Clarke on Who Was Behind the Stuxnet Attack," *Smithsonian Magazine*, April 2012, https://www.smithsonianmag.com/history/richard-clarke-on-who-was-behind-the-stuxnet-attack-160630516/

65. Falliere, Murchu, and Chien, "W32.Stuxnet Dossier," 6–10.

66. David E. Sanger and Mark Mazzetti, "U.S. Had Cyberattack Plan If Iran Nuclear Dispute Led to Conflict," *The New York Times*, 16 February 2016, sec. World, https://www.nytimes.com/2016/02/17/world/middleeast/us-had-cyberattack-planned-if-iran-nuclear-negotiations-failed.html

67. Ellen Nakashima, "Trump Approved Cyber-Strikes against Iranian Computer Database Used to Plan Attacks on Oil Tankers," *Washington Post*, 22 June 2019.

68. Eric Schmitt, Farnaz Fassihi, and David D. Kirkpatrick, "Saudi Oil Attack Photos Implicate Iran, U.S. Says; Trump Hints at Military Action," *The New York Times*, 15 September 2019, sec. World, https://www.nytimes.com/2019/09/15/world/middleeast/iran-us-saudi-arabia-attack.html

69. Schmitt, Fassihi, and Kirkpatrick.

70. Chris Bing and Patrick Howell O'Neill, "Kaspersky's 'Slingshot' Report Burned an ISIS-Focused Intelligence Operation," *CyberScoop* (blog), 20 March 2018, https://www.cyberscoop.com/kaspersky-slingshot-isis-operation-socom-five-eyes/

71. NSA, "ANT Product Catalog," 2009.

72. GReAT, "Equation: The Death Star of Malware Galaxy," *Securelist* (blog),

16 February 2015, https://securelist.com/equation-the-death-star-of-malware-galaxy/68750/

73. Resilience methods allowing a network intrusion tool to survive device restarts and attempts to remove it.

74. Comae, "The Shadow Brokers: Cyber Fear Game-Changers."

75. Left of launch references a timeline which places all actions prior to the launch on the left-hand, while post-launch mitigation attempts are right of launch.

76. Herbert C. Kemp, "Left of Launch: Countering Theater Ballistic Missiles," Issue Brief, Atlantic Council, July 2017, 2.

77. US Department of Defense, "Declaratory Policy, Concept of Operations, and Employment Guidelines for Left-of-Launch Capability," 10 May 2017, 2.

78. David E. Sanger and William J. Broad, "Trump Inherits a Secret Cyberwar Against North Korean Missiles," *The New York Times*, 20 January 2018, sec. World, https://www.nytimes.com/2017/03/04/world/asia/north-korea-missile-program-sabotage.html

79. Jeffrey Lewis, "Is the United States Really Blowing Up North Korea's Missiles?," *Foreign Policy* (blog), 19 April 2017, https://foreignpolicy.com/2017/04/19/the-united-states-isnt-hacking-north-koreas-missile-launches/

80. Kenneth Todorov et al., Panel on Full Spectrum Missile Defense, Center for Strategic and International Studies 4 December 2015.

81. Futter, "The Dangers of Using Cyberattacks to Counter Nuclear Threats," *Arms Control Today* 46, no. 6 (2016): 8–14.

82. US Joint Chiefs of Staff, "Joint Publication 3–12: Cyberspace Operations," 8 June 2018.

83. This is echoed strongly in the words of the commander of the 24th Air Force, defining cyberspace as "a warfighting domain of operations where cyber operators generate effects differently than the ways we generate effects in Air, Space, Land and Sea.", see US 24th Air Force, "Commander's Strategic Vision,"8 March 2017, 2.

6. THE RUSSIAN SPECTRUM OF CONFLICT

1. Stephen Blank, "Cyber War and Information War à La Russe," in *Understanding Cyber Conflict: 14 Analogies*, ed. George Perkovich and Ariel E. Levite, (Washington D.C.: Georgetown University Press, 2017), https://carnegieendowment.org/2017/10/16/cyber-war-and-information-war-la-russe-pub-73399

2. Valery Gerasimov, "The Value of Science Is in the Foresight," *Military Review* 96, no. 1 (2016): 24.

3. Chekinov and Bogdanov, "The Nature and Content of a New-Generation War."

4. Bilyana Lilly and Joe Cheravitch, "The Past, Present, and Future of Russia's Cyber Strategy and Forces," in *2020 12th International Conference on Cyber Conflict (CyCon)* (Tallinn, Estonia: IEEE, 2020), 129–55, https://doi.org/10.23919/CyCon 49761.2020.9131723

5. Dmitry Adamsky, "Cross-Domain Coercion: The Current Russian Art of Strategy,"

Institut Français des Relations Internationales, 2015, 21–23, http://www.ifri.org/sites/default/files/atoms/files/pp54adamsky.pdf

6. Chekinov and Bogdanov, "The Nature and Content of a New-Generation War," 16–17.

7. This includes Mark Galeotti, who inadvertently suggested the Gerasimov Doctrine concept and then railed against its use. See Mark Galeotti, "The 'Gerasimov Doctrine' and Russian Non-Linear War," *In Moscow's Shadows* (blog), 6 July 2014, https://inmoscowsshadows.wordpress.com/2014/07/06/the-gerasimov-doctrine-and-russian-non-linear-war/. Galeotti later dedicated a polemic to the accidentally minted term, see Mark Galeotti, "The Mythical 'Gerasimov Doctrine' and the Language of Threat," *Critical Studies on Security* 7, no. 2 (May 4, 2019): 157–61, https://doi.org/10.1080/21624887.2018.1441623

8. Frank G. Hoffman, "Hybrid Warfare & Challenges," *Joint Forces Quarterly*, no. 52 (2009): 34–47.

9. Giles, "Russia's 'New' Tools for Confronting the West," 5.

10. Charles K. Bartles, "Getting Gerasimov Right," *Military Review* 96, no. 1 (2016): 34; Giles, "Russia's 'New' Tools for Confronting the West," 9.

11. Diane Chotikul, "The Soviet Theory of Reflexive Control in Historical and Psychocultural Perspective: A Preliminary Study," Fort Belvoir, VA: Defense Technical Information Center, 1 July 1986.

12. Timothy Thomas, "Russia's Reflexive Control Theory and the Military," *The Journal of Slavic Military Studies* 17, no. 2 (June 2004): 237.

13. Thomas, 242.

14. Chotikul, "The Soviet Theory of Reflexive Control in Historical and Psychocultural Perspective: A Preliminary Study:," 73–74.

15. Maria Snegovaya, "Putin's Information Warfare in Ukraine: Soviet Origins Of Russia's Hybrid Warfare," Washington: Institute for the Study of War, 2015, 10; Franklin D. Kramer et al., *Meeting the Russian Hybrid Challenge: A Comprehensive Strategic Framework*, 2017, 11.

16. Snegovaya, "Putin's Information Warfare in Ukraine," 10; Kramer et al., *Meeting the Russian Hybrid Challenge*, 11.

17. Janne Hakala and Jazlyn Melnychuk, "Russia's Strategy in Cyberspace," Riga, Latvia: NATO Strategic Communications Centre of Excellence, June 2021, 5–7.

18. Valeriy Akimenko and Keir Giles, "Russia's Cyber and Information Warfare," *Asia Policy* 27, no. 2 (2020): 67.

19. Dr Lester W. Grau and Charles K. Bartles, *The Russian Way of War: Force Structure, Tactics, and Modernization of the Russian Ground Forces* (Fort Leavenworth: Foreign Military Studies Office, 2016), 289–90.

20. Franke, "War by Non-Military Means," 14–15.

21. Adamsky, *Cross-Domain Coercion*, 9.

22. Lilly and Cheravitch, "The Past, Present, and Future of Russia's Cyber Strategy and Forces," 135.

23. A. Yu. Maruyev, "Russia and the U.S.A in Confrontation: Military and Political Aspects," *Military Thought* 18, no. 3 (1 July 2009): 2.

24. Maruyev, 3.

25. Bartles, "Getting Gerasimov Right," 32–33.

26. Kramer et al., *Meeting the Russian Hybrid Challenge*, 4.

27. Snegovaya, "Putin's Information Warfare in Ukraine," 9.

28. Chekinov and Bogdanov, "The Nature and Content of a New-Generation War," 13.

29. Col. E. A. Perov and Col. A. V. Pereverzev, "On the Prospective Digital Communication Network of the RF Armed Forces," *Military Thought* 17, no. 2 (2008): 89–95.

30. Col. S. I. Baylev and Col. I. N. Dylevsky, "The Russian Armed Forces in the Information Environment: Principles, Rules, and Confidence-Building Measures," n.d., 12–13; Giles, "Russia's 'New' Tools for Confronting the West," 25–27.

31. Franke, "War by Non-Military Means," 14–15.

32. Giles, "Russia's 'New' Tools for Confronting the West," 27.

33. Chekinov and Bogdanov, "The Nature and Content of a New-Generation War," 16.

34. Lt. Gen. V. I. Kuznetsov, Col. Yu. Ye. Donskov, and Lt. Col. O. G. Nikitin, "Cyberspace in Military Operations Today," *Military Thought* 23, no. 1 (2014): 22.

35. Bellingcat, "FSB Team of Chemical Weapon Experts Implicated in Alexey Navalny Novichok Poisoning," 14 December 2020, https://www.bellingcat.com/news/uk-and-europe/2020/12/14/fsb-team-of-chemical-weapon-experts-implicated-in-alexey-navalny-novichok-poisoning/

36. Adamsky, *Cross-Domain Coercion*, 19.

37. Giles, "Russia's 'New' Tools for Confronting the West," 25.

38. Russian Federation, "The Military Doctrine of the Russian Federation."

39. It remains indeterminate how actively this sentiment was stoked by Russian influence operations.

40. Maruyev, "Russia and the U.S.A in Confrontation: Military and Political Aspects," 11.

41. Snegovaya, "Putin's Information Warfare in Ukraine," 9–10.

42. Rod Thornton, "The Changing Nature of Modern Warfare: Responding to Russian Information Warfare," *The RUSI Journal* 160, no. 4 (4 July 2015): 40.

43. Thornton, 42.

44. Giles, "Russia's 'New' Tools for Confronting the West," 28.

45. See Chapter Four for the analysis on strategic principals within cyber-warfare.

46. Michael Connell and Sarah Vogler, "Russia's Approach to Cyber Warfare," Arlington, VA: CNA, 24 March 2017, 27.

47. Adamsky, *Cross-Domain Coercion*, 23–24.

48. Jānis Bērziš, "Russian New Generation Warfare Is Not Hybrid Warfare," in *The War in Ukraine: Lessons for Europe*, ed. Artis Pabriks and Andis Kudors (Rīga: The Centre for East European Policy Studies: University of Latvia Press, 2015), 45.

49. Thornton, "The Changing Nature of Modern Warfare," 42–43.

50. Clausewitz, *On War*, 1:36.

51. Franke, "War by Non-Military Means," 24.

52. Activation entails reaching the effects phase of the operation. This can be seen in their operations against Ukrainian critical infrastructure, against which they have triggered several disruptive and destructive payloads. As for premature detection, Russian operations attributed to GRU, FSB, SVR and other agencies are comparatively exposed with alarming efficiency and consistency. This is in part due to reuse of network infrastructure, malicious code, and offensive techniques.

53. Grau and Bartles, *The Russian Way of War: Force Structure, Tactics, and Modernization of the Russian Ground Forces*, 289–91.

54. Estonian Foreign Intelligence Service, "International Security and Estonia," 2018, 36–39.

55. US Press Secretary, "Statement from the Press Secretary."

56. Maynor et al., "The MeDoc Connection."

57. NotPetya can be contrasted with the North Korean WannaCry, which had ostensibly intended to infect without limit, and Western-made Stuxnet, which infected broadly but only attacked specific targets.

58. US Department of Justice, "U.S. Charges Russian GRU Officers with International Hacking and Related Influence and Disinformation Operations," 4 October 2018, https://www.justice.gov/opa/pr/us-charges-russian-gru-officers-international-hacking-and-related-influence-and

59. For an extensive accounting of Unit 74555's activity, see Andy Greenberg, *Sandworm: A New Era of Cyberwar and the Hunt for the Kremlin's Most Dangerous Hackers*, 1st ed. (New York: Doubleday, 2019).

60. For example, this reliance on external operations has resulted in a 2018 mass-casualty incident in Syria, where US forces opened fire and killed an estimated 200 Russian mercenaries deployed to assist Bash al-Assad's forces. See Neil Hauer, "Russia's Mercenary Debacle in Syria," *Foreign Affairs*, 26 February 2018, https://www.foreignaffairs.com/articles/syria/2018-02-26/russias-mercenary-debacle-syria

61. Eneken Tikk et al., *International Cyber Incidents: Legal Considerations* (Tallinn: Cooperative Cyber Defence Centre of Excellence, 2010), 15.

62. Tikk et al., 16.

63. Tikk et al., 18.

64. Connell and Vogler, "Russia's Approach to Cyber Warfare," 13.

65. This process had several key milestones, including the formation of the CCDCOE in Tallinn in 2008, the publishing of the Tallinn Manual in 2013, and the recognition by NATO of cyberspace as a distinct "domain of operations" in 2016.

66. Government of Georgia, "Russian Cyberwar on Georgia," 10 November 2008.

67. Evgeny Morozov, "How I Became a Soldier in the Georgia–Russia Cyberwar," *Slate*, 14 August 2008, http://www.slate.com/articles/technology/technology/2008/08/an_army_of_ones_and_zeroes.html

68. Deibert, Rohozinski, and Crete-Nishihata, "Cyclones in Cyberspace," 12; Government of Georgia, "Russian Cyberwar on Georgia," 6.

69. Deibert, Rohozinski, and Crete-Nishihata, "Cyclones in Cyberspace," 13.

70. Connell and Vogler, "Russia's Approach to Cyber Warfare," 18.

71. E Lincoln Bonner III, "Cyber Power in 21st-Century Joint Warfare," *Joint Forces Quarterly* 74, no. 3 (2014): 107.

72. Giles, "Russia's 'New' Tools for Confronting the West," 4.

73. This is made eminent by the deployment of numerous state-of-the-art Russian war-fighting platforms, including the Sukhoi Su-35 fighter, The S-400 Triumf air defence system, and numerous variants of the T-90 main battle tank.

74. Walter Russell Mead, "The Return of Geopolitics: The Revenge of the Revisionist Powers," *Foreign Affairs* 93 (2014): 74–75.

75. Chekinov and Bogdanov, "The Nature and Content of a New-Generation War," 20.

76. Giannangeli, "Russians forcing RAF to abort missions in Syria by 'hacking into' their systems."

77. Ministry of Defense of the Russian Federation, "Head of the Russian General Staff's Office for UAV Development Major General Alexander Novikov Holds Briefing for Domestic and Foreign Reporters: Ministry of Defense of the Russian Federation," 11 January 2018, http://eng.mil.ru/en/news_page/country/more.htm?id=12157872@egNews

78. Raf Sanchez, "Russia Uses Missiles and Cyber Warfare to Fight off 'swarm of Drones' Attacking Military Bases in Syria," *The Telegraph*, 9 January 2018, https://www.telegraph.co.uk/news/2018/01/09/russia-fought-swarm-drones-attacking-military-bases-syria/

79. The technology has become so abundant as to be made available as a commercial product for use in non-kinetic corporate perimeter defence.

80. Maynor et al., "The MeDoc Connection"; Karan Sood and Shaun Hurley, "NotPetya Ransomware Attack," *CrowdStrike Blog* (blog), 29 June 2017.

81. See for example US Press Secretary, "Statement from the Press Secretary."

82. US-CERT, "Petya Ransomware," 15 February 2018, https://www.us-cert.gov/ncas/alerts/TA17-181A

83. Andy Greenberg, "Ukrainians Say Petya Ransomware Hides State-Sponsored Attacks," *Wired*, 28 June 2017, https://www.wired.com/story/petya-ransomware-ukraine/

84. Bob Drogin, "Russians Seem To Be Hacking Into Pentagon / Sensitive Information Taken—But Nothing Top Secret," *Los Angeles Times*, 7 October 1999, https://www.sfgate.com/news/article/Russians-Seem-To-Be-Hacking-Into-Pentagon-2903309.php

85. Costin Raiu et al., "Penquin's Moonlit Maze," *Securelist* (blog), 3 April 2017, https://securelist.com/penquins-moonlit-maze/77883/

86. CrowdStrike, "Danger Close: Fancy Bear Tracking of Ukrainian Field Artillery Units," *CrowdStrike Blog* (blog), 22 December 2016, https://www.crowdstrike.com/blog/danger-close-fancy-bear-tracking-ukrainian-field-artillery-units/

87. The extent of evidence pertaining to Russian use of intelligence garnered from this malware is limited and contentious. CrowdStrike initially claimed that the

compromised app led to an 80% attrition rate for Ukrainian D-30s. After challenges to these figures, including from the Ukrainian ministry of defense (see Ukraine Ministry of Defense, "Інформація Про 'Втрати у ЗС України 80% Гаубиць Д-30' Не Відповідає Дійсності," 6 January 2017, http://www.mil.gov.ua/news/2017/01/06/informacziya-po-vtrati-u-zs-ukraini-80-gaubicz-d-30%E2%80%9D-ne-vidpovidae-dijsnosti/), CrowdStrike amended their figures to reflect a 15–20% attrition rate. Yet as it stands, it is unclear how much of this is attributable to information gleaned from the app itself.

88. This campaign and associated malware is also known as HAVEX.
89. Symantec, "Dragonfly: Cyberespionage Attacks Against Energy Suppliers," 7 July 2014.
90. Dragos, "CRASHOVERRIDE: Threat to the Electric Grid Operations," 9.
91. Dragos, 16.
92. Slowik, "CRASHOVERRIDE: Reassessing the 2016 Ukraine Electric Power Event as a Protection-Focused Attack."
93. Slowik, 4–5.
94. Sheera Frenkel, "Experts Say Russians May Have Posed As ISIS To Hack French TV Channel," BuzzFeed, 10 June 2015, https://www.buzzfeed.com/sheerafrenkel/experts-say-russians-may-have-posed-as-isis-to-hack-french-t
95. Frenkel.
96. Trend Micro indirectly received technical data from the hack provided by the French National Cybersecurity Agency (ANSSI) which indicated infections by malware associated with APT28, commonly associated with the Russian GRU. See Rik Ferguson, "TV5 Monde, Russia and the CyberCaliphate," Trend Micro (blog), 10 June 2015, 5, http://blog.trendmicro.co.uk/tv5-monde-russia-and-the-cybercaliphate/
97. Nicole Perlroth, "Cyberattack Caused Olympic Opening Ceremony Disruption," The New York Times, 13 February 2018, sec. Technology, https://www.nytimes.com/2018/02/12/technology/winter-olympic-games-hack.html
98. Jay Rosenberg, "2018 Winter Cyber Olympics: Code Similarities with Cyber Attacks in Pyeongchang," Intezer (blog), 12 February 2018, https://www.intezer.com/2018-winter-cyber-olympics-code-similarities-cyber-attacks-pyeongchang/
99. GReAT, "The Devil's in the Rich Header," Securelist (blog), 8 March 2018, https://securelist.com/the-devils-in-the-rich-header/84348/
100. Warren Mercer, Paul Rascagneres, and Matthew Molyett, "Olympic Destroyer Takes Aim At Winter Olympics," Cisco's Talos Intelligence (blog), 12 February 2018, http://blog.talosintelligence.com/2018/02/olympic-destroyer.html
101. Perlroth, "Cyberattack Caused Olympic Opening Ceremony Disruption."
102. Warren P. Strobel, "Pompeo Blames Russia for Hack as Trump Casts Doubt on Widespread Conclusion," Wall Street Journal, 19 December 2020, sec. Politics, https://www.wsj.com/articles/pompeo-blames-russia-for-solarwinds-hack-11608391515
103. Cimpanu, "CISA Updates SolarWinds Guidance, Tells US Govt Agencies to Update Right Away."

104. Raphael Satter and Joseph Menn, "SolarWinds Hackers Accessed Microsoft Source Code, the Company Says," *Reuters*, 1 January 2021, https://www.reuters.com/article/global-cyber-microsoft-idINKBN29620C

105. Brad Smith, "A Moment of Reckoning: The Need for a Strong and Global Cybersecurity Response," *Microsoft On the Issues* (blog), 17 December 2020, https://blogs.microsoft.com/on-the-issues/2020/12/17/cyberattacks-cyber-security-solarwinds-fireeye/

106. FireEye, "Unauthorized Access of FireEye Red Team Tools"; FireEye, "Highly Evasive Attacker Leverages SolarWinds Supply Chain to Compromise Multiple Global Victims With SUNBURST Backdoor."

107. Mark Clayton, "Ukraine Election Narrowly Avoided 'Wanton Destruction' from Hackers," *Christian Science Monitor*, 17 June 2014, https://www.csmonitor.com/World/Passcode/2014/0617/Ukraine-election-narrowly-avoided-wanton-destruction-from-hackers

108. This is assuming the stated goal was to facilitate a Trump-led Republican victory in the 2016 elections, as he was perhaps perceived as being more pliable with regard to Russian grand-strategy and less hawkish in its geopolitical disposition.

109. CrowdStrike, the security company solicited to assist in the investigation and cleanup of the DNC hack, was fairly straightforward in its attribution, see Dmitri Alperovitch, "Bears in the Midst: Intrusion into the Democratic National Committee»," *CrowdStrike Blog* (blog), 15 June 2016, https://www.crowdstrike.com/blog/bears-midst-intrusion-democratic-national-committee/. The US government publicly decried the Russian government as responsible on numerous official occasions, see for example DHS Press Office, "Joint Statement from the Department Of Homeland Security and Office of the Director of National Intelligence on Election Security."

110. Franceschi-Bicchierai, "We Spoke to DNC Hacker 'Guccifer 2.0.'"

111. Grau and Bartles, *The Russian Way of War: Force Structure, Tactics, and Modernization of the Russian Ground Forces.*

7. ASSERTING CHINESE DOMINANCE

1. Jyrki Kallio and Julie Chen, "Taiwan's 2020 Election and Its Implications: Dark Clouds Looming for Already Strained Cross-Strait Relations," FIIA Comment, Finnish Institute of International Affairs, January 2020.

2. Gao Xu, "说透了！'武统'台湾什么时候开始？解放军专家权威解读_两岸快评_中国台湾网" China Taiwan Network, April 15, 2020, http://www.taiwan.cn/plzhx/plyzl/202004/t20200415_12265753.htm

3. Trevor Hunnicutt, "Biden Says United States Would Come to Taiwan's Defense," *Reuters*, 22 October 2021, sec. Asia Pacific, https://www.reuters.com/world/asia-pacific/biden-says-united-states-would-come-taiwans-defense-2021-10-22/

4. It is crucial to note that the United States and Taiwan do not have a true defensive pact. As a result of the United States recognising the PRC as China, the US does not

officially acknowledge Taiwan's independence as a political entity. Instead, the Taiwan Relations Act, signed in 1979, establishes ambiguous military support to Taiwan. Considering mounting geopolitical tensions with the PRC, it is likely that a destabilising move to claim Taiwan would result in a US response or risk losing much of its freedom to operate and influence in the entire region.

5. Ian Easton, *The Chinese Invasion Threat: Taiwan's Defense and American Strategy in Asia*, 1st ed. (Arlington, Virgina: Project 2049 Institute, 2017); Drew Thompson, "China Is Still Wary of Invading Taiwan," *Foreign Policy* (blog), 11 May 2020, https://foreignpolicy.com/2020/05/11/china-taiwan-reunification-invasion-coronavirus-pandemic/

6. Costello, John, "The Strategic Support Force: China's Information Warfare Service," *China Brief* (blog), 8 February 2016, https://jamestown.org/program/the-strategic-support-force-chinas-information-warfare-service/

7. Ji-Jen Hwang, "China's Military Reform: The Strategic Support Force, Non-Traditional Warfare, and the Impact on Cross-Strait Security," *Issues & Studies* 53, no. 3 (September 2017): 4.

8. The PRC has aggressively pursued diplomatic isolation as a means of weakening Taiwan's overall standing. As of 2021, only 14 countries (plus the Vatican) have formal relations with Taiwan, see Thomas J. Shattuck, "The Race to Zero?: China's Poaching of Taiwan's Diplomatic Allies," *Orbis* 64, no. 2 (February 2020): 334–52, https://doi.org/10.1016/j.orbis.2020.02.003.

9. Jan Van Tol et al., "AirSea Battle: A Point-of-Departure Operational Concept," DTIC Document, 2010, 3.

10. Michael McDevitt, "The PLA Navy's Antiaccess Role in a Taiwan Contingency," in *The Chinese Navy*, ed. Phillip C. Saunders et al. (Washington DC: Institute for National Strategic Studies, 2011), 204.

11. Oriana Skylar Maestro, "Military Confrontation in the South China Sea," Council on Foreign Relations, 21 May 2020, https://www.cfr.org/report/military-confrontation-south-china-sea

12. These include an incident in which a PLA J-11 fighter buzzed a US P-8A patrol plane, various incidents in which PLA Navy ships rammed or otherwise harassed fishing vessels, and a near collision between a PLA navy warship and the US Missile Destroyer Cowpens in 2014.

13. As of 2021, this includes 20 outposts in the Paracel Islands and 7 in the Spartlys, see continuous tracking via "China Tracker," Asia Maritime Transparency Initiative, accessed 26 December 2020, https://amti.csis.org/island-tracker/china/

14. Gregory Poling, "New Imagery Release," Asia Maritime Transparency Initiative, 10 September 2015, http://amti.csis.org/new-imagery-release/

15. Frances Mangosing, "New Photos Show China Is Nearly Done with Its Militarization of South China Sea," Inquirer.Net, 5 February 2018, http://www.inquirer.net/specials/exclusive-china-militarization-south-china-sea.

16. The Active Defence strategy has existed in the PLA in some forms since the 1980s, and was first incepted as a means of countering a Soviet invasion. The 1980s ver-

sion of the strategy bares minimal resemblance to its modern incarnation, perhaps only in the lack of supposed desire to initiate conflict. See Fravel, *Active Defense*, 141.

17. The State Council Information Office, China's Military Strategy.

18. Michael Green, Ernest Bower, and Center for Strategic and International Studies, *Asia-Pacific Rebalance 2025: Capabilities, Presence, and Partnerships: An Independent Review of U.S. Defense Strategy in the Asia-Pacific* (Washington D.C.: Center for Strategic and International Studies, 2016).

19. US Department of Defense, "Military and Security Developments Involving the People's Republic of China," Office of the Secretary of Defense, 2020, https://media.defense.gov/2020/Sep/01/2002488689/-1/-1/1/2020-DOD-CHINA-MILITARY-POWER-REPORT-FINAL.PDF

20. The State Council Information Office, China's Military Strategy.

21. Larry M. Wortzel, "PLA 'Joint' Operational Contingencies in South Asia, Central Asia, and Korea," in *Beyond The Strait: PLA Missions Other Than Taiwan*, ed. Roy Kamphausen, David Lai, and Andrew Scobell (Carlisle, PA: Strategic Studies Institute, 2009), 328.

22. Lindsay Maizland, "China's Modernizing Military," *Council on Foreign Relations* (blog), 5 February 2020, https://www.cfr.org/backgrounder/chinas-moderniz-ing-military

23. Larry Wortzel, "PLA Command, Control and Targeting Architectures," 191–93; Vincent Wei-cheng Wang, "The Chinese Military and the" Taiwan Issue": How China Assesses Its Security Environment," *Tamkang Journal of International Affairs* 10, no. 4 (2007): 109.

24. US Department of Defense, "Annual Report to Congress: Military and Security Developments Involving the People's Republic of China," 15 May 2017, 1–2.

25. Nigel Inkster, "Conflict Foretold: America and China," *Survival* 55, no. 5 (October 2013): 12–20.

26. At least not initially. In the advanced stages of a Taiwan invasion, combat operations between land forces on both sides are widely indicated as crucial to China's success.

27. The State Council Information Office, China's Military Strategy.

28. This notion is supported by statements from PRC high ranking officials, who publicly remarked on Taiwanese ploys towards independence as reasonable cause for military action. See Taiwan Affairs Office of the State Council, "国台办新闻发布会辑录（2018-05-16）中共中央台湾工作办公室、国务院台湾事务办公室," 16 May 2018, http://www.gwytb.gov.cn/xwfbh/201805/t20180516_11955430.htm

29. Easton, *The Chinese Invasion Threat*, 22.

30. US Department of Defense, "Annual Report to Congress: Military and Security Developments Involving the People's Republic of China."

31. Ian Easton, "Able Archers: Taiwan Defense Strategy in an Age of Precision Strike," Project 2049 Institute, n.d., 8.

32. McReynolds, "China's Evolving Perspectives on Network Warfare: Lessons from the Science of Military Strategy," 4.

33. People's Liberation Army News, "Cracking the Winning Source Code of Information Warfare-Xinhuanet.Com," Xinhuanet, November 7, 2017.

34. Kania and Costello, "Seizing the Commanding Heights."

35. Costello, John, "The Strategic Support Force."

36. Elsa Kania and Costello, John, "The Strategic Support Force and the Future of Chinese Information Operations," *Cyber Defense Review* 3, no. 1 (Spring 2018): 105.

37. Hwang, "China's Military Reform," 8.

38. Ministry of National Defense of the People's Republic of China, "China's National Defense in the New Era," Xinhuanet, 24 July 2019, http://eng.mod.gov.cn/publications/2019–07/24/content_4846452.htm; US Department of Defense, "Military and Security Developments Involving the People's Republic of China."

39. Examples are abundant, but perhaps the most notorious is that of Unit 61398, unmasked by private US security company Mandiant in 2013. See Mandiant, "APT1—Exposing One of China's Cyber Espionage Units."

40. Easton, *The Chinese Invasion Threat*, 85.

41. Roger Cliff, "Anti-Access Measures in Chinese Defense Strategy." Testimony before the US–China Economic and Security Review Commission, 2011, 3.

42. Van Tol et al., "AirSea Battle," 14.

43. Rather than a single network in the technical sense, it is reasonable to view DoDIN as a shared-topology logical superstructure that includes the numerous networks operated by the Department of Defense for its requirements.

44. US Army, "Army Field Manual 3–38—Cyber Electromagnetic Activities."

45. US Army, "Deployed Tactical Network Guidance," US Army Chief Information Office, 31 May 2012, 2.

46. Carlo Kopp, "JTIDS/MIDS—Network Centric Warfare Fundamentals," *DefenceTODAY*, n.d.

47. David Bennett, "An Analysis of the China's Offshore Active Defense and the People's Liberation Army Navy," *Global Security Studies* 1, no. 1 (2010): 129–30.

48. Cliff, "Anti-Access Measures in Chinese Defense Strategy," 2.

49. Ronald O'Rourke, "China Naval Modernization: Implications for U.S. Navy Capabilities—Background and Issues for Congress," Congressional Research Service, 17 December 2020, 57–59.

50. A2AD zones are usually neatly defined by the range of corresponding military hardware within the region, to include anti-aircraft missiles and radar, anti-ship cruise missiles, and other range-bound capabilities.

51. This varies with the unit type, but generally entails operation within a limited mission area. Examples include infantry companies, tank battalions or warplane sorties.

52. Vinod Anand, "Chinese Concepts and Capabilities of Information Warfare," *Strategic Analysis* 30 (2006): 789; Brian M. Mazanec, "The Art of (Cyber) War," *The Journal of International Security Affairs* 16 (Spring 2009): 6.

53. McReynolds, "China's Evolving Perspectives on Network Warfare: Lessons from the Science of Military Strategy," 6.

54. Garamone, "Esper Describes DOD's Increased Cyber Offensive Strategy."

55. A key advantage of cyber-warfare includes the capacity to attack below the threshold of inflicting damage; subtle manipulative attacks can often be as detrimental to the opponent and far more attainable.

56. Fraser et al., "Double Dragon: APT41, a Dual Espionage and Cyber Crime Operation."

57. These suspicions have resulted in blocking communications giants Huawei and ZTE from entering the US market altogether. Similar concerns have now surfaced in other countries, who are seeking to block or remove Huawei from participation in mobile 5G network construction.

58. Most prominently by bombing runways with special munitions developed to thoroughly crater the tarmac, effectively preventing all take-off and landing capabilities.

59. Easton, "Able Archers: Taiwan Defense Strategy in an Age of Precision Strike," 37.

60. Michael Lostumbo et al., *Air Defense Options for Taiwan: An Assessment of Relative Costs and Operational Benefits*, Research Report, RR-1051-OSD (Santa Monica, California: Rand Corporation, 2016), 20–22.

61. See for example the aforementioned Link-16 and its sibling protocol Link-11, which are also used across Taiwanese forces.

62. Arms trade to Taiwan has fluctuated in correlation to the relationship with the PRC. In 2020, The US Trump administration has authorised a significant arms sale package that specifically focused on hardware used to deter or defeat an amphibious campaign. See Sidharth Kaushal, "US Weapons Sales to Taiwan: Upholding the Porcupine Strategy," RUSI, 8 December 2020, https://rusi.org/commentary/us-weapons-sales-taiwan-upholding-porcupine-strategy

63. It is not unusual to continuously upgrade reliable platforms such as the F-16. Many countries, including the US and Israel, have done the same. As of 2021, Taiwan has upgraded 30% of its F-16s to the latest standard, F-16V. This standard includes essential updates including improved electronic countermeasures and onboarding computers. Once completed, the updates will alleviate some long-standing vulnerabilities but likely create new ones due to the additional software involved. See Taiwan News, "Taiwan on Schedule to Complete F-16 Upgrades by 2023," Taiwan News, 17 March 2021, https://www.taiwannews.com.tw/en/news/4153117.

64. McReynolds, "China's Evolving Perspectives on Network Warfare: Lessons from the Science of Military Strategy," 5.

65. Andre Toonk, "Chinese ISP Hijacks the Internet," *BGPMon* (blog), 8 April 2010, https://bgpmon.net/chinese-isp-hijacked-10-of-the-internet/

66. Dan Goodin, "Citing BGP Hijacks and Hack Attacks, Feds Want China Telecom out of the US," *Ars Technica*, 10 April 2020, https://arstechnica.com/tech-policy/2020/04/citing-bgp-hijacks-and-hack-attacks-feds-want-china-telecom-out-of-the-us/

67. Successfully undermining attack attribution is not easy to achieve; attribution is not a binary process, but rather thresholds of confidence. It is entirely possible that

even a misleading attack on civilian infrastructure would be sufficiently attributed to Chinese authorities as to incur the anger of other nations.

68. See, for example, the Stuxnet campaign.

69. This is to contrast other threat actors such as Russian intelligence agencies, that often rely on OCOs and other information operations to pre-emptively achieve their strategic-political objectives.

70. Airborne Warning and Control Systems.

71. Air Land Sea Application Center, "TADIL J: Introduction to Tactical Digital Information Link J and Quick Reference Guide," June 2000.

72. Understanding Voice and Data Link Networking, 2–19.

73. Hura et al., "Tactical Data Links," 109–10.

74. Link-16 enabled platforms include the F15/16 strike fighters, guided missile destroyers and AWACS early warning planes, to name a few.

75. Air Land Sea Application Center, "TADIL J: Introduction to Tactical Digital Information Link J and Quick Reference Guide."

76. Siobahn Gorman, August Cole, and Yochi Dreazen, "Computer Spies Breach Fighter-Jet Project," *The Wall Street Journal*, 21 April 2009, http://www.wsj.com/articles/SB124027491029837401

77. ViaSat, "Link-16 Message Card," October 2012, https://www.viasat.com/files/assets/assets/Link16_NPG_Message_Card_100112a.pdf; Air Land Sea Application Center, "TADIL J: Introduction to Tactical Digital Information Link J and Quick Reference Guide," 19–20.

78. US Navy, "Electronics Technician Volume 03-Communications Systems," July 1997, 144, http://electronicstechnician.tpub.com/14088/css/14088_144.htm

79. Global Security, "AEGIS Combat System," accessed October 2, 2015, http://www.globalsecurity.org/military/systems/ship/systems/aegis.htm.

80. Lockheed Martin, "Lockheed Martin to Enhance U.S. Navy's C4ISR Capabilities," Naval Today, 1 July 2014, http://navaltoday.com/2014/07/01/lockheed-martin-to-enhance-u-s-navys-c4isr-capabilities/

81. US Department of Defense, "Fiscal Year 2016 Navy Programs—Surface Electronic Warfare Improvement Program (SEWIP) Block 2," December 2016, https://www.dote.osd.mil/Portals/97/pub/reports/FY2016/navy/2016sewip.pdf?ver=2019-08-22-105305-400

82. Matthew J. Schwarz, "Lockheed Martin Suffers Massive Cyberattack," *Dark Reading* (blog), 30 May 2011, http://www.darkreading.com/risk-management/lockheed-martin-suffers-massive-cyberattack/d/d-id/1098013?

83. John McHale, "Record Number of Cyber Attacks Hit Lockheed Martin in 2014," Military Embedded Systems, 18 February 2015, http://mil-embedded.com/3499-record-number-of-cyber-attacks-hit-lockheed-martin-in-2014/

84. A high-level component architecture of said ACS is freely available online; see for example Global Security, "AEGIS Combat System."

85. PR Newswire, "Lockheed Martin and DRS Technologies Deliver 4000th AN/UYQ-70 Ship Display System to the U.S. Navy," 11 May 2012.

86. Global Security, "AEGIS Combat System"; PR Newswire, "Lockheed Martin and DRS Technologies Deliver 4000th AN/UYQ-70 Ship Display System to the U.S. Navy."

87. As a counterpoint, open-source software may be better tested due to its broad reach, so vulnerabilities disclosed by researchers are far more likely to have been submitted and remediated. In comparison, "closed" platforms can only rely on whatever testing processes are internally in place.

88. US Navy, "The US Navy Fact File: Tomahawk Cruise Missile," US Navy Official Website, accessed October 1, 2015.

89. Jane's International Defence Review, "Exploiting The Network For Smarter Weapon Effects," *Jane's International Defence Review*, August 2015, 2.

90. USCYBERCOM devotes ever-increasing budgets and attention to both safeguarding and recuperating from attacks against critical networks.

91. US Homeland Security, "Blueprint for a Secure Cyber Future," November 2011, 16.

92. US Army, "Deployed Tactical Network Guidance," 1; Coile, "WIN-T SATCOM Overview Briefing," 8; Epperson, "Satellite Communications Within the Army's WIN-T Architecture," 9.

93. Fravel, *Active Defense*.

94. Qiao Liang and Wang Xiangsui, *Unrestricted Warfare* (Beijing: PLA Literature and Arts Publishing House, 1999).

95. 寿晓松, 军事科学院, and 军事战略研究部, *The Science of Military Strategy* (北京: 军事科学出版社, 2013).

8. APPROXIMATING THE IRANIAN THREAT

1. J. Matthew McInnis, "Iranian Concepts of Warfare: Understanding Tehran's Evolving Military Doctrine," American Enterprise Institutes, February 2017, 4–5.

2. Hadi Ajili and Masha Rough, "Iran's Military Strategy," *Survival* 61, no. 6 (November 2019): 139–52.

3. McInnis, "Iranian Concepts of Warfare: Understanding Tehran's Evolving Military Doctrine," 19.

4. Kenneth Katzman, "Iran's Foreign and Defense Policies," Congressional Research Service, 11 January 2021, 3–4.

5. McInnis, "Iranian Concepts of Warfare: Understanding Tehran's Evolving Military Doctrine," 17.

6. Langner, "Stuxnet—Dissecting a Cyberwarfare Weapon."

7. See Kim Zetter's comprehensive book on Stuxnet as a resource on the campaign: Zetter, *Countdown to Zero Day: Stuxnet and the Launch of the World's First Digital Weapon*.

8. Unknown, "Untitled Paste from 'Cutting Sword of Justice,'" Paste Site, Pastebin, 15 August 2012, https://pastebin.com/HqAgaQRj

9. GReAT, "Shamoon The Wiper: Further Details (Part II)," *Securelist* (blog), 11 September 2012, https://securelist.com/shamoon-the-wiper-further-details-part-ii/57784/

10. GReAT, "Shamoon the Wiper—Copycats at Work," *Securelist* (blog), 16 August 2012, https://securelist.com/shamoon-the-wiper-copycats-at-work/57854/

11. Itamar Rabinovich, "How Iran's Regional Ambitions Have Developed since 1979," Iran's Revolution, 40 Years on—What It Has Meant for Iran, America, and the Region (series), Brookings, 24 January 2019, https://www.brookings.edu/blog/order-from-chaos/2019/01/24/how-irans-regional-ambitions-have-developed-since-1979/

12. Brian Katz, "Axis Rising: Iran's Evolving Regional Strategy and Non-State Partnerships in the Middle East," CSIS Briefs, Center for Strategic and International Studies, 11 October 2018, https://www.csis.org/analysis/axis-rising-irans-evolving-regional-strategy-and-non-state-partnerships-middle-east.

13. Siobahn Gorman and Julian E. Barnes, "Iranian Hacking to Test NSA Nominee Michael Rogers," *The Wall Street Journal*, 18 February 2014.

14. US Department of Justice, "Former U.S. Counterintelligence Agent Charged With Espionage on Behalf of Iran; Four Iranians Charged With a Cyber Campaign Targeting Her Former Colleagues," 13 February 2019, https://www.justice.gov/opa/pr/former-us-counterintelligence-agent-charged-espionage-behalf-iran-four-iranians-charged-cyber

15. The phishing attempts themselves against Witt's former colleagues were crude. This was, in fact, a golden opportunity missed by Iranian intelligence. Were they to use Witt's familiarity to craft truly compelling lures highly tailored to their targets, the odds of success—and avoiding detection—would have been immeasurably higher.

16. See for example Iran's fluctuating support to regional Jihadist groups, or the Sunni Palestinian Hamas.

17. One notable example is the three-way axis cemented between Syria, Iran, and Lebanon's Hezbollah. On this topic, see Rola El Husseini, "Hezbollah and the Axis of Refusal: Hamas, Iran and Syria," *Third World Quarterly* 31, no. 5 (July 2010): 803–15, https://doi.org/10.1080/01436597.2010.502695

18. Ajili and Rough, "Iran's Military Strategy."

19. Insikt Group, "The History of Ashiyane: Iran's First Security Forum," *Recorded Future* (blog), 16 January 2019, https://www.recordedfuture.com/ashiyane-forum-history/

20. Assistance included warplanes sold by the US to Iran, and collaboration on advanced missile systems with Israel. On the latter, see Dalia Dassa Kaye, Alireza Nader, and Parisa Roshan, "A Brief History of Israeli-Iranian Cooperation and Confrontation," in *Israel and Iran: A Dangerous Rivalry* (Santa Monica, CA: RAND, 2011), 13.

21. This was not an immediate process; Iranian, Israeli, and US pragmatism sustained their relationship for a limited period post-revolution. See Kaye, Nader, and Roshan, 14–16.

22. Eva Patricia Rakel, "Iranian Foreign Policy since the Iranian Islamic Revolution: 1979–2006," *Perspectives on Global Development and Technology* 6, no. 1–3 (2007): 169–70, https://doi.org/10.1163/156914907X207711

23. Throughout the 1960s and 1970s, Israel had been embroiled in several bouts of high and low-intensity conflict with all of its neighbours, including Syria, Lebanon, Jordan, and Egypt. While its performance varied throughout these wars, it had eventually emerged with invigorated US support and swaths of new territory under Israeli control. Some, including the Sinai region, would later negotiate back to their original owners through peace treaties. Other territories including Palestinian lands and the Golan Heights remained under Israeli control.

24. Martin van Bruinessen, "The Kurds between Iran and Iraq," *MERIP Middle East Report*, no. 141 (July 1986): 14–27, https://doi.org/10.2307/3011925

25. IISS, "Chapter One: Tehran's Strategic Intent," in *Iran's Networks of Influence in the Middle East* (IISS, 2019), 11–38, https://www.iiss.org/publications/strategic-dossiers/iran-dossier/iran-19-03-ch-1-tehrans-strategic-intent

26. Casey L Addis and Christopher M Blanchard, "Hezbollah: Background and Issues for Congress," Congressional Research Service, 3 January 2011.

27. Afshon Ostovar, "Sectarian Dilemmas in Iranian Foreign Policy: When Strategy and Identity Politics Collide," Carnegie Endowment for International Peace, November 2016, 11–14.

28. Paul Hastert, "Al Qaeda and Iran: Friends or Foes, or Somewhere in Between?," *Studies in Conflict & Terrorism* 30, no. 4 (April 1, 2007): 327–36, https://doi.org/10.1080/10576100701200132

29. IISS, "Chapter One."

30. Volz and Finkle, "U.S. Indicts Iranians for Hacking Dozens of Banks, New York Dam."

31. Volz and Finkle.

32. Volz and Finkle.

33. Nalani Fraser et al., "Double Dragon: APT41, a Dual Espionage and Cyber Crime Operation" FireEye, 7 August 2019.

34. Catherine A Theohary, "Iranian Offensive Cyber Attack Capabilities," In Focus, Congressional Research Service, 13 January 2020.

35. IISS, "Chapter One."

36. Insikt Group, "The History of Ashiyane."

37. Even Ashiyane Forum originated with Ashiyane Digital Security Team, a network security company established in 2002.

38. Yossi Mansharof, "Iran's Cyber War: Hackers In Service Of The Regime; IRGC Claims Iran Can Hack Enemy's Advanced Weapons Systems; Iranian Army Official: 'The Cyber Arena Is Actually The Arena Of The Hidden Imam,'" MEMRI, 25 August 2013, https://www.memri.org/reports/irans-cyber-war-hackers-service-regime-irgc-claims-iran-can-hack-enemys-advanced-weapons

39. Defacements are perhaps best viewed as online graffiti; acts that overwrite or replace visible content in a website with content chosen by the perpetrator. This often includes ideological imagery and text acknowledging the hack, assuming responsibility on behalf of a group, and disparaging the victim. Such attacks scale

well, especially against small websites, are often easily accomplished by automated tools, and rarely involve a significant level of skill to execute.

40. Collin Anderson and Karim Sadjadpour, "Iran's Cyber Threat: Espionage, Sabotage, and Revenge," Carnegie Endowment for International Peace, 2018, https://carnegieendowment.org/files/Iran_Cyber_Final_Full_v2.pdf

41. Mansharof, "Iran's Cyber War."

42. Biz Stone, "DNS Disruption," *Twitter Official Blog* (blog), 18 December 2009, https://blog.twitter.com/en_us/a/2009/dns-disruption.html

43. Michael Connell, "Deterring Iran's Use of Offensive Cyber: A Case Study," Fort Belvoir, VA: Defense Technical Information Center, 1 October 2014, 11.

44. The IRGC is often assessed to be the more significant military threat than its conventional Iranian military counterpart. The organisation enjoys substantial funding and deploys comparatively advanced equipment.

45. See for example the US indictment against Arabi, Reza, and Espargham in 2020. The three were an IRGC officer, the leader of the Iranian Dark Coders Team, and a civilian malware developer, respectively. US Department of Justice, "Iranian Hackers Indicted for Stealing Data from Aerospace and Satellite Tracking Companies."

46. Tasnim News Agency, "Commander Reiterates Iran's Preparedness to Confront Enemies in Cyber Warfare—Defense News," 18 February 2014, https://www.tasnimnews.com/en/news/2014/02/18/287797/commander-reiterates-iran-s-preparedness-to-confront-enemies-in-cyber-warfare

47. Shane Harris, "Forget China: Iran's Hackers Are America's Newest Cyber Threat," *Foreign Policy* (blog), 18 February 2014, https://foreignpolicy.com/2014/02/18/forget-china-irans-hackers-are-americas-newest-cyber-threat/

48. Natasha Bertrand, "Iran Is Building a Non-Nuclear Threat Faster than Experts 'Would Have Ever Imagined,'" Business Insider, 27 March 2015.

49. Perlroth and Hardy, "Online Banking Attacks Were Work of Iran, U.S. Officials Say."

50. Michael Mimoso, "Automated Toolkits Named in Massive DDoS Attacks Against U.S. Banks," ThreatPost, 2 October 2012, https://threatpost.com/automated-toolkits-named-massive-ddos-attacks-against-us-banks-100212/77068/

51. Antone Gonsalves, "Expert Fingers DDoS Toolkit Used in Bank Cyberattacks," CSO Online, 1 October 2012, https://www.csoonline.com/article/2138382/expert-fingers-ddos-toolkit-used-in-bank-cyberattacks.html

52. Gonsalves.

53. Alex Kirk, "Itsoknoproblembro, the VRT Has You Covered," *Cisco's Talos Intelligence* (blog), 16 October 2012, http://blog.talosintelligence.com/2012/10/itsoknoproblembro-vrt-has-you-covered.html

54. Mimoso, "Automated Toolkits Named in Massive DDoS Attacks Against U.S. Banks."

55. Connor Simpson, "Sheldon Adelson Has an Idea: Lob a Nuclear Bomb into the

Iranian Desert," The Atlantic, 23 October 2013, https://www.theatlantic.com/international/archive/2013/10/sheldon-adelson-has-idea-lob-nuclear-bomb-iranian-desert/309657/

56. Benjamin Elgin and Michael Riley, "Now at the Sands Casino: An Iranian Hacker in Every Server," Bloomberg, 12 December 2014, https://web.archive.org/web/20160412171928/https://www.bloomberg.com/news/articles/2014-12-11/iranian-hackers-hit-sheldon-adelsons-sands-casino-in-las-vegas

57. FireEye, "FireEye Responds to Wave of Destructive Cyber Attacks in Gulf Region," FireEye, 1 December 2016, https://www.fireeye.com/blog/threat-research/2016/11/fireeye_respondsto.html

58. Robert Falcone and Bryan Lee, "Shamoon 2: Delivering Disttrack," *Unit42* (blog), 26 March 2017, https://unit42.paloaltonetworks.com/unit42-shamoon-2-delivering-disttrack/

59. Robert Falcone, "Shamoon 3 Targets Oil and Gas Organization," *Palo Alto Networks Unit 42* (blog), 13 December 2018, https://unit42.paloaltonetworks.com/shamoon-3-targets-oil-gas-organization/

60. Michael Connell, "Iran's Military Doctrine," US Institute of Peace, 11 October 2010, https://iranprimer.usip.org/resource/irans-military-doctrine

61. Cimpanu, "Source Code of Iranian Cyber-Espionage Tools Leaked on Telegram."

62. The March 2019 OilRig leak would not be the last by Lab Dookhtegan, they would proceed to share additional information and documents pertaining to alleged state-aligned Iranian network operations. There is a distinct possibility that this was in fact an information operation by an adversary of Iran, meant to diminish its capacities while signaling coercive intent.

63. Bryan Lee and Robert Falcone, "Behind the Scenes with OilRig," *Unit42* (blog), 30 April 2019, https://unit42.paloaltonetworks.com/behind-the-scenes-with-oilrig/

64. Allison Wikoff and Richard Emerson, "New Research Exposes Iranian Threat Group Operations," Security Intelligence, 16 July 2020, https://securityintelligence.com/posts/new-research-exposes-iranian-threat-group-operations/

65. Gabriel Weimann, "Cyberterrorism: How Real Is the Threat?" Special Report, US Institute of Peace, December 2004.

66. Gwen Ackerman, "Iranian Hackers Drew Worryingly Close to Israel's Missile Alarm," Bloomberg, 24 February 2019.

67. Ahiya Raved, "Cyber Attack Targeted Israel's Water Supply, Internal Report Claims," *Ynet News*, 26 April 2020, https://www.ynetnews.com/article/HJX1mWmF8; T. O. I. staff, "Iran Cyberattack on Israel's Water Supply Could Have Sickened Hundreds," The Times of Israel, 1 June 2020, https://www.timesofisrael.com/iran-cyberattack-on-israels-water-supply-could-have-sickened-hundreds-report/

68. Ahiya Raved, "שוב: מתקפת סייבר על מתקני מים בישראל," *Ynet News*, 17 July 2020, https://www.ynet.co.il/article/rJrCqmAkw

9. A REVOLUTION IN CYBER AFFAIRS?

1. Ingvild Bode and Hendrik Huelss, "The Future of Remote Warfare? Artificial Intelligence, Weapons Systems and Human Control," in *Remote Warfare: Interdisciplinary Perspectives*, ed. Alasdair McKay, Abigail Watson and Megan Karlshøj-Pedersen, (Bristol, England: E-International Relations Publishing, 2021), 219, https://www.e-ir.info/publication/remote-warfare-interdisciplinary-perspectives/

2. Reviewed later in greater detail, the idea of AI-supported network defence has already shown to have limited success. This was recently proven by the DARPA-supported "Cyber Grand Challenge", in which multiple AI "Cyber Reasoning Systems" competed against one another in autonomously detecting threats, thwarting attacks, and mitigating vulnerabilities. See Teresa Nicole Brooks, "Survey of Automated Vulnerability Detection and Exploit Generation Techniques in Cyber Reasoning Systems," *ArXiv:1702.06162 [Cs]*, 20 February 2017, http://arxiv.org/abs/1702.06162

3. Commonly called the "internet of things" (IoT), this refers to the explosive rise of so-called "smart" devices, ranging from previously simple house appliances to door locks, thermostats and many others. These are often made by less security-aware vendors, who repurpose generic hardware and software for various household functions. This architecture along with poor secure development practices and limited support for subsequent patching means the devices are often internet-enabled and highly vulnerable, even to minimally-resourced adversaries.

4. Patrick Howell O'Neill, "2021 Has Broken the Record for Zero-Day Hacking Attacks," *MIT Technology Review*, 23 September 2021, https://www.technologyreview.com/2021/09/23/1036140/2021-record-zero-day-hacks-reasons/

5. Adrienne Porter Felt et al., "Measuring HTTPS Adoption on the Web," in *26th USENIX Security Symposium*, 2017, 1323–38; Google, "HTTPS Encryption on the Web—Google Transparency Report," 2020, https://transparencyreport.google.com/https/overview?hl=en

6. Meta, "Introducing Ray-Ban Stories: First-Generation Smart Glasses," *Meta* (blog), 9 September 2021, https://about.fb.com/news/2021/09/introducing-ray-ban-stories-smart-glasses/

7. Erik Lin-Greenberg, "Allies and Artificial Intelligence: Obstacles to Operations and Decision-Making," *Texas National Security Review* 3, no. 2 (Spring 2020): 65.

8. Libicki, "Why Cyber Will Not and Should Not Have Its Grand Strategist," 31.

9. Bode and Huelss, "The Future of Remote Warfare? Artificial Intelligence, Weapons Systems and Human Control," 221.

10. James Johnson, "Artificial Intelligence in Nuclear Warfare: A Perfect Storm of Instability?," *The Washington Quarterly* 43, no. 2 (2 April 2020): 197–98, https://doi.org/10.1080/0163660X.2020.1770968

11. "AI Enhancing Land Forces' Weapons and Platforms," *Defense Update:* (blog), 27 December 2020, https://defense-update.com/20201227_fire-weaver-ot.html

12. Miles Brundage et al., "The Malicious Use of Artificial Intelligence: Forecasting, Prevention, and Mitigation," *ArXiv:1802.07228*, February 2018, 58–59.

13. When dissecting its attack surface, Microsoft's Windows operating system has come a long way since its early days. Numerous mechanisms, configurations, and capabilities have been introduced that serve to limit the attack surface and afford greater flexibility to network administrators looking to safeguard their deployments. Microsoft's own defensive tools, including Microsoft Defender, have also improved drastically in their ability to detect and mitigate malicious software.

14. Brittany Day, "Linux: An OS Capable of Effectively Meeting the US Government's," Linux Security, accessed 29 December 2020, https://linuxsecurity.com/features/features/linux-an-os-capable-of-effectively-meeting-the-us-government-s-security-needs-heading-into-2020

15. Examples include Qubes, a fully open-source secure Linux variant pertaining to compartmentalise running processes, and Trusted End Node Security (TENS), a US Department of Defense lightweight Linux variant.

16. To varying levels of success. Vulnerabilities, including full remote code execution (RCE), remain possible even if they are complex to identify and exploit.

17. This process entails capitalising on multiple vulnerabilities within the targeted system in order to gain administrative access and achieve persistence. The initial exploit used to gain foothold may be insufficient for the task and require additional exploitation of the system itself, including through local privilege escalation exploits.

18. For example, as of 2021, private exploit company Zerodium offers 2.5 million US dollars for a full exploit chain with persistence for Android devices, and 2.0 million US dollars for equivalent Apple iOS devices. See https://zerodium.com/program.html

19. See for example the Spectre and Meltdown CPU vulnerabilities, which led to widespread panic as they were originally deemed unpatchable. While the vulnerabilities were eventually addressed via manufacturer patches and operating system kernel fixes, it indicated how complex mitigation was for some hardware vulnerabilities. See CERT.org, "CPU Hardware Vulnerable to Side-Channel Attacks," Carnegie Mellon University CERT, 3 January 2018, https://www.kb.cert.org/vuls/id/584653

20. One apt anecdote describes the exponential increase in code complexity deployed within military aircraft. See Robert N. Charette, "This Car Runs on Code," IEEE Spectrum: Technology, Engineering, and Science News, 1 February 2009, https://spectrum.ieee.org/transportation/systems/this-car-runs-on-code

21. Catalin Cimpanu, "Windows 10, IOS 15, Ubuntu, Chrome Fall at China's Tianfu Hacking Contest," The Record by Recorded Future, 17 October 2021, https://therecord.media/windows-10-ios-15-ubuntu-chrome-fall-at-chinas-tianfu-hacking-contest/

22. Ministry of Industry and Information Technology Network Security, "Notice of the Ministry of Industry and Information Technology and the State Internet Information Office of the Ministry of Public Security on Issuing the Regulations on the

Administration of Network Product Security Vulnerabilities," Cyberspace Administration of China, 13 July 2021, http://www.cac.gov.cn/2021–07/13/c_1627761607640342.htm

23. Verizon, "2020 Data Breach Investigations Report," 2020, 13.

24. Drew Robb, "Building the Global Heatmap," *Strava Engineering* (blog), 1 November 2017, https://medium.com/strava-engineering/the-global-heatmap-now-6x-hotter-23fc01d301de

25. Jeremy Hsu, "The Strava Heat Map Shows Even Militaries Can't Keep Secrets from Social Data," *Wired*, 30 January 2018, https://www.wired.com/story/strava-heat-map-military-bases-fitness-trackers-privacy/

26. Michael C Horowitz, "Artificial Intelligence, International Competition, and the Balance of Power," *Texas National Security Review* 1, no. 3 (May 2018): 40.

27. M L Cummings, "Artificial Intelligence and the Future of Warfare," London: Chatham House, January 2017, 2.

28. Kareem Ayoub and Kenneth Payne, "Strategy in the Age of Artificial Intelligence," *Journal of Strategic Studies* 39, no. 5–6 (18 September 2016): 794.

29. Deep Blue was an IBM supercomputer purpose-built to excel at chess. For a detailed analysis of the system and its development, see Murray Campbell, A. Joseph Hoane Jr, and Feng-hsiung Hsu, "Deep Blue," *Artificial Intelligence* 134, no. 1–2 (2002): 57–83.

30. Brooks, "Survey of Automated Vulnerability Detection and Exploit Generation Techniques in Cyber Reasoning Systems," 8.

31. Ben Goertzel and Cassio Pennachin, eds., *Artificial General Intelligence (Cognitive Technologies)* (Berlin; New York: Springer, 2007), 15.

32. The list of those concerned by the potential impact of AGI includes physicist Stephen Hawking and technology industrialist Elon Musk, among many others.

33. Brundage et al., "The Malicious Use of Artificial Intelligence: Forecasting, Prevention, and Mitigation," 61.

34. Kania, Elsa, "Battlefield Singularity: Artificial Intelligence, Military Revolution, and China's Future Military Power," Center for a New American Security, November 2017.

35. Huw Roberts et al., "The Chinese Approach to Artificial Intelligence: An Analysis of Policy, Ethics, and Regulation," *AI & SOCIETY* 36, no. 1 (March 2021): 60.

36. Robert Work, "Remarks by Deputy Secretary Work on Third Offset Strategy," US Department of Defense, 28 April 2016, https://www.defense.gov/News/Speeches/Speech-View/Article/753482/remarks-by-deputy-secretary-work-on-third-offset-strategy/

37. Fryer-Biggs, "Secretive Pentagon Research Program Looks to Replace Human Hackers with AI."

38. Horowitz, "Artificial Intelligence, International Competition, and the Balance of Power," 41.

39. Johnson, "Artificial Intelligence in Nuclear Warfare," 200–201.

40. "AI Enhancing Land Forces' Weapons and Platforms."

41. Johnson, "Artificial Intelligence in Nuclear Warfare," 200; Bode and Huelss, "The

Future of Remote Warfare? Artificial Intelligence, Weapons Systems and Human Control," 221.

42. Examples include the 2014 paper demonstrating fooling classification neural networks, see Christian Szegedy et al., "Intriguing Properties of Neural Networks," *ArXiv:1312.6199 [Cs]*, 20 December 2013, http://arxiv.org/abs/1312.6199

43. Fryer-Biggs, "Secretive Pentagon Research Program Looks to Replace Human Hackers with AI."

44. Johnson, "Artificial Intelligence in Nuclear Warfare," 201.

45. See Chapter 2: "Offensive Network Operations".

46. Patents suggesting this course of action have already been approved, though their method of implementation is unclear. See for example Hershey, Chapa, and Umberger, Methods and apparatuses for eliminating a missile threat.

47. One such automated method is called *fuzzing*, in which software attempts to automatically find faults in other software by providing it with a variety of unexpected inputs in hopes of creating unexpected, exploitable behaviour. Discussion of AI-enabled fuzzing is well underway, with some claiming it could radically advance vulnerability hunting and exploitation. See for example Derek Manky, "Using Fuzzing to Mine for Zero-Days," ThreatPost, 7 December 2018, https://threatpost.com/using-fuzzing-to-mine-for-zero-days/139683/

48. Wade Shen, "The Information Domain and the Future of Conflict" in *2017 9th International Conference on Cyber Conflict (CyCon)* (Tallinn, Estonia: IEEE, 1 June 2017).

49. Many such examples have been mentioned throughout this work including internal threats such as Edward Snowden's exfiltration of broad NSA data, and external threats such as the Shadow Brokers' theft of NSA TAO network operation capabilities.

50. One example of an unmanned space vehicle is the reusable Boeing X-37, used by the US Air Force for undisclosed mission types.

51. Martti Lehto and Bill Hutchinson, "Mini-Drones Swarms and Their Potential in Conflict Situations" in *2020 International Conference on Cyber Warfare and Security* (Reading, UK, 2020), 328.

52. TRADOC, "The Operational Environment and the Changing Character of Future Warfare," US Army Training and Doctrine Command, 31 May 2017, 11.

53. Chekinov and Bogdanov, "The Nature and Content of a New-Generation War."

54. People's Liberation Army, "顺应军事变革潮流把握改革主动——中国军网-军报记者," 5 January 2016.

55. Kim Hartmann and Christoph Steup, "The Vulnerability of UAVs to Cyber Attacks—An Approach to the Risk Assessment," n.d., 2.

56. Hartmann and Steup, 7.

57. Lehto and Hutchinson, "Mini-Drones Swarms and Their Potential in Conflict Situations"; Bode and Huelss, "The Future of Remote Warfare? Artificial Intelligence, Weapons Systems and Human Control."

58. Sean J. A. Edwards, *Swarming on the Battlefield: Past, Present, and Future* (Santa Monica, CA: Rand, 2000).

59. Brett Davis, "Learning Curve: Iranian Asymmetrical Warfare and Millennium Challenge 2002," Center for International Maritime Security, 14 August 2014.

60. Kania, Elsa, "Swarms at War: Chinese Advances in Swarm Intelligence," Jamestown, 6 July 2017, https://jamestown.org/program/swarms-war-chinese-advances-swarm-intelligence/

61. Sanchez, "Russia Uses Missiles and Cyber Warfare to Fight off 'swarm of Drones' Attacking Military Bases in Syria."

62. Kyle Rempfer, "DARPA Hopes to Swarm Drones out of C-130s in 2019 Test," *Air Force Times*, 19 December 2017, https://www.airforcetimes.com/newsletters/daily-news-roundup/2017/12/18/darpa-hopes-to-swarm-drones-out-of-c-130s-in-2019-test/

63. Lehto and Hutchinson, "Mini-Drones Swarms and Their Potential in Conflict Situations," 330.

64. Buchanan, *The Cybersecurity Dilemma Hacking, Trust and Fear Between Nations.*

65. Rid, *Rise of the Machines: The Lost History of Cybernetics*, 47–49.

CONCLUSIONS

1. In one curious example, Israel chose to counter Hamas network operations through an airstrike against their facilities, subsequently tweeting that "HamasCyberHQ.exe has been removed". See Zak Doffman, "Israel Responds To Cyber Attack With Air Strike On Cyber Attackers In World First," Forbes, 6 May 2019.

2. Elad Popovich, "A Classical Analysis of the 2014 Israel–Hamas Conflict," *CTC Sentinel* 7, no. 11 (November 2014): 20–24.

3. Easton, *The Chinese Invasion Threat*, 132.

4. One notable but indeterminate exception is the alleged efforts to curtail North Korean ballistic missile tests by way of OCOs against the associated infrastructure, see Sanger and Broad, "Trump Inherits a Secret Cyberwar Against North Korean Missiles."

5. See for example Ben Schaefer, "The Cyber Party of God: How Hezbollah Could Transform Cyberterrorism," *Georgetown Security Studies Review* (blog), 11 March 2018, http://georgetownsecuritystudiesreview.org/2018/03/11/the-cyber-party-of-god-how-hezbollah-could-transform-cyberterrorism/

6. Debates are ongoing as to how networked should it be after a much-needed refurbish. For a comprehensive report on the overall attack surface for the arsenal, see Bezya Unal and Patricia Lewis, "Cybersecurity of Nuclear Weapons Systems: Threats, Vulnerabilities, and Consequences," London: Chatham House, January 2018.

BIBLIOGRAPHY

Ablon, Lillian. *Zero Days, Thousands of Nights: The Life and Times of Zero-Day Vulnerabilities and Their Exploits.* Santa Monica: RAND, 2017.

Ackerman, Gwen. "Iranian Hackers Drew Worryingly Close to Israel's Missile Alarm." Bloomberg, 24 February 2019.

Adamsky, Dmitry. "Cross-Domain Coercion: The Current Russian Art of Strategy." Institut Français des Relations Internationales, 2015. http://www.ifri.org/sites/default/files/atoms/files/pp54adamsky.pdf

Addis, Casey L, and Christopher M Blanchard. "Hezbollah: Background and Issues for Congress." Congressional Research Service, 3 January 2011.

Defense Update: "AI Enhancing Land Forces' Weapons and Platforms," 27 December 2020. https://defense-update.com/20201227_fire-weaver-ot.html

Air Combat Command Public Affairs. "ACC Discusses 16th Air Force as New Information Warfare NAF." US Air Force, 18 September 2019. https://www.af.mil/News/Article-Display/Article/1964795/acc-discusses-16th-air-force-as-new-information-warfare-naf/

Air Land Sea Application Center. "TADIL J: Introduction to Tactical Digital Information Link J and Quick Reference Guide," June 2000.

Ajili, Hadi, and Masha Rough. "Iran's Military Strategy." *Survival* 61, no. 6 (November 2019): 139–52.

Akimenko, Valeriy, and Keir Giles. "Russia's Cyber and Information Warfare." *Asia Policy* 27, no. 2 (2020): 67–75. https://doi.org/10.1353/asp. 2020.0014

Alberts, David S. "Agility, Focus, and Convergence: The Future of Command and Control." Office of the Assistant Secretary of Defense for Networks and Information Integration, 2007.

Alberts, David S., John Garstka, and Frederick P. Stein. *Network Centric Warfare: Developing and Leveraging Information Superiority.* CCRP Publication Series. Washington, DC: National Defense University Press, 1999.

BIBLIOGRAPHY

Alperovitch, Dmitri. "Bears in the Midst: Intrusion into the Democratic National Committee," *CrowdStrike* (blog), 15 June 2016. https://www.crowdstrike.com/blog/bears-midst-intrusion-democratic-national-committee/

Anderson, Collin, and Karim Sadjadpour. "Iran's Cyber Threat: Espionage, Sabotage, and Revenge." Carnegie Endowment for International Peace, 2018. https://carnegieendowment.org/files/Iran_Cyber_Final_Full_v2.pdf

Army Headquarters. "US Army Field Manual 3–13—Information Operations," November 2003.

Arquilla, John. "Ethics and Information Warfare." In *Strategic Appraisal: The Changing Role of Information in Warfare*, edited by Zalmay Khalilzad, John P. White, and Andrew Marshall, 379–401. Santa Monica, CA: RAND, 1999.

Arquilla, John, and David Ronfeldt. "Cyberwar Is Coming!" *Comparative Strategy* 12, no. 2 (1993): 141–65.

Ayoub, Kareem, and Kenneth Payne. "Strategy in the Age of Artificial Intelligence." *Journal of Strategic Studies* 39, no. 5–6 (18 September 2016): 793–819.

Barnes, Julian E. "U.S. Cyber Command Bolsters Allied Defenses to Impose Cost on Moscow." *The New York Times*, 7 May 2019, sec. US https://www.nytimes.com/2019/05/07/us/politics/cyber-command-russian-interference.html

Barnes, Julian E., and Thomas Gibbons-Neff. "U.S. Carried Out Cyberattacks on Iran." *The New York Times*, 22 June 2019. https://www.nytimes.com/2019/06/22/us/politics/us-iran-cyber-attacks.html

Barrett, Neil. "Penetration Testing and Social Engineering: Hacking the Weakest Link." *Information Security Technical Report* 8, no. 4 (2003): 56–64.

Bartles, Charles K. "Getting Gerasimov Right." *Military Review* 96, no. 1 (2016): 30–38.

Baylev, Col. S. I., and Col. I. N. Dylevsky. "The Russian Armed Forces in the Information Environment: Principles, Rules, and Confidence-Building Measures," n.d., 6.

Bellingcat. "FSB Team of Chemical Weapon Experts Implicated in Alexey Navalny Novichok Poisoning." Bellingcat, 14 December 2020. https://www.bellingcat.com/news/uk-and-europe/2020/12/14/fsb-team-of-chemical-weapon-experts-implicated-in-alexey-navalny-novichok-poisoning/

Bennett, David. "China's Offshore Active Defense and the People's Liberation Army Navy." *Global Security Studies* 1, no. 1 (2010).

Berman, Ilan. The Iranian Cyber Threat, Revisited. US House of Represen-

BIBLIOGRAPHY

tatives Committee on Homeland Security Subcommittee on Cyber-security, Infrastructure Protection, and Security Technologies (2013).

Bertrand, Natasha. "Iran Is Building a Non-Nuclear Threat Faster than Experts 'Would Have Ever Imagined.'" Business Insider, 27 March 2015.

Bērziņš, Jānis. "Russian New Generation Warfare Is Not Hybrid Warfare." In *The War in Ukraine: Lessons for Europe*, edited by Artis Pabriks and Andis Kudors, 40–51. Riga: The Centre for East European Policy Studies: University of Latvia Press, 2015.

Bing, Chris, and Patrick Howell O'Neill. "Kaspersky's 'Slingshot' Report Burned an ISIS-Focused Intelligence Operation." *CyberScoop* (blog), 20 March 2018. https://www.cyberscoop.com/kaspersky-slingshot-isis-operation-socom-five-eyes/

Bisson, David. "Hacker Admits to Stealing Military Data from U.S. Department of Defense." Tripwire, 16 June 2017. https://www.tripwire.com/state-of-security/latest-security-news/hacker-admits-to-stealing-military-data-from-u-s-department-of-defense/

Blank, Stephen. "Cyber War and Information War à La Russe." In *Understanding Cyber Conflict: 14 Analogies*, 81–98. Washington D.C.: Georgetown University Press, 2017. https://carnegieendowment.org/2017/10/16/cyber-war-and-information-war-la-russe-pub-73399

Bode, Ingvild, and Hendrik Huelss. "The Future of Remote Warfare? Artificial Intelligence, Weapons Systems and Human Control." In *Remote Warfare: Interdisciplinary Perspectives*, 218–33. Bristol, England: E-International Relations Publishing, 2021. https://www.e-ir.info/publication/remote-warfare-interdisciplinary-perspectives/

Bolia, Robert S. "Overreliance on Technology in Warfare: The Yom Kippur War as a Case Study." DTIC Document, 2004.

Borland, John. "Analyzing the Internet Collapse." MIT Technology Review. Accessed 27 January 2018. https://www.technologyreview.com/s/409491/analyzing-the-internet-collapse/

Boyd, John. "The Essence of Winning and Losing." Edited by Chet Richards and Chuck Spinney. Pogo Archives. 8 June 1995. http://pogoarchives.org/m/dni/john_boyd_compendium/essence_of_winning_losing.pdf

Bradbury, Danny. "FSB Hackers Drop Files Online." *Naked Security* (blog), 23 July 2019. https://nakedsecurity.sophos.com/2019/07/23/fsb-hackers-drop-files-online/

Brady, James S. "Remarks by the President to the White House Press Corps." The White House, 20 August 2012. https://www.whitehouse.gov/the-press-office/2012/08/20/remarks-president-white-house-press-corps

Bronk, Christopher, and Eneken Tikk-Ringas. "The Cyber Attack on Saudi Aramco." *Survival* 55, no. 2 (May 2013): 81–96.

BIBLIOGRAPHY

Brooks, Teresa Nicole. "Survey of Automated Vulnerability Detection and Exploit Generation Techniques in Cyber Reasoning Systems." *ArXiv:1702. 06162 [Cs]*, 20 February 2017. http://arxiv.org/abs/1702.06162

Bruinessen, Martin van. "The Kurds between Iran and Iraq." *MERIP Middle East Report*, no. 141 (July 1986): 14–27. https://doi.org/10.2307/301 1925

Brundage, Miles, Shahar Avin, Jack Clark, Helen Toner, Peter Eckersley, Ben Garfinkel, Allan Dafoe, et al. "The Malicious Use of Artificial Intelligence: Forecasting, Prevention, and Mitigation." *ArXiv:1802.07228*, February 2018.

Buchanan, Ben. *The Cybersecurity Dilemma Hacking, Trust and Fear Between Nations*. Oxford: Oxford University Press, 2017.

Burbank, Jack L., Philip F. Chimento, Brian K. Haberman, and William T. Kasch. "Key Challenges of Military Tactical Networking and the Elusive Promise of MANET Technology." *IEEE Communications Magazine* 44, no. 11 (2006): 39–42.

Burton, Robert W., Frank L. Cloutier, Clarence S. Summers, Elliott R. Brown, and John A. Zingg. *The Strategy of Electronic Warfare*. US Air Force Academy, 1979.

Campbell, Murray, A Joseph Hoane Jr, and Feng-hsiung Hsu. "Deep Blue." *Artificial Intelligence* 134, no. 1–2 (2002): 57–83.

Carter, Ash. "A Lasting Defeat: The Campaign to Destroy ISIS." Cambridge, Mass: The Belfer Center, October 2017.

Caton, Jeffrey, L. "Army Support of Military Cyberspace Operations." Strategic Studies Institute, January 2015.

Cebrowski, Arthur K., and John J. Garstka. "Network-Centric Warfare: Its Origin and Future." In *US Naval Institute Proceedings* 124 (1998):28–35.

Cenciotti, David. "The Amazing Growler Ball 2020 Video Teases An EA-18G Cyber Attack Capability That Is Yet To Come." *The Aviationist* (blog), 2 September 2020. https://theaviationist.com/2020/09/02/the-amazing-growler-ball-2020-video-teases-an-ea-18g-cyber-attack-ca-pability-that-is-yet-to-come/

CERT.org. "CPU Hardware Vulnerable to Side-Channel Attacks." Carnegie Mellon University CERT, 3 January 2018. https://www.kb.cert.org/vuls/id/584653

Chairman of the Joint Chiefs of Staff. "Memorandum of Policy No. 30: Command and Control Warfare," 8 March 1993.

———. "National Military Strategy for Cyberspace Operations," December 2006.

Charette, Robert N. "This Car Runs on Code." IEEE Spectrum: Technology, Engineering, and Science News, February 1, 2009. https://spectrum.ieee.org/transportation/systems/this-car-runs-on-code.

BIBLIOGRAPHY

Chekinov, Sergey G., and Sergey A. Bogdanov. "The Nature and Content of a New-Generation War." *Military Thought* 4 (2013): 12–23.

Asia Maritime Transparency Initiative. "China Tracker." Accessed 26 December 2020. https://amti.csis.org/island-tracker/china/

Chirgwin, Richard. "IT 'heroes' Saved Maersk from NotPetya with Ten-Day Reinstallation Bliz." *The Register*, 25 January 2018. https://www.theregister.co.uk/2018/01/25/after_notpetya_maersk_replaced_everything/

Chotikul, Diane. "The Soviet Theory of Reflexive Control in Historical and Psychocultural Perspective: A Preliminary Study." Fort Belvoir, VA: Defense Technical Information Center, 1 July 1986.

Cimpanu, Catalin. "CISA Updates SolarWinds Guidance, Tells US Govt Agencies to Update Right Away." ZDNet, 30 December 2020. https://www.zdnet.com/article/cisa-updates-solarwinds-guidance-tells-us-govt-agencies-to-update-right-away/

——. "Source Code of Iranian Cyber-Espionage Tools Leaked on Telegram." ZDNet, 19 April 2019. https://www.zdnet.com/article/source-code-of-iranian-cyber-espionage-tools-leaked-on-telegram/

——. "Windows 10, IOS 15, Ubuntu, Chrome Fall at China's Tianfu Hacking Contest." The Record by Recorded Future, 17 October 2021. https://therecord.media/windows-10-ios-15-ubuntu-chrome-fall-at-chinas-tianfu-hacking-contest/

Clapper Jr, James R. "Challenging Joint Military Intelligence." DTIC Document, 1994.

Clapper Jr, James R., Michael S. Rogers, and Marcel Lettre. Statement on Foreign Cyber Threats to the United States, US Senate Armed Services Committee (2017).

Clarke, Richard A. "War From Cyberspace." *The National Interest* 104 (2009): 31–36.

Clarke, Richard A., and Robert K. Knake. *Cyber War: The Next Threat to National Security and What to Do About It*. 1st ed. New York: Ecco, 2010.

Clausewitz, Carl Von. *On War*. 3rd ed. Vol. 1. London: N. Trubner & Co, 1873.

Clayton, Mark. "Ukraine Election Narrowly Avoided 'wanton Destruction' from Hackers." *Christian Science Monitor*, 17 June 2014. https://www.csmonitor.com/World/Passcode/2014/0617/Ukraine-election-narrowly-avoided-wanton-destruction-from-hackers

Cliff, Roger. "Anti-Access Measures in Chinese Defense Strategy." *Testimony before the US-China Economic and Security Review Commission*, 2011.

Cockburn, Robert. "The Radio War." *IEE Proceedings A-Physical Science, Measurement and Instrumentation, Management and Education-Reviews* 132, no. 6 (1985): 423–34.

BIBLIOGRAPHY

Cohen, Eliot A. "A Revolution in Warfare." *Foreign Affairs* 75, no. 2 (1996): 37–54.

Coile, Gregory. "WIN-T SATCOM Overview Briefing." Program Executive Office Command Control Communications-Tactical, 2009.

Coker, Margaret, and Paul Sonne. "Ukraine: Cyberwar's Hottest Front." *Wall Street Journal*, 10 November 2015, sec. World. http://www.wsj.com/articles/ukraine-cyberwars-hottest-front-1447121671

Comae. "The Shadow Brokers: Cyber Fear Game-Changers." Comae Technologies, July 2017.

Comey, James B. "Addressing the Cyber Security Threat." Speech. Federal Bureau of Investigation, 7 January 2015. https://www.fbi.gov/news/speeches/addressing-the-cyber-security-threat

Tasnim News Agency. "Commander Reiterates Iran's Preparedness to Confront Enemies in Cyber Warfare—Defense News," 18 February 2014. https://www.tasnimnews.com/en/news/2014/02/18/287797/commander-reiterates-iran-s-preparedness-to-confront-enemies-in-cyber-warfare

Connell, Michael. "Deterring Iran's Use of Offensive Cyber: A Case Study:" Fort Belvoir, VA: Defense Technical Information Center, 1 October 2014.

———. "Iran's Military Doctrine." US Institute of Peace, 11 October 2010. https://iranprimer.usip.org/resource/irans-military-doctrine

Connell, Michael, and Sarah Vogler. "Russia's Approach to Cyber Warfare." Occasional Paper Series. Arlington, VA: CNA, 24 March 2017.

Conti, Gregory, and David Raymond. *On Cyber: Towards an Operational Art for Cyber Conflict*. New York: Kopidion Press, 2017.

Cordesman, Anthony H. "An Open Strategic Challenge to the United States, But One Which Does Not Have to Lead to Conflict." Center for Strategic and International Studies 24 July 2019.

———. "Chinese Strategy and Military Power in 2014." Washington D. C.: Center for Strategic and International Studies, November 2014.

Cordesman, Anthony H, Ashley Hess, and Nicholas S Yarosh. *Chinese Military Modernization and Force Development: A Western Perspective*. Washington D.C.: Center for Strategic and International Studies, 2013.

Costello, John. "The Strategic Support Force: China's Information Warfare Service." *China Brief* (blog), 8 February 2016. https://jamestown.org/program/the-strategic-support-force-chinas-information-warfare-service/

CPNI. "Critical National Infrastructure." UK Centre for the Protection of National Infrastructure, 2020. https://www.cpni.gov.uk/critical-national-infrastructure-0

Crail, Peter. "IAEA Sends Syria Nuclear Case to UN." Arms Control

Association, 7 July 2011. https://www.armscontrol.org/act/2011_%20
07–08/%20IAEA_Sends_Syria_Nuclear_Case_to_UN.

Crawford, Bruce T., James J. Mingus, and Gary P. Martin. The United
States Army Network Modernization Strategy, Committee on the Armed
Services (2017). http://docs.house.gov/meetings/AS/AS25/20170
927/106451/HHRG-115-AS25-Wstate-CrawfordB-20170927.pdf

Crow, Jason. "The Situation Is Developing, but the More I Learn This
Could Be Our Modern Day, Cyber Equivalent of Pearl Harbor." Tweet.
@RepJasonCrow (blog), 18 December 2020. https://twitter.com/
RepJasonCrow/status/1339950863801061377

CrowdStrike. "Danger Close: Fancy Bear Tracking of Ukrainian Field
Artillery Units." *CrowdStrike Blog* (blog), 22 December 2016. https://
www.crowdstrike.com/blog/danger-close-fancy-bear-tracking-
ukrainian-field-artillery-units/

Crowdy, Terry. *The Enemy Within: A History of Spies, Spymasters and Espionage.*
London: Bloomsbury Publishing, 2011.

Cummings, M. L. "Artificial Intelligence and the Future of Warfare."
London: Chatham House, January 2017.

Cybereason. "Paying the Price of Destructive Cyber Attacks," 2017.

Davis, Brett. "Learning Curve: Iranian Asymmetrical Warfare and
Millennium Challenge 2002." Center for International Maritime Security,
14 August 2014. http://cimsec.org/learning-curve-iranian-asymmetrical-
warfare-millennium-challenge-2002–2/11640

Day, Brittany. "Linux: An OS Capable of Effectively Meeting the US
Government's." Linux Security. Accessed 29 December 2020. https://
linuxsecurity.com/features/features/linux-an-os-capable-of-effectively-
meeting-the-us-government-s-security-needs-heading-into-2020

Deeks, Ashley. "The Geography of Cyber Conflict: Through a Glass
Darkly," 2013.

Defense Science Board. "Task Force on Cyber Deterrence." Department of
Defense, February 2017.

Deibert, R. J., R. Rohozinski, and M. Crete-Nishihata. "Cyclones in
Cyberspace: Information Shaping and Denial in the 2008 Russia-Georgia
War." *Security Dialogue* 43, no. 1 (1 February 2012): 3–24.

Dekker, Anthony H. "Measuring the Agility of Networked Military Forces."
Journal of Battlefield Technology 9, no. 1 (2006): 19.

Denning, Dorothy E. *Information Warfare and Security.* 4th ed. Reading:
Addison-Wesley, 1999.

DHS Press Office. "Joint Statement from the Department Of Homeland
Security and Office of the Director of National Intelligence on Election
Security." Department of Homeland Security, 7 October 2016. https://
www.dhs.gov/node/23199

BIBLIOGRAPHY

Doffman, Zak. "Israel Responds To Cyber Attack With Air Strike On Cyber Attackers In World First." Forbes, 6 May 2019. https://www.forbes.com/sites/zakdoffman/2019/05/06/israeli-military-strikes-and-destroys-hamas-cyber-hq-in-world-first/

Dragos. "CRASHOVERRIDE: Threat to the Electric Grid Operations." 12 June 2017.

Drew, Christopher. "Stolen Data Is Tracked to Hacking at Lockheed." *The New York Times*, 3 June 2011, sec. Technology. https://www.nytimes.com/2011/06/04/technology/04security.html

Drogin, Bob. "Russians Seem To Be Hacking Into Pentagon / Sensitive Information Taken—But Nothing Top Secret." *Los Angeles Times*, 7 October 1999. https://www.sfgate.com/news/article/Russians-Seem-To-Be-Hacking-Into-Pentagon-2903309.php

Dyer, Geoff. "US Launches Online Assault against Isis." *Financial Times*, 6 April 2016. https://www.ft.com/content/4d98edd0-fba5-11e5-b3f6-11d5706b613b.

Easton, Ian. "Able Archers: Taiwan Defense Strategy in an Age of Precision Strike." Project 2049 Institute, n.d.

———. *The Chinese Invasion Threat: Taiwan's Defense and American Strategy in Asia*. 1st ed. Arlington, Virgina: Project 2049 Institute, 2017.

Edwards, Sean J. A. *Swarming on the Battlefield: Past, Present, and Future*. Santa Monica, CA: Rand, 2000.

El Husseini, Rola. "Hezbollah and the Axis of Refusal: Hamas, Iran and Syria." *Third World Quarterly* 31, no. 5 (July 2010): 803–15. https://doi.org/10.1080/01436597.2010.502695

Elgin, Benjamin, and Michael Riley. "Now at the Sands Casino: An Iranian Hacker in Every Server." Bloomberg, 12 December 2014. https://www.bloomberg.com/news/articles/2014-12-11/iranian-hackers-hit-sheldon-adelsons-sands-casino-in-las-vegas

Elkind, Peter. "Sony Pictures: Inside the Hack of the Century." *Fortune* (blog), 1 July 2015. http://fortune.com/sony-hack-part-1/

Epperson, Lynn. "Satellite Communications Within the Army's WIN-T Architecture." Program Executive Office Command Control Communications-Tactical, 6 February 2014.

Estonian Foreign Intelligence Service. "International Security and Estonia," 2018.

Falcone, Robert. "Shamoon 3 Targets Oil and Gas Organization." *Palo Alto Networks Unit 42* (blog), 13 December 2018. https://unit42.paloaltonetworks.com/shamoon-3-targets-oil-gas-organization/

Falcone, Robert, and Bryan Lee. "Shamoon 2: Delivering Disttrack." *Unit42* (blog), 26 March 2017. https://unit42.paloaltonetworks.com/unit42-shamoon-2-delivering-disttrack/

BIBLIOGRAPHY

Falliere, Nicolas, Liam O Murchu, and Eric Chien. "W32.Stuxnet Dossier." Symantec, February 2011.

Farwell, James P., and Rafal Rohozinski. "Stuxnet and the Future of Cyber War." *Survival* 53, no. 1 (February 2011): 23–40.

Felt, Adrienne Porter, Richard Barnes, April King, Chris Palmer, Chris Bentzel, and Parisa Tabriz. "Measuring HTTPS Adoption on the Web." In *26th USENIX Security Symposium*, 1323–38, 2017.

Ferguson, Rik. "TV5 Monde, Russia and the CyberCaliphate." *Trend Micro* (blog), 10 June 2015. http://blog.trendmicro.co.uk/tv5-monde-russia-and-the-cybercaliphate/

FireEye. "FireEye Responds to Wave of Destructive Cyber Attacks in Gulf Region." FireEye, 1 December 2016.

———. "Highly Evasive Attacker Leverages SolarWinds Supply Chain to Compromise Multiple Global Victims With SUNBURST Backdoor." FireEye, 13 December 2020. https://www.fireeye.com/blog/threat-research/2020/12/evasive-attacker-leverages-solarwinds-supply-chain-compromises-with-sunburst-backdoor.html

———. "Unauthorized Access of FireEye Red Team Tools." FireEye, 8 December 2020. https://www.fireeye.com/blog/threat-research/2020/12/unauthorized-access-of-fireeye-red-team-tools.html

Fleming, Jeremy. "GCHQ Director's Speech at CYBERUK 2018." CYBERUK, 12 April 2018.

Forcese, Craig. "Spies Without Borders: International Law and Intelligence Collection." *Journal of National Security Law and Policy* 5 (2011): 179–210.

Franceschi-Bicchierai, Lorenzo. "We Spoke to DNC Hacker 'Guccifer 2.0.'" *Motherboard*, 21 June 2016. https://motherboard.vice.com/en_us/article/aek7ea/dnc-hacker-guccifer-20-interview

Franke, Ulrik. "War by Non-Military Means." Stockholm, Sweden: FOI, 2015.

Fraser, Nalani, Fred Plan, Jacqueline O'Leary, Vincent Cannon, Raymond Leong, Dan Perez, and Chi-en Shen. "Double Dragon: APT41, a Dual Espionage and Cyber Crime Operation." FireEye, 7 August 2019.

Fravel, M. Taylor. *Active Defense: China's Military Strategy since 1949*. Princeton Studies in International History and Politics. Princeton, New Jersey: Princeton University Press, 2019.

Freedberg Jr., Sydney J. "ALIS Glitch Grounds Marine F-35Bs." *Breaking Defense* (blog), 22 June 2017. http://breakingdefense.com/2017/06/breaking-alis-glitch-grounds-marine-f-35bs/

Freedberg Jr., Sydney J. "Wireless Hacking In Flight: Air Force Demos Cyber EC-130." *Breaking Defense* (blog), 15 September 2015. https://

breakingdefense.com/2015/09/wireless-hacking-in-flight-air-force-demos-cyber-ec-130/

Freedman, Lawrence. *Strategy: A History.* Oxford: Oxford University Press, 2015.

Frenkel, Sheera. "Experts Say Russians May Have Posed As ISIS To Hack French TV Channel." BuzzFeed, 10 June 2015. https://www.buzzfeed.com/sheerafrenkel/experts-say-russians-may-have-posed-as-isis-to-hack-french-t

Fryer-Biggs, Zachary. "Secretive Pentagon Research Program Looks to Replace Human Hackers with AI." Yahoo! News, 13 September 2020. https://news.yahoo.com/secretive-pentagon-research-program-looks-to-replace-human-hackers-with-ai-090032920.html

Fuller, J.F.C. *The Foundations of the Science of War.* London: Hutchinson & Company, 1926.

Futter. "The Dangers of Using Cyberattacks to Counter Nuclear Threats." *Arms Control Today* 46, no. 6 (2016): 8–14.

Galeotti, Mark. "The 'Gerasimov Doctrine' and Russian Non-Linear War." *In Moscow's Shadows* (blog), 6 July 2014. https://inmoscowsshadows.wordpress.com/2014/07/06/the-gerasimov-doctrine-and-russian-non-linear-war/

————. "The Mythical 'Gerasimov Doctrine' and the Language of Threat." *Critical Studies on Security* 7, no. 2 (4 May 2019): 157–61. https://doi.org/10.1080/21624887.2018.1441623

Gallagher, Ryan. "The Inside Story of How British Spies Hacked Belgium's Largest Telco." *The Intercept* (blog), 13 December 2014. https://theintercept.com/2014/12/13/belgacom-hack-gchq-inside-story/

Gallagher, Sean. "F-35 Radar System Has Bug That Requires Hard Reboot in Flight." *Ars Technica*, 10 March 2016. https://arstechnica.com/information-technology/2016/03/f-35-radar-system-has-bug-that-requires-hard-reboot-in-flight/

Government Accountability Office. "DOD Needs a Strategy for Re-Designing the F-35's Central Logistics System." Government Accountability Office, March 2020. https://www.gao.gov/assets/710/705154.pdf

Garamone, Jim. "Esper Describes DOD's Increased Cyber Offensive Strategy." US Department of Defense, 20 September 2019. https://www.defense.gov/Explore/News/Article/Article/1966758/esper-describes-dods-increased-cyber-offensive-strategy/

GCHQ. "Full-Spectrum Cyber Effects." GCHQ, 2012.

Geller, Eric. "Trump Scraps Obama Rules on Cyberattacks, Giving Military Freer Hand." *POLITICO*, 16 August 2018. https://politi.co/2MSWCnS

Gerasimov, Valery. "The Value of Science Is in the Foresight." *Military Review* 96, no. 1 (2016): 23.

Gerges, Fawaz A. "The Obama Approach to the Middle East: The End of America's Moment?" *International Affairs* 89, no. 2 (2013): 299–323.

Gettle, Mitch. "Air Force Releases New Mission Statement." United States Air Force, 8 December 2005. http://www.af.mil/News/ArticleDisplay/tabid/223/Article/132526/air-force-releases-new-mission-statement.aspx

Giannangeli, Marco. "Russians forcing RAF to abort missions in Syria by 'hacking into' their systems." *The Daily Express*, 15 January 2017. http://www.express.co.uk/news/world/754236/russia-raf-bombers-syria-hacking-missions-military-army

Giles, Keir. "'Information Troops'-A Russian Cyber Command?" In *2011 3rd International Conference on Cyber Conflict*, 45–60. Tallinn, Estonia, 2011.

———. "Russia's 'New' Tools for Confronting the West." London: Chatham House, March 2016.

Global Security. "AEGIS Combat System." Global Security. Accessed 2 October 2015. http://www.globalsecurity.org/military/systems/ship/systems/aegis.htm

Goertzel, Ben, and Cassio Pennachin, eds. *Artificial General Intelligence (Cognitive Technologies)*. Berlin; New York: Springer, 2007.

Goldman, Adam. "New Charges in Huge C.I.A. Breach Known as Vault 7." *The New York Times*, 19 June 2018, sec. US https://www.nytimes.com/2018/06/18/us/politics/charges-cia-breach-vault-7.html

Gonsalves, Antone. "Expert Fingers DDoS Toolkit Used in Bank Cyberattacks." CSO Online, 1 October 2012. https://www.csoonline.com/article/2138382/expert-fingers-ddos-toolkit-used-in-bank-cyber-attacks.html

Goodin, Dan. "Citing BGP Hijacks and Hack Attacks, Feds Want China Telecom out of the US." *Ars Technica*, 10 April 10, 2020. https://arstechnica.com/tech-policy/2020/04/citing-bgp-hijacks-and-hack-attacks-feds-want-china-telecom-out-of-the-us/

———. "Group Claims to Hack NSA-Tied Hackers, Posts Exploits as Proof." *Ars Technica*, 16 August 2016. https://arstechnica.com/information-technology/2016/08/group-claims-to-hack-nsa-tied-hackers-posts-exploits-as-proof/

Gordon, Michael R. "Despite Cold War's End, Russia Keeps Building a Secret Complex." *The New York Times*, 16 April 1996, sec. World. https://www.nytimes.com/1996/04/16/world/despite-cold-war-s-end-russia-keeps-building-a-secret-complex.html

Gorman, Siobahn, and Julian E. Barnes. "Iranian Hacking to Test NSA

Nominee Michael Rogers." *The Wall Street Journal*, 18 February 2014. https://web.archive.org/web/20200303081105/https://www. wsj.com/articles/iranian-hacking-to-test-nsa-nominee-michael-rogers-1392694544?tesla=y

Gorman, Siobahn, August Cole, and Yochi Dreazen. "Computer Spies Breach Fighter-Jet Project." *The Wall Street Journal*, 21 April 2009. http://www.wsj.com/articles/SB124027491029837401

Government of Georgia. "Russian Cyberwar on Georgia," November 10, 2008.

Graham, Bradley. "Military Grappling With Rules For Cyber Warfare." *Washington Post*, 8 November 1999.

Grau, Dr Lester W, and Charles K Bartles. *The Russian Way of War: Force Structure, Tactics, and Modernization of the Russian Ground Forces*. Fort Leavenworth: Foreign Military Studies Office, 2016.

GReAT. "Equation: The Death Star of Malware Galaxy." *Securelist* (blog), 16 February 2015. https://securelist.com/equation-the-death-star-of-malware-galaxy/68750/

―――. "Shamoon the Wiper—Copycats at Work." *Securelist* (blog), 16 August 2012. https://securelist.com/shamoon-the-wiper-copycats-at-work/57854/

―――. "Shamoon The Wiper: Further Details (Part II)." *Securelist* (blog), 11 September 2012. https://securelist.com/shamoon-the-wiper-further-details-part-ii/57784/

―――. "The Devil's in the Rich Header." *Securelist* (blog), 8 March 2018. https://securelist.com/the-devils-in-the-rich-header/84348/

―――. "The Duqu 2.0: Technical Details." Kaspersky Lab, 11 June 2015. https://securelist.com/the-mystery-of-duqu-2-0-a-sophisticated-cyberespionage-actor-returns/70504/

―――. "The ProjectSauron APT." Kaspersky Lab, 9 August 2016. https://securelist.com/faq-the-projectsauron-apt/75533/

Green, Michael, Ernest Bower, and Center for Strategic and International Studies. *Asia-Pacific Rebalance 2025: Capabilities, Presence, and Partnerships: An Independent Review of U.S. Defense Strategy in the Asia-Pacific*. Washington D.C.: Center for Strategic and International Studies, 2016.

Greenberg, Andy. *Sandworm: A New Era of Cyberwar and the Hunt for the Kremlin's Most Dangerous Hackers*. 1st ed. New York: Doubleday, 2019.

―――. "Ukrainians Say Petya Ransomware Hides State-Sponsored Attacks." *Wired*, 28 June 2017. https://www.wired.com/story/petya-ransomware-ukraine/

Hakala, Janne, and Jazlyn Melnychuk. "Russia's Strategy in Cyberspace." Riga, Latvia: NATO Strategic Communications Centre of Excellence, June 2021.

BIBLIOGRAPHY

Hamill, Jasper. "Bank-Busting Jihadi Botnet Comes Back To Life. But Who Is Controlling It This Time?" Forbes, 30 June 2014. https://www.forbes.com/sites/jasperhamill/2014/06/30/bank-busting-jihadi-botnet-comes-back-to-life-but-who-is-controlling-it-this-time/#3df4bb0f6f07

Hammond, Phillip. "Chancellor Speech: Launching the National Cyber Security Strategy." GOV.UK, 1 November 2016. https://www.gov.uk/government/speeches/chancellor-speech-launching-the-national-cyber-security-strategy

Harel, Amos, and Aluf Benn. "No Longer a Secret: How Israel Destroyed Syria's Nuclear Reactor." Haaretz, 23 March 2018. https://web.archive.org/web/20201223133510/https://www.haaretz.com/middle-east-news/syria/MAGAZINE-no-longer-a-secret-how-israel-destroyed-syria-s-nuclear-reactor-1.5914407

Harris, Eimi. "NATO Adds Cyber to Operational Domain." NATO Association of Canada, 4 July 2016. http://natoassociation.ca/nato-adds-cyber-to-operational-domain/

Harris, Shane. "Forget China: Iran's Hackers Are America's Newest Cyber Threat." Foreign Policy (blog), 18 February 2014. https://foreignpolicy.com/2014/02/18/forget-china-irans-hackers-are-americas-newest-cyber-threat/

Hartmann, Kim, and Christoph Steup. "The Vulnerability of UAVs to Cyber Attacks—An Approach to the Risk Assessment," In 5th International Conference on Cyber Conflict (CYCON 2013), 2013, 23.

Hastert, Paul. "Al Qaeda and Iran: Friends or Foes, or Somewhere in Between?" Studies in Conflict & Terrorism 30, no. 4 (1 April 2007): 327–36. https://doi.org/10.1080/10576100701200132

Hauer, Neil. "Russia's Mercenary Debacle in Syria." Foreign Affairs, 26 February 2018.

Heickero, Roland. "Emerging Cyber Threats and Russian Views on Information Warfare and Information Operations." Swedish Defence Research Agency, March 2010.

Hershey, Paul C., Joseph O. Chapa, and Elizabeth Umberger. Methods and apparatuses for eliminating a missile threat. United States US201600 70674A1, filed 9 September 2014, and issued 10 March 2016.

Hershey, Paul C., Robert E. Dehnert JR, and John J. Williams. Digital weapons factory and digital operations center for producing, deploying, assessing, and managing digital defects. United States US9544326B2, filed 20 January 2015, and issued 10 January 2017.

Hershey, Paul Christian, Marilyn Winklareth Zett, Angelo Cianciosi II Michael, Brianne Rene-Martinek Hoppes, Roland Dige Chang, Andrew Arnold, and John Zolper JR. System and method for integrated and synchronized planning and response to defeat disparate threats over the threat kill chain with combined cyber, electronic warfare and kinetic effects. United States US20180038669A1, filed 28 February 2017, and issued 8 February 2018.

Hoffman, Frank G. "Hybrid Warfare & Challenges." *Joint Forces Quarterly*, no. 52 (2009): 34–47.

Horowitz, Michael C. "Artificial Intelligence, International Competition, and the Balance of Power." *Texas National Security Review* 1, no. 3 (May 2018): 37–57.

HP Security Research. "Profiling an Enigma: The Mystery of North Korea's Cyber Threat Landscape." Hewlett Packard, 16 August 2014.

Hsu, Jeremy. "The Strava Heat Map Shows Even Militaries Can't Keep Secrets from Social Data." *Wired*, 30 January 2018. https://www.wired.com/story/strava-heat-map-military-bases-fitness-trackers-privacy/

Google. "HTTPS Encryption on the Web—Google Transparency Report," 2020. https://transparencyreport.google.com/https/overview?hl=en

Hunnicutt, Trevor. "Biden Says United States Would Come to Taiwan's Defense." *Reuters*, 22 October 2021, sec. Asia Pacific.

Hura, Myron, Gary McLeod, James Schneider, Daniel Gonzales, Daniel M. Norton, Jody Jacobs, Kevin M. O'Connel, William Little, Richard Mesic, and Lewis Jamison. "Chapter 9—Tactical Data Links." In *Interoperability: A Continuing Challenge*, 107–21. Santa Monica, CA: RAND, 2000.

Hutcherson, Norman B. "Command & Control Warfare: Putting Another Tool in the War-Fighter's Data Base." Alabama, United States: Air University Press, 1994.

Hutchins, Eric M., Michael J. Cloppert, and Rohan M. Amin. "Intelligence-Driven Computer Network Defense Informed by Analysis of Adversary Campaigns and Intrusion Kill Chains." *Leading Issues in Information Warfare & Security Research* 1 (2011): 80.

Hwang, Ji-Jen. "China's Military Reform: The Strategic Support Force, Non-Traditional Warfare, and the Impact on Cross-Strait Security." *Issues & Studies* 53, no. 3 (September 2017): 1–25.

IISS. "Chapter One: Tehran's Strategic Intent." In *Iran's Networks of Influence in the Middle East*, 11–38. IISS, 2019. https://www.iiss.org/publications/strategic-dossiers/iran-dossier/iran-19-03-ch-1-tehrans-strategic-intent

Inkster, Nigel. "Conflict Foretold: America and China." *Survival* 55, no. 5 (October 2013): 7–28.

Insikt Group. "The History of Ashiyane: Iran's First Security Forum." *Recorded Future* (blog), 16 January 2019. https://www.recordedfuture.com/ashiyane-forum-history/.

BIBLIOGRAPHY

Israeli Navy. "אתר חיל הים". 1973—מלחמת יום הכיפורים. Accessed 23 January 2017. http://www.navy.idf.il/1274-he/Navy.aspx

Jahne, Seth L., Blake Jeffrey Harnden, Eric R. Van Alst, James M. Chan, James M. Kalasky, and Andrew Paul Riha. Techniques Deployment System. United States US20150369569A1, filed 24 June 2014, and issued 24 December 2015.

Jane's. "Exploiting The Network For Smarter Weapon Effects." Jane's International Defence Review, August 2015.

Johnson, James. "Artificial Intelligence in Nuclear Warfare: A Perfect Storm of Instability?" The Washington Quarterly 43, no. 2 (2 April 2020): 197–211. https://doi.org/10.1080/0163660X.2020.1770968

Jones, Reginald V. "Scientific Intelligence." Journal of the Royal United Service Institution 92 (1956): 352–69.

Jones, Sam. "Ministry of Defence Fends Off 'Thousands' of Daily Cyber Attacks." Financial Times, 25 June 2015.

Jun, Jenny, Scott LaFoy, and Ethan Sohn. "North Korea's Cyber Operations." Center for Strategic and International Studies, December 2015.

Kallio, Jyrki, and Julie Chen. "Taiwan's 2020 Election and Its Implications: Dark Clouds Looming for Already Strained Cross-Strait Relations." FIIA Comment. Finnish Institute of International Affairs, January 2020.

Kania, Elsa. "Battlefield Singularity: Artificial Intelligence, Military Revolution, and China's Future Military Power." Center for a New American Security, November 2017.

———. "Swarms at War: Chinese Advances in Swarm Intelligence." Jamestown, 6 July 2017. https://jamestown.org/program/swarms-war-chinese-advances-swarm-intelligence/

Kania, Elsa B., and John Costello. "Seizing the Commanding Heights: The PLA Strategic Support Force in Chinese Military Power." Journal of Strategic Studies, 12 May 2020, 1–47. https://doi.org/10.1080/01402390.2020.1747444

Kania, Elsa, and Costello, John. "The Strategic Support Force and the Future of Chinese Information Operations." Cyber Defense Review 3, no. 1 (Spring 2018): 105–21.

Kaspersky Lab. "Taste of Topinambour: Turla Hacking Group Hides Malware in Anti-Internet Censorship Software." 15 July 2019. https://www.kaspersky.com/about/press-releases/2019_taste-of-topinambour

———. "The Regin Platform: Nation State Ownage of GSM Networks," 24 November 2014.

Katz, Brian. "Axis Rising: Iran's Evolving Regional Strategy and Non-State Partnerships in the Middle East." CSIS Briefs. Center for Strategic and International Studies, 11 October 2018. https://www.csis.org/analysis/

axis-rising-irans-evolving-regional-strategy-and-non-state-partnerships-middle-east

Katzman, Kenneth. "Iran's Foreign and Defense Policies." Congressional Research Service, 11 January 2021.

Katzman, Kenneth, and Paul K. Kerr. "Iran Nuclear Agreement." Congressional Research Service, 2015.

Kaushal, Sidharth. "US Weapons Sales to Taiwan: Upholding the Porcupine Strategy." RUSI, 8 December 2020. https://rusi.org/commentary/us-weapons-sales-taiwan-upholding-porcupine-strategy

Kaye, Dalia Dassa, Alireza Nader, and Parisa Roshan. "A Brief History of Israeli–Iranian Cooperation and Confrontation." In *Israel and Iran: A Dangerous Rivalry*, 9–18. Santa Monica, CA: RAND, 2011.

Keegan, Matthew, and Stephen Leonard Engelson Wyatt. Method and system for a small unmanned aerial system for delivering electronic warfare and cyber effects. United States US20180009525A1, filed 15 March 2016, and issued 11 January 2018.

Keller, John. "Navy and Air Force Choose DRFM Jammers from Mercury Systems to Help Spoof Enemy Radar." Military & Aerospace Electronics, 18 June 2014. https://www.militaryaerospace.com/articles/2014/06/mercury-drfm-jammer.html

Kemp, Herbert C. "Left of Launch: Countering Theater Ballistic Missiles." Issue Brief. Atlantic Council, July 2017.

Kimmons, Sean. "Cyber Teams Throw Virtual Effects, Defend Networks against ISIS." US Army, 15 February 2017. http://www.army.mil/article/182400/cyber_teams_throw_virtual_effects_defend_networks_against_isil

King, David R., and Joseph D. Massey. "History of the F-15 Program: A Silver Anniversary First Flight Remembrance." 1997, *Air Force Journal of Logistics* (Winter 1997) pp.10–16.

Kirk, Alex. "Itsoknoproblembro, the VRT Has You Covered," 16 October 2012. http://blog.talosintelligence.com/2012/10/itsoknoproblembro-vrt-has-you-covered.html

Kopan, Tal. "DNC Hack: What You Need to Know." *CNN*, 21 June 2016. http://www.cnn.com/2016/06/21/politics/dnc-hack-russians-guccifer-claims/index.html

Kopp, Carlo. "JTIDS/MIDS—Network Centric Warfare Fundamentals." *DefenceTODAY*, n.d.

Korns, Stephen W., and Joshua E. Kastenberg. "Georgia's Cyber Left Hook." *Parameters* 38, no. 4 (2008): 60.

Kozy, Adam. "Two Birds, One STONE PANDA." *CrowdStrike Blog* (blog), 30 August 2018. https://www.crowdstrike.com/blog/two-birds-one-stone-panda/

BIBLIOGRAPHY

Kramer, Franklin D, Lauren Speranza M, Atlantic Council of the United States, and Brent Scowcroft Center on International Security. *Meeting the Russian Hybrid Challenge: A Comprehensive Strategic Framework*, 2017.

Kuehl, Dan, and Leigh Armistead. "Information Operations: The Policy and Organizational Evaluation." In *Information Operations*. Washington D.C.: Potomac Books, 2007.

Kuznetsov, Lt. Gen. V. I., Col. Yu. Ye. Donskov, and Lt. Col. O. G. Nikitin. "Cyberspace in Military Operations Today." *Military Thought* 23, no. 1 (2014): 20–25.

Langner, Ralph. "Stuxnet—Dissecting a Cyberwarfare Weapon." *IEEE Security and Privacy* 9, no. 3 (June 2011): 49–51.

Lee, Bryan, and Robert Falcone. "Behind the Scenes with OilRig." *Unit42* (blog), 30 April 2019. https://unit42.paloaltonetworks.com/behind-the-scenes-with-oilrig/

Lee, Robert M. "Potential Sample of Malware from the Ukrainian Cyber Attack Uncovered." *SANS Industrial Control Systems Security Blog* (blog), 1 January 2016. https://ics.sans.org/blog/2016/01/01/potential-sample-of-malware-from-the-ukrainian-cyber-attack-uncovered

Lee, Robert M., Michael J. Assante, and Tim Conway. "German Steel Mill Cyber Attack." SANS ICS, 2014.

Lehto, Martti, and Bill Hutchinson. "Mini-Drones Swarms and Their Potential in Conflict Situations." In *2020 12th International Conference on Cyber Warfare and Security*, 326–34. Reading, UK, 2020.

Leibunguth, Jonathon P. Command and Control Systems for Cyber Warfare. United States US20090249483A1, filed 30 March 2009, and issued 1 October 2009.

Lewis, James A. "Thresholds for Cyberwar." Center for Strategic and International Studies, 2010.

Lewis, Jeffrey. "Is the United States Really Blowing Up North Korea's Missiles?" *Foreign Policy* (blog), 19 April 2017. https://foreignpolicy.com/2017/04/19/the-united-states-isnt-hacking-north-koreas-missile-launches/

Liang, Qiao, and Wang Xiangsui. *Unrestricted Warfare*. PLA Literature and Arts Publishing House, 1999.

Libicki, Martin C. *Cyberdeterrence and Cyberwar*. Santa Monica, CA: RAND, 2009.

———. *What Is Information Warfare?* 3rd ed. Washington DC: National Defense University, 1995.

———. "Why Cyber Will Not and Should Not Have Its Grand Strategist." 2014, Strategic Studies Quarterly (Spring 2014), pp. 23–39.

Liddell Hart, B.H. *Strategy*. 2nd rev. ed. New York: Meridian, 1991.

Lilly, Bilyana, and Joe Cheravitch. "The Past, Present, and Future of

BIBLIOGRAPHY

Russia's Cyber Strategy and Forces." In *2020 12th International Conference on Cyber Conflict (CyCon)*, 129–55. Tallinn, Estonia: IEEE, 2020. https://doi.org/10.23919/CyCon49761.2020.9131723

Lin, Herbert S. "Offensive Cyber Operations and the Use of Force." *Journal of National Security Law and Policy* 4 (2010): 63–86.

Lincoln Bonner III, E. "Cyber Power in 21st-Century Joint Warfare." *Joint Forces Quarterly* 74, no. 3 (2014): 102–9.

Lin-Greenberg, Erik. "Allies and Artificial Intelligence: Obstacles to Operations and Decision-Making." *Texas National Security Review* 3, no. 2 (Spring 2020): 57–76.

Lockheed Martin. "Autonomic Logistics Information System (ALIS)," November 2009. https://www.lockheedmartin.com/content/dam/lockheed-martin/rms/documents/alis/CS00086-55%20(ALIS%20Product%20Card).pdf

————. "Lockheed Martin to Enhance U.S. Navy's C4ISR Capabilities." NavalToday, 1 July 2014. http://navaltoday.com/2014/07/01/lockheed-martin-to-enhance-u-s-navys-c4isr-capabilities/

Lostumbo, Michael, David R. Frelinger, James Williams, and Barry Wilson. *Air Defense Options for Taiwan: An Assessment of Relative Costs and Operational Benefits*. Research Report, RR-1051-OSD. Santa Monica, California: Rand Corporation, 2016.

Lynn, William. "Defending a New Domain—The Pentagon's Cyber-strategy." *Foreign Affairs* 89, no. 5 (October 2010): 97–108.

Maestro, Oriana Skylar. "Military Confrontation in the South China Sea." Council on Foreign Relations, May 21, 2020. https://www.cfr.org/report/military-confrontation-south-china-sea

Mahan, Alfred Thayer. *The Influence of Sea Power upon History, 1660–1783*. Read Books Ltd, 2013.

Maizland, Lindsay. "China's Modernizing Military." *Council on Foreign Relations* (blog), 5 February 2020. https://www.cfr.org/backgrounder/chinas-modernizing-military

Makovsky, David. "The Silent Strike." *The New Yorker*, 17 September 2012. http://www.newyorker.com/magazine/2012/09/17/the-silent-strike

Malone, Jeff. "Intelligence Support Requirements for Offensive CNO." In *Cyber Warfare and Nation States Conference*. Canberra, Australia, 23 August 2010.

Mandiant. "APT1—Exposing One of China's Cyber Espionage Units," 2013.

Mangosing, Frances. "New Photos Show China Is Nearly Done with Its Militarization of South China Sea." *Inquirer.Net*, 5 February 2018. http://

www.inquirer.net/specials/exclusive-china-militarization-south-china-sea

Manky, Derek. "Using Fuzzing to Mine for Zero-Days." ThreatPost, December7,2018.https://threatpost.com/using-fuzzing-to-mine-for-zero-days/139683/

Mansharof, Yossi. "Iran's Cyber War: Hackers In Service Of The Regime; IRGC Claims Iran Can Hack Enemy's Advanced Weapons Systems; Iranian Army Official: 'The Cyber Arena Is Actually The Arena Of The Hidden Imam.'" MEMRI, 25 August 2013. https://www.memri.org/reports/irans-cyber-war-hackers-service-regime-irgc-claims-iran-can-hack-enemys-advanced-weapons

Markoff, John. "SecurID Company Suffers Security Breach." *The New York Times*, 17 March 2011, sec. Technology. https://www.nytimes.com/2011/03/18/technology/18secure.html

Maruyev, A. Yu. "Russia and the U.S.A in Confrontation: Military and Political Aspects." *Military Thought* 18, no. 3 (1 July 2009): 1–8.

Maurer, Diana. F-35 Sustainment: DOD Needs to Address Key Uncertainties as It Re-Designs the Aircraft's Logistics System, Committee on Oversight and Reform (2020).

Maynor, David, Aleksander Nikolic, Matt Olney, and Yves Younan. "The MeDoc Connection." *Cisco Talos* (blog), 5 July 2017. http://blog.talosintelligence.com/2017/07/the-medoc-connection.html

Mazanec, Brian M. "The Art of (Cyber) War." *The Journal of International Security Affairs* 16 (Spring 2009): 81–90.

McDevitt, Michael. "The PLA Navy's Antiaccess Role in a Taiwan Contingency." In *The Chinese Navy*, edited by Phillip C. Saunders, Christopher Yung, Michael Swaine, and Andrew Nien-Dzu Yang. Washington DC: Institute for National Strategic Studies, 2011.

McHale, John. "Record Number of Cyber Attacks Hit Lockheed Martin in 2014." Military Embedded Systems, 18 February 2015. http://mil-embedded.com/3499-record-number-of-cyber-attacks-hit-lockheed-martin-in-2014/

McInnis, J Matthew. "Iranian Concepts of Warfare: Understanding Tehran's Evolving Military Doctrine." American Enterprise Institutes, February 2017.

McMaster, H. R. "The Human Element: When Gadgetry Becomes Strategy." *World Affairs* 171, no. 3 (2009): 31–43.

McReynolds, Joe. "China's Evolving Perspectives on Network Warfare: Lessons from the Science of Military Strategy." *China Brief* 15, no. 8 (17 April 2015): 3–7.

Mead, Walter Russell. "The Return of Geopolitics: The Revenge of the Revisionist Powers." *Foreign Aff.* 93 (2014): 69.

Mercer, Warren, Paul Rascagneres, and Matthew Molyett. "Olympic Destroyer Takes Aim At Winter Olympics." *Cisco's Talos Intelligence* (blog), 12 February 2018. http://blog.talosintelligence.com/2018/02/olympic-destroyer.html

Meta. "Introducing Ray-Ban Stories: First-Generation Smart Glasses." *Meta* (blog), 9 September 2021. https://about.fb.com/news/2021/09/introducing-ray-ban-stories-smart-glasses/

Microsoft. "Microsoft Security Bulletin MS17–010—Critical." 14 July 2017. https://docs.microsoft.com/en-us/security-updates/securitybulletins/2017/ms17-010.

Midson, David. "Geography, Territory and Sovereignty in Cyber Warfare." In *New Technologies and the Law of Armed Conflict*, edited by Hitoshi Nasu and Robert McLaughlin, 75–93. The Hague: T.M.C. Asser Press, 2014.

Mimoso, Michael. "Automated Toolkits Named in Massive DDoS Attacks Against U.S. Banks." ThreatPost, 2 October 2012. https://threatpost.com/automated-toolkits-named-massive-ddos-attacks-against-us-banks-100212/77068/

Ministry of Defense of the Russian Federation. "Head of the Russian General Staff's Office for UAV Development Major General Alexander Novikov Holds Briefing for Domestic and Foreign Reporters: Ministry of Defence of the Russian Federation." 11 January 2018. http://eng.mil.ru/en/news_page/country/more.htm?id=12157872@egNews

Ministry of Industry and Information Technology Network Security. "Notice of the Ministry of Industry and Information Technology and the State Internet Information Office of the Ministry of Public Security on Issuing the Regulations on the Administration of Network Product Security Vulnerabilities." Cyberspace Administration of China, 13 July 2021.

Ministry of National Defense of the People's Republic of China. "China's National Defense in the New Era." Xinhuanet, 24 July 2019.

Mizokami, Kyle. "Why the Air Force Is Buying a Bunch of F-15s Even Though the F-35 Is Coming." Popular Mechanics, 19 February 2019. https://www.popularmechanics.com/military/aviation/a26413900/air-force-buying-new-f-15/

Monte, Matthew. *Network Attacks & Exploitation: A Framework*. Indianapolis, IN: John Wiley & Sons, Inc, 2015.

Moore, Daniel. "Targeting Technology: Mapping Military Offensive Network Operations." In *2018 10th International Conference on Cyber Conflict (CyCon)*, 89–108. Tallinn, Estonia: IEEE, 2018.

Morozov, Evgeny. "How I Became a Soldier in the Georgia-Russia Cyberwar." *Slate*, 14 August 2008. http://www.slate.com/articles/technology/technology/2008/08/an_army_of_ones_and_zeroes.html

BIBLIOGRAPHY

Nakashima, Ellen. "Trump Approved Cyber-Strikes against Iranian Computer Database Used to Plan Attacks on Oil Tankers." *Washington Post*, 22 June 2019.

———. "U.S. Cybercom Contemplates Information Warfare to Counter Russian Interference in 2020 Election." *Washington Post*, 25 December 2019. https://www.washingtonpost.com/national-security/us-cyber-com-contemplates-information-warfare-to-counter-russian-interference-in-the-2020-election/2019/12/25/21bb246e-20e8-11ea-bed5-880264cc91a9_story.html

Nakashima, Ellen, and Craig Timberg. "NSA Officials Worried about the Day Its Potent Hacking Tool Would Get Loose. Then It Did." *Washington Post*, 16 May 2017, sec. Technology.

Nakasone, Paul M. "A Cyber Force for Persistent Operations." *Joint Forces Quarterly* 92 (2019): 10–14.

National Audit Office. "Investigation: WannaCry Cyber Attack and the NHS." Accessed 1 January 2018. https://www.nao.org.uk/report/investigation-wannacry-cyber-attack-and-the-nhs/

NATO. "Allied Joint Publication 3.20: Allied Joint Doctrine for Cyberspace Operations (Edition A)." NATO Standardization Office, January 2020.

NCSC. "Russian Military 'Almost Certainly' Responsible for Destructive 2017 Cyber Attack." UK National Cyber Security Centre, 15 February 2018. https://www.ncsc.gov.uk/news/russian-military-almost-certainly-responsible-destructive-2017-cyber-attack

NIST. "NVD—CVSS Severity Distribution Over Time."2020. https://nvd.nist.gov/general/visualizations/vulnerability-visualizations/cvss-severity-distribution-over-time

NSA. "ANT Product Catalog," 2009.

———. "Case Studies of Integrated Cyber Operation Techniques." 2011.

———. "Computer Network Operations—GENIE," 2013. https://www.eff.org/files/2015/02/03/20150117-spiegel-excerpt_from_the_secret_nsa_budget_on_computer_network_operations_-_code_word_genie.pdf

———. "Getting Close to the Adversary: Forward-Based Defense with QFIRE." 3 June 2011.

———. "SID and DIA Collaborate Virtually on Russian Targets." 18 May 2004.

———. "The ROC: NSA's Epicenter for Computer Network Operations," 6 September 2006.

NSA, and USSTRATCOM. "National Initiative Protection Program—Sentry Eagle," November 23, 2004.

Nye, Joseph S. "Cyber Power." DTIC Document, 2010.

BIBLIOGRAPHY

O'Conner, Sean. "Access Denial—Syria's Air Defence Network." Jane's International Defence Review, 2014.

O'Neill, Patrick Howell. "2021 Has Broken the Record for Zero-Day Hacking Attacks." MIT Technology Review, 23 September 2021. https://www.technologyreview.com/2021/09/23/1036140/ 2021-record-zero-day-hacks-reasons/

MuckRock. "Operation Aurora," 3 July 2014. https://www.muckrock. com/foi/united-states-of-america-10/operation-aurora-11765/#files

O'Rourke, Ronald. "China Naval Modernization: Implications for U.S. Navy Capabilities—Background and Issues for Congress." Congressional Research Service, 17 December 2020.

Ostovar, Afshon. "Sectarian Dilemmas in Iranian Foreign Policy: When Strategy and Identity Politics Collide." Carnegie Endowment for International Peace, November 2016.

Parks, Raymond C. and Duggan, David P. "Principles of Cyber-Warfare." In *Proceedings from the Second Annual IEEE SMC Information Assurance Workshop*, 122–26. New York: West Point, 2001.

People's Liberation Army. "顺应军事变革潮流把握改革主动—中国军网-军报记者," 5 January 2016.

People's Liberation Army News. "Cracking the Winning Source Code of Information Warfare-Xinhuanet.Com." Xinhuanet, 7 November 2017. http://m.xinhuanet.com/mil/2017–11/07/c_129734404.htm

Perlroth, Nicole. "Cyberattack Caused Olympic Opening Ceremony Disruption." *The New York Times*, 13 February 2018, sec. Technology. https://www.nytimes.com/2018/02/12/technology/winter-olympic-games-hack.html

Perlroth, Nicole, and Quentin Hardy. "Online Banking Attacks Were Work of Iran, U.S. Officials Say." *The New York Times*, 8 January 2013, sec. Technology. http://www.nytimes.com/2013/01/09/technology/ online-banking-attacks-were-work-of-iran-us-officials-say.html

Perlroth, Nicole, and Savage, Charlie. "Is D.N.C. Email Hacker a Person or a Russian Front? Experts Aren't Sure." *The New York Times*, 27 July 2016, sec US. http://www.nytimes.com/2016/07/28/us/politics/is-dnc-email-hacker-a-person-or-a-russian-front-experts-arent-sure.html

Perlroth, Nicole, Mark Scott, and Sheera Frenkel. "Cyberattack Hits Ukraine Then Spreads Internationally." *The New York Times*, 27 June 2017, sec. Technology. https://www.nytimes.com/2017/06/27/technol-ogy/ransomware-hackers.html

Perov, Col E A, and Col A V Pereverzev. "On the Prospective Digital Communication Network of the RF Armed Forces." *Military Thought* 17, no. 2 (2008): 89–95.

Peterson, Dale. "Offensive Cyber Weapons: Construction, Development, and Employment." *Journal of Strategic Studies* 36, no. 1 (February 2013): 120–24.

Pierson, Brendan, and Jan Wolfe. "Explainer-U.S. Government Hack: Espionage or Act of War?" *Reuters*, 21 December 2020. https://www.reuters.com/article/global-cyber-legal-idUSKBN28T0HH

Plucker, Ron C. "Command and Control Warfare—A New Concept for the Joint Operational Commander." DTIC Document, 1993.

Poling, Gregory. "New Imagery Release." Asia Maritime Transparency Initiative, 10 September 2015. http://amti.csis.org/new-imagery-release/

Pomerleau, Mark. "US is 'Outgunned' in Electronic Warfare, Says Cyber Commander." C4ISRNET, 10 August 2017. https://www.c4isrnet.com/show-reporter/technet-augusta/2017/08/10/us-is-outgunned-in-electronic-warfare-says-cyber-commander/

———. "Cyber Command Shifts Counterterrorism Task Force to Focus on Higher-Priority Threats," *Army Times*, 4 May 2021, https://www.armytimes.com/cyber/2021/05/04/cyber-command-shifts-counterterrorism-task-force-to-focus-on-higher-priority-threats/

Popovich, Elad. "A Classical Analysis of the 2014 Israel-Hamas Conflict." *CTC Sentinel* 7, no. 11 (November 2014): 20–24.

PR Newswire. "Lockheed Martin and DRS Technologies Deliver 4000th AN/UYQ-70 Ship Display System to the U.S. Navy." PR Newswire, 11 May 2012.

Price, Alfred. *Instruments of Darkness*. Barnsley, UK: Frontline Books, 2017.

Rabinovich, Itamar. "How Iran's Regional Ambitions Have Developed since 1979." Iran's Revolution, 40 Years on—What It Has Meant for Iran, America, and the Region (series). Brookings, 24 January 2019. https://www.brookings.edu/blog/order-from-chaos/2019/01/24/how-irans-regional-ambitions-have-developed-since-1979/

Rainey, James W. "Ambivalent Warfare: Tactical Doctrine of the AEF in World War I." *Parameters* 13, no. 3 (1983): 34–46.

Raiu, Costin, Daniel Moore, Juan Andres Guerrero-Saade, and Thomas Rid. "Penquin's Moonlit Maze." *Securelist* (blog), 3 April 2017. https://securelist.com/penquins-moonlit-maze/77883/

Rakel, Eva Patricia. "Iranian Foreign Policy since the Iranian Islamic Revolution: 1979–2006." *Perspectives on Global Development and Technology* 6, no. 1–3 (2007): 159–87. https://doi.org/10.1163/156914907X207711

Ranger, Steve. "NATO Updates Cyber Defence Policy as Digital Attacks Become a Standard Part of Conflict." *ZDNet*, 30 June 2014. http://www.zdnet.com/article/nato-updates-cyber-defence-policy-as-digital-attacks-become-a-standard-part-of-conflict/

BIBLIOGRAPHY

Rattray, Gregory J. *Strategic Warfare in Cyberspace*. Cambridge, Mass: MIT Press, 2001.

Rattray, Gregory J., and Jason Healey. "Categorizing and Understanding Offensive Cyber Capabilities and Their Use." In *Proceedings of a Workshop on Deterring Cyberattacks: Informing Strategies and Developing Options for U.S. Policy*. Washington D.C.: National Academic Press, 2010.

Raved, Ahiya. "Cyber Attack Targeted Israel's Water Supply, Internal Report Claims." *Ynet News*, 26 April 2020. https://www.ynetnews.com/article/HJX1mWmF8

———. "שוב: מתקפת סייבר על מתקני מים בישראל." *Ynet News*, 17 July 2020. https://www.ynet.co.il/article/rJrCqmAkw

Rempfer, Kyle. "Army Cyber Lobbies for Name Change This Year, as Information Warfare Grows in Importance." *Army Times*, 16 October 2019. https://www.armytimes.com/news/your-army/2019/10/16/ausa-army-cyber-lobbies-for-name-change-this-year-as-information-warfare-grows-in-importance/

———. "DARPA Hopes to Swarm Drones out of C-130s in 2019 Test." *Air Force Times*, 19 December 2017. https://www.airforcetimes.com/newsletters/daily-news-roundup/2017/12/18/darpa-hopes-to-swarm-drones-out-of-c-130s-in-2019-test/

Rid, Thomas. "All Signs Point to Russia Being Behind the DNC Hack." *Motherboard*, 25 July 2016. http://motherboard.vice.com/read/all-signs-point-to-russia-being-behind-the-dnc-hack.

———. *Rise of the Machines: The Lost History of Cybernetics*. Scribe Publications, 2016.

Rid, Thomas, and Ben Buchanan. "Attributing Cyber Attacks." *Journal of Strategic Studies* 38, no. 1–2 (2 January 2015): 4–37.

Rid, Thomas, and Peter McBurney. "Cyber-Weapons." *The RUSI Journal* 157, no. 1 (February 2012): 6–13.

Risen, James. "U.S. Secretly Negotiated With Russians to Buy Stolen NSA Documents—and the Russians Offered Trump-Related Material, Too." *The Intercept* (blog), 9 February 2018. https://theintercept.com/2018/02/09/donald-trump-russia-election-nsa/

Rivner, Uri. "Anatomy of an Attack." *Speaking of Security—The RSA Blog* (blog), 1 April 2011. https://blogs.rsa.com/anatomy-of-an-attack/

Robb, Drew. "Building the Global Heatmap." *Strava Engineering* (blog), 1 November 2017. https://medium.com/strava-engineering/the-global-heatmap-now-6x-hotter-23fc01d301de

Roberts, Huw, Josh Cowls, Jessica Morley, Mariarosaria Taddeo, Vincent Wang, and Luciano Floridi. "The Chinese Approach to Artificial Intelligence: An Analysis of Policy, Ethics, and Regulation." *AI & SOCIETY* 36, no. 1 (March 2021): 59–77.

BIBLIOGRAPHY

Rosenbach, Marcel, Hilmar Schmundt, and Christian Stöcker. "Source Code Similarities: Experts Unmask 'Regin' Trojan as NSA Tool." *Spiegel Online*, 27 January 2015, sec. International. http://www.spiegel.de/international/world/regin-malware-unmasked-as-nsa-tool-after-spiegel-publishes-source-code-a-1015255.html

Rosenbaum, Ron. "Richard Clarke on Who Was Behind the Stuxnet Attack." *Smithsonian Magazine*, April 2012. https://www.smithsonianmag.com/history/richard-clarke-on-who-was-behind-the-stuxnet-attack-160630516/

Rosenberg, Jay. "2018 Winter Cyber Olympics: Code Similarities with Cyber Attacks in Pyeongchang." *Intezer* (blog), 12 February 2018. https://www.intezer.com/2018-winter-cyber-olympics-code-similarities-cyber-attacks-pyeongchang/

Russian Federation. "The Military Doctrine of the Russian Federation," 25 December 2014. http://rusemb.org.uk/press/2029

Safran, Nadav. "Trial by Ordeal: The Yom Kippur War, October 1973." *International Security* 2, no. 2 (1977): 133–70.

Samit, Sarkar. "Massive DDoS Attack Affecting PSN, Some Xbox Live Apps." *Polygon*, October 21, 2016. https://www.polygon.com/2016/10/21/13361014/psn-xbox-live-down-ddos-attack-dyn

Sanchez, Raf. "Russia Uses Missiles and Cyber Warfare to Fight off 'swarm of Drones' Attacking Military Bases in Syria." *The Telegraph*, 9 January 2018. https://www.telegraph.co.uk/news/2018/01/09/russia-fought-swarm-drones-attacking-military-bases-syria/

Sanger, David E. "Obama Ordered Wave of Cyberattacks Against Iran." *The New York Times*, 1 June 2012, sec. World. http://www.nytimes.com/2012/06/01/world/middleeast/obama-ordered-wave-of-cyberattacks-against-iran.html

———. "U.S. Cyberattacks Target ISIS in a New Line of Combat." *The New York Times*, 24 April 2016, sec. US. http://www.nytimes.com/2016/04/25/us/politics/us-directs-cyberweapons-at-isis-for-first-time.html

Sanger, David E., and William J. Broad. "Trump Inherits a Secret Cyberwar Against North Korean Missiles." *The New York Times*, 20 January 2018, sec. World. https://www.nytimes.com/2017/03/04/world/asia/north-korea-missile-program-sabotage.html

Sanger, David E., and Mark Mazzetti. "Analysts Find Israel Struck a Syrian Nuclear Project." *The New York Times*, 14 October 2007.

———. "U.S. Had Cyberattack Plan If Iran Nuclear Dispute Led to Conflict." *The New York Times*, 16 February 2016, sec. World. https://www.nytimes.com/2016/02/17/world/middleeast/us-had-cyberattack-planned-if-iran-nuclear-negotiations-failed.html

Sanger, David E., and Eric Schmitt. "Russian Ships Near Data Cables Are Too Close for U.S. Comfort." *The New York Times*, 25 October 2015, sec. Europe. https://www.nytimes.com/2015/10/26/world/europe/russian-presence-near-undersea-cables-concerns-us.html

Satter, Raphael, and Joseph Menn. "SolarWinds Hackers Accessed Microsoft Source Code, the Company Says." *Reuters*, 1 January 2021. https://www.reuters.com/article/global-cyber-microsoft-idINKBN29620C

Schaefer, Ben. "The Cyber Party of God: How Hezbollah Could Transform Cyberterrorism." *Georgetown Security Studies Review* (blog), 11 March 2018. http://georgetownsecuritystudiesreview.org/2018/03/11/the-cyber-party-of-god-how-hezbollah-could-transform-cyberterrorism/

Schmitt, Eric, Farnaz Fassihi, and David D. Kirkpatrick. "Saudi Oil Attack Photos Implicate Iran, U.S. Says; Trump Hints at Military Action." *The New York Times*, 15 September 2019, sec. World. https://www.nytimes.com/2019/09/15/world/middleeast/iran-us-saudi-arabia-attack.html

Schmitt, Michael N. *Tallinn Manual on the International Law Applicable to Cyber Warfare*. New York: Cambridge University Press, 2013.

Schmitt, Michael N., and NATO Cooperative Cyber Defence Centre of Excellence, eds. *Tallinn Manual 2.0 on the International Law Applicable to Cyber Operations*. 2nd ed. Cambridge, UK; New York: Cambridge University Press, 2017.

Schneider, James, and Lawrence L. Izzo. "Clausewitz's Elusive Center of Gravity." *Parameters* (September 1987): 46–57.

Schwarz, Matthew J. "Lockheed Martin Suffers Massive Cyberattack." *Dark Reading* (blog), 30 May 2011. http://www.darkreading.com/risk-management/lockheed-martin-suffers-massive-cyberattack/d/d-id/1098013?

Scott, Roger D. "Territorially Intrusive Intelligence Collection and International Law." *The Air Force Law Review* 46 (1999): 217–24.

Shane, Scott. "Ex-N.S.A. Worker Accused of Stealing Trove of Secrets Offers to Plead Guilty." *The New York Times*, 1 January 2018, sec. US https://www.nytimes.com/2018/01/03/us/politics/harold-martin-nsa-guilty-plea-offer.html

Shattuck, Thomas J. "The Race to Zero?: China's Poaching of Taiwan's Diplomatic Allies." *Orbis* 64, no. 2 (February 2020): 334–52. https://doi.org/10.1016/j.orbis.2020.02.003

Shen, Wade. "The Information Domain and the Future of Conflict." In *2017 9th International Conference on Cyber Conflict*. Tallinn, Estonia, 1 June 2017.

Shields, Nathan P. United States of America V. Park Jin Hyok, No. MJ 18–1479 (United States District Court, 8 June 2018).

Shimshoni, Jonathan. "Technology, Military Advantage, and World War I:

BIBLIOGRAPHY

A Case for Military Entrepreneurship." *International Security* 15, no. 3 (1990): 187.

Simpson, Connor. "Sheldon Adelson Has an Idea: Lob a Nuclear Bomb into the Iranian Desert." The Atlantic, 23 October 2013. https://www.the-atlantic.com/international/archive/2013/10/sheldon-adelson-has-idea-lob-nuclear-bomb-iranian-desert/309657/

Slowik, Joe. "CRASHOVERRIDE: Reassessing the 2016 Ukraine Electric Power Event as a Protection-Focused Attack." Dragos, 15 August 2019.

Smeets, Max. "A Matter of Time: On the Transitory Nature of Cyber-weapons." *Journal of Strategic Studies* (16 February 2017): 1–28.

————. "The Strategic Promise of Offensive Cyber Operations." *Strategic Studies Quarterly* (Fall 2018): 90–113.

Smith, Brad. "A Moment of Reckoning: The Need for a Strong and Global Cybersecurity Response." *Microsoft On the Issues* (blog), 17 December 2020. https://blogs.microsoft.com/on-the-issues/2020/12/17/cyber-attacks-cybersecurity-solarwinds-fireeye/

Smith, Edward A. "Effects Based Operations." *Applying Network Centric Warfare in Peace*, 2005.

Snegovaya, Maria. "Putin's Information Warfare in Ukraine: *Soviet Origins Of Russia's Hybrid* Warfare." Washington: Institute for the Study of War, 2015.

Sood, Karan, and Shaun Hurley. "NotPetya Ransomware Attack." *CrowdStrike Blog* (blog), 29 June 2017. https://www.crowdstrike.com/blog/petrwrap-ransomware-technical-analysis-triple-threat-file-encryption-mft-encryption-credential-theft/

Space and Missile Systems Center Public Affairs. "Counter Communications System Block 10.2 Achieves IOC, Ready for the Warfighter." United States Space Force, 13 March 2020. https://www.spaceforce.mil/News/Article/2113447/counter-communications-system-block-102-achieves-ioc-ready-for-the-warfighter

Stone, Biz. "DNS Disruption." *Twitter Official Blog* (blog), 18 December 2009. https://blog.twitter.com/en_us/a/2009/dns-disruption.html

Strachan, Hew. "The Battle of the Somme and British Strategy." *Journal of Strategic Studies* 21, no. 1 (March 1998): 79–95.

Strategic Comments. "Syria: Foreign Intervention Still Debated, but Distant." *Strategic Comments* 18, no. 6 (August 2012): 1–5.

Strobel, Warren P. "Pompeo Blames Russia for Hack as Trump Casts Doubt on Widespread Conclusion." *Wall Street Journal*, 19 December 2020, sec. Politics. https://www.wsj.com/articles/pompeo-blames-russia-for-solarwinds-hack-11608391515

Symantec. "Dragonfly: Cyberespionage Attacks Against Energy Suppliers." 7 July 2014.

————. "Internet Security Threat Report." February 2019.

————. "Petya Ransomware Outbreak: Here's What You Need to Know," 24 October 2017. https://www.symantec.com/blogs/threat-intelligence/petya-ransomware-wiper

————. "What You Need to Know about the WannaCry Ransomware." 23 October 2017. https://www.symantec.com/blogs/threat-intelligence/wannacry-ransomware-attack

Szegedy, Christian, Wojciech Zaremba, Ilya Sutskever, Joan Bruna, Dumitru Erhan, Ian Goodfellow, and Rob Fergus. "Intriguing Properties of Neural Networks." *ArXiv:1312.6199 [Cs]*, 20 December 2013. http://arxiv.org/abs/1312.6199

Taiwan Affairs Office of the State Council. "国台办新闻发布会辑录（2018-05-16）中共中央台湾工作办公室、国务院台湾事务办公室." 16 May 2018. http://www.gwytb.gov.cn/xwfbh/201805/t20180516_11955430.htm

Taiwan News. "Taiwan on Schedule to Complete F-16 Upgrades by 2023." Taiwan News, 17 March 2021. https://www.taiwannews.com.tw/en/news/4153117

Temple-Raston, Dina. "How The U.S. Hacked ISIS." NPR, 26 September 2019. https://www.npr.org/2019/09/26/763545811/how-the-u-s-hacked-isis

The State Council Information Office. China's Military Strategy (2015).

Theohary, Catherine A. "Iranian Offensive Cyber Attack Capabilities." In Focus. Congressional Research Service, 13 January 2020.

Thomas, Timothy. "Russia's Reflexive Control Theory and the Military." *The Journal of Slavic Military Studies* 17, no. 2 (June 2004): 237–56.

Thompson, Drew. "China Is Still Wary of Invading Taiwan." *Foreign Policy* (blog), 11 May 2020. https://foreignpolicy.com/2020/05/11/china-taiwan-reunification-invasion-coronavirus-pandemic/

Thornton, Rod. "The Changing Nature of Modern Warfare: Responding to Russian Information Warfare." *The RUSI Journal* 160, no. 4 (4 July 2015): 40–48.

ThreatConnect Research. "Shiny Object? Guccifer 2.0 and the DNC Breach." ThreatConnect, 29 June 2016.

Tikk, Eneken, Kadri Kaska, Liis Vihul, and NATO Cooperative Cyber Defence Centre of Excellence. *International Cyber Incidents: Legal Considerations*. Tallinn: Cooperative Cyber Defence Centre of Excellence, 2010.

Tirpak, John A. "Making the Best of the Fighter Force." *Air Force Magazine* 90, no. 3 (2007): 40.

Todorov, Kenneth, Archer Macy, Richard Formica, Joseph Horn, and Thomas Karako. Panel on Full Spectrum Missile Defense. Center for Strategic and International Studies, 4 December 2015.

Times Of Israel. "Iran Cyberattack on Israel's Water Supply Could Have Sickened Hundreds." The Times of Israel, 1 June 2020. https://www.timesofisrael.com/iran-cyberattack-on-israels-water-supply-could-have-sickened-hundreds-report/

Toonk, Andre. "Chinese ISP Hijacks the Internet." *BGPMon* (blog), 8 April 2010. https://bgpmon.net/chinese-isp-hijacked-10-of-the-internet/

Townsend, Stephen. "Accelerating Multi-Domain Operations: Evolution of an Idea." US Army Training and Doctrine Command, 2018.

TRADOC. "Multi-Domain Battle: Evolution of Combined Arms for the 21st Century 2025–2040." US Army Training and Doctrine Command, October 2017.

———. "The Operational Environment and the Changing Character of Future Warfare." US Army Training and Doctrine Command, 31 May 2017.

Tzu, Sun. *The Art of War*. (Brooklyn, NY: Sheba Blake Publishing, 2017).

UK Cabinet Office. "The UK Cyber Security Strategy: Report on Progress and Forward Plans," 2014.

UK Foreign Ministry. "Foreign Office Minister Condemns North Korean Actor for WannaCry Attacks." GOV.UK, 19 December 2017. https://www.gov.uk/government/news/foreign-office-minister-condemns-north-korean-actor-for-wannacry-attacks

UK Ministry of Defense. "Joint Doctrine Note 1/18: Cyber and Electromagnetic Activities," February 2018.

Ukraine Ministry of Defense. "Втрати у ЗС України 80% Гаубиць Д-30," Не Відповідає Дійсності," 6 January 2017. http://www.mil.gov.ua/news/2017/01/06/informaciya-po-vtrati-u-zs-ukraini-80-gaubicz-d-30%E2%80%9D-ne-vidpovidae-dijsnosti/

Unal, Bezya, and Patricia Lewis. "Cybersecurity of Nuclear Weapons Systems: Threats, Vulnerabilities, and Consequences." London: Chatham House, January 2018.

Underwood, Kimberly. "The Army Evolves Its Formations for Cyber and Electronic Warfare." SIGNAL Magazine, 21 October 2020. https://www.afcea.org/content/army-evolves-its-formations-cyber-and-electronic-warfare

United Kingdom Government. "National Cyber Force Transforms Country's Cyber Capabilities to Protect UK." GOV.UK, 19 November 2020. https://www.gov.uk/government/news/national-cyber-force-transforms-countrys-cyber-capabilities-to-protect-uk

United States Government. "Presidential Policy Directive 20—U.S. Cyber Operations Policy," October 2012.

United States Industrial Control Systems Cyber Emergency Response

BIBLIOGRAPHY

Team. "Cyber-Attack Against Ukrainian Critical Infrastructure | ICS-CERT." ICS-CERT, February 25, 2016.

Unknown. "Untitled Paste from 'Cutting Sword of Justice.'" Paste Site. Pastebin, 15 August 2012. https://pastebin.com/HqAgaQRj

US 24th Air Force. "Commander's Strategic Vision." US Air Force, 8 March 2017.

US Air Force. "Air Force Doctrine Document 3–12," 15 July 2010.

——. "Department of Defense Fiscal Year 2021 Budget Estimates: Air Force, Research, Development, Test & Evaluation, Space Force," February 2020. https://www.saffm.hq.af.mil/Portals/84/documents/FY21/RDTE_/FY21%20Space%20Force%20Research%20Development%20Test%20and%20Evaluation.pdf?ver=202002-11-083608–887

US Army. "Army Field Manual 3–12: Cyberspace Operations and Electromagnetic Warfare." August 2021.

——. "Army Field Manual 3–38—Cyber Electromagnetic Activities," 12 February 2014.

——. "Army Regulation 525–20: Command & Control Countermeasures (C2CM)." US Army Headquarters, 31 July 1992.

——. "Deployed Tactical Network Guidance." US Army Chief Information Office, 31 May 2012.

——. "U.S. Army Field Manual 100–2–1: Soviet Forces." Headquarters of the Department of the US Army, 16 July 1984.

——. "U.S. Army Field Manual 100–6: Information Operations." Headquarters of the Department of the US Army, August 1996.

US Army Headquarters. "Army Doctrine Publication 3–0: Operations." US Army Headquarters, October 2011.

US Cyber Command. "Achieve and Maintain Cyberspace Superiority: Command Vision for US Cyber Command," 23 March 2018.

——. "JFHQ-C Certification: Framework to Operationalize the JFHQ-C." October 2013.

——. "USCYBERCOM 120-Day Assessment of Operation GLOWING SYMPHONY," 15 June 2016.

US CYBERCOM JIOC. "Improving Targeting Support to Cyber Operations," 30 November 2016.

US Department of Defense. "Aegis Modernization Report Program—Fiscal Year 2016," 2017.

——. "Chinese Exfiltrate Sensitive Military Technology," 2011.

——. "Cyber Strategy Summary," 2018. https://media.defense.gov/2018/Sep/18/2002041658/-1/-1/1/CYBER_STRATEGY_SUMMARY_FINAL.PDF

——. "Cybercom to Elevate to Combatant Command." Accessed

10 June 2018. https://www.defense.gov/News/Article/Article/151 1959/cybercom-to-elevate-to-combatant-command/

————. "Declaratory Policy, Concept of Operations, and Employment Guidelines for Left-of-Launch Capability," 10 May 2017.

————. "Fiscal Year 2015 DoD Programs—F-35 Joint Strike Fighter (JSF)," January 2016.

————. "Fiscal Year 2016 DoD Programs—F-35 Joint Strike Fighter (JSF)," January 2017.

————. "Fiscal Year 2016 Navy Programs—Surface Electronic Warfare Improvement Program (SEWIP) Block 2," December 2016. https://www.dote.osd.mil/Portals/97/pub/reports/FY2016/navy/2016 sewip.pdf?ver=2019–08–22–105305–400

————. "Fiscal Year 2019 DoD Programs—F-35 Joint Strike Fighter (JSF)," January 2020. https://www.documentcloud.org/documents/ 6780413-FY2019-DOT-E-F-35-Annual-Report.html

————. "Fiscal Year 2020 DoD Programs—F-35 Joint Strike Fighter (JSF)," December 2020. https://www.dote.osd.mil/Portals/97/pub/ reports/FY2020/other/2020DOTEAnnualReport.pdf

————. "Information Operations Roadmap," 30 October 2003.

————. "Military and Security Developments Involving the People's Republic of China." Office of the Secretary of Defense, 2020. https:// media.defense.gov/2020/Sep/01/2002488689/-1/-1/1/2020-DOD-CHINA-MILITARY-POWER-REPORT-FINAL.PDF

————. The Department of Defense Cyber Strategy (2015).

————. "U.S. Department of Defense Directive 3600.01—Information Operations," December 1996.

————. "U.S. Department of Defense Directive 3600.01—Information Warfare," November 1992.

US Department of Justice. "Former U.S. Counterintelligence Agent Charged With Espionage on Behalf of Iran; Four Iranians Charged With a Cyber Campaign Targeting Her Former Colleagues," 13 February 2019. https://www.justice.gov/opa/pr/former-us-counterintelligence-agent-charged-espionage-behalf-iran-four-iranians-charged-cyber

————. "Iranian Hackers Indicted for Stealing Data from Aerospace and Satellite Tracking Companies," 17 September 2020. https://www.justice.gov/usao-edva/pr/iranian-hackers-indicted-stealing-data-aerospace-and-satellite-tracking-companies

————. "Six Russian GRU Officers Charged in Connection with Worldwide Deployment of Destructive Malware and Other Disruptive Actions in Cyberspace." US Department of Justice, October 19, 2020. https://www.justice.gov/opa/pr/six-russian-gru-officers-charged-connection-worldwide-deployment-destructive-malware-and

BIBLIOGRAPHY

————. "Two Chinese Hackers Working with the Ministry of State Security Charged with Global Computer Intrusion Campaign Targeting Intellectual Property and Confidential Business Information, Including COVID-19 Research." US Department of Justice, 21 July 2020. https://www.justice.gov/opa/pr/two-chinese-hackers-working-ministry-state-security-charged-global-computer-intrusion

————. "U.S. Charges Russian GRU Officers with International Hacking and Related Influence and Disinformation Operations." US Department of Justice, 4 October 2018. https://www.justice.gov/opa/pr/us-charges-russian-gru-officers-international-hacking-and-related-influence-and

US, DNI. "Cyber Threat Framework Lexicon." Office of the Director of National Intelligence, 2013.

US DoD. "Annual Report to Congress: Military and Security Developments Involving the People's Republic of China." US Department of Defense, 15 May 2017.

US Homeland Security. "Blueprint for a Secure Cyber Future." US Homeland Security, November 2011.

US Joint Chiefs of Staff. "Joint Publication 1–02: DoD Dictionary." US Department of Defense, May 2017.

————. "Joint Publication 2–0: Joint Intelligence." US Department of Defense, 22 October 2013.

————. "Joint Publication 3–0: Operations," 11 August 2011.

————. "Joint Publication 3–12: Cyberspace Operations," 2 May 2013.

————. "Joint Publication 3–12: Cyberspace Operations," 8 June 2018.

————. "Joint Publication 3–13: Command and Control Warfare (C2W)," 7 February 1996.

————. "Joint Publication 3–13: Information Operations," 20 November 2014.

————. "Joint Publication 3–13.1: Electronic Warfare," 8 February 2012.

————. "Joint Publication 5–0: Joint Planning," 16 June 2017.

US Navy. "Electronics Technician Volume 03-Communications Systems," July 1997. http://electronicstechnician.tpub.com/14088/css/14088_144.htm

————. "The US Navy Fact File: Tomahawk Cruise Missile." US Navy Official Website. Accessed 1 October 2015.

US Press Secretary. "Statement from the Press Secretary." The White House, February 15, 2018.

US Strategic Command. "U.S. Cyber Command (USCYBERCOM)," 30 September 2016. http://www.stratcom.mil/Media/Factsheets/Factsheet-View/Article/960492/us-cyber-command-uscybercom/

BIBLIOGRAPHY

US-CERT. "Petya Ransomware," 15 February 2018. https://www.us-cert. gov/ncas/alerts/TA17–181A

USSR Exercise Control Staff. "Task for the Operational Command Staff Exercise Soyuz-75 for the 4th Army." Cold War International History Project, Polish Institute of National Remembrance, March 1975. http:// digitalarchive.wilsoncenter.org/document/113511

———. "The Operational-Tactical Exercise of Allied Fleets in the Baltic Sea, Codenamed VAL-77." Cold War International History Project, Polish Institute of National Remembrance, 1977. http://digitalarchive. wilsoncenter.org/document/114599

Van Tol, Jan, Mark Gunzinger, Andrew Krepinevich, and Jim Thomas. "AirSea Battle: A Point-of-Departure Operational Concept." DTIC Document, 2010.

Verizon. "2020 Data Breach Investigations Report." Verizon, 2020.

ViaSat. "Link-16 Message Card," October 2012.

Vinod Anand. "Chinese Concepts and Capabilities of Information Warfare." *Strategic Analysis* 30 (2006): 781–97.

Volz, Duston, and Jim Finkle. "U.S. Indicts Iranians for Hacking Dozens of Banks, New York Dam." *Reuters*, March 25, 2016. http://www.reuters. com/article/us-usa-iran-cyber-idUSKCN0WQ1JF

Waltz, Kenneth. *Man, the State, and War: A Theoretical Analysis*. Columbia University Press, 2013.

Waltz, Kenneth N. "The Origins of War in Neorealist Theory." *Journal of Interdisciplinary History* 18, no. 4 (1988): 615.

Wang, Vincent Wei-cheng. "The Chinese Military and the" Taiwan Issue": How China Assesses Its Security Environment." *Tamkang Journal of International Affairs* 10, no. 4 (2007): 89.

Watson-Watt, Robert. "Battle Scars of Military Electronics—The Scharnhorst Break-Through." *IRE Transactions on Military Electronics* 1, no. 1 (March 1957): 19–25.

Weimann, Gabriel. "Cyberterrorism: How Real Is the Threat?" Special Report. United States Institute of Peace, December 2004.

Weinberger, Sharon. "Is This the Start of Cyberwarfare?" *Nature* 474, no. 7350 (2011): 142.

Whetten, Lawrence, and Michael Johnson. "Military Lessons of the Yom Kippur War." *The World Today* 30, no. 3 (March 1974): 101–10.

White House. "International Strategy for Cyberspace," May 6, 2011.

Wikileaks. "Vault 7: CIA Hacking Tools Revealed." Wikileaks, March 7, 2017. https://wikileaks.org/ciav7p1/

Wikoff, Allison, and Richard Emerson. "New Research Exposes Iranian Threat Group Operations." Security Intelligence, July 16, 2020. https://

securityintelligence.com/posts/new-research-exposes-iranian-threat-group-operations/

Wilson, Clay. "Information Operations, Electronic Warfare, and Cyberwar: Capabilities and Related Policy Issues." DTIC Document, 2007.

Work, Robert. "Remarks by Deputy Secretary Work on Third Offset Strategy." US Department of Defense, 28 April 2016. https://www.defense.gov/News/Speeches/Speech-View/Article/753482/remarks-by-deputy-secretary-work-on-third-offset-strategy/

Wortzel, Larry M. "PLA Command, Control and Targeting Architectures: Theory, Doctrine, and Warfighting Applications." In *Right-Sizing the People's Liberation Army: Exploring the Contours of China's Military*, edited by Roy Kamphausen and Andrew Scobell, 191–235. Carlisle, PA: Strategic Studies Institute, 2007.

———. "PLA 'Joint' Operational Contingencies in South Asia, Central Asia, and Korea." In *Beyond The Strait: PLA Missions Other Than Taiwan*, edited by Roy Kamphausen, David Lai, and Andrew Scobell. Carlisle, PA: Strategic Studies Institute, 2009.

Xu, Gao. "说透了！'武统'台湾什么时候开始？解放军专家权威解读_两岸快评_中国台湾网." China Taiwan Network, 15 April 2020. http://www.taiwan.cn/plzhx/plyzl/202004/t20200415_12265753.htm

Zakheim, Dov S. "The United States Navy and Israeli Navy: Background, Current Issues, Scenarios, and Prospects." Center for Naval Assessment, 2012.

Zetter, Kim. *Countdown to Zero Day: Stuxnet and the Launch of the World's First Digital Weapon*. Broadway Books, 2014.

———. "How Digital Detectives Deciphered Stuxnet, the Most Menacing Malware in History." *Wired*, 11 July 2011. https://www.wired.com/2011/07/how-digital-detectives-deciphered-stuxnet/

寿晓松, 军事科学院 and 军事战略研究部. *The Science of Military Strategy*. 北京: 军事科学出版社, 2013.

INDEX

INDEX

information warfare, 59
Iraq, relations with, 208,
   209–10
Islamic Revolution (1979), 201,
   204, 207
Israel, relations with, 203, 207,
   208, 209, 210, 216, 217,
   225
Ministry of Intelligence leaks
   (2019), 221
Nitro Zeus plan (2015), 136
Operation Ababil (2012–13),
   29–30, 211, 210, 214,
   218–19
operational trajectory, 213–22
presence-based operations,
   204, 211, 213, 217, 219,
   220, 222, 225
proxies, use of, 204–5,
   206–13, 215–16, 217, 221,
   224, 226
Quds Force, 208, 214
Revolutionary Guard Corps,
   211, 216–17
Saudi Arabia, relations with,
   203–4, 208, 215, 220
Sentinel UAV interception
   (2011), 240
Shamoon attacks, 203–4,
   220–21
Shi'a Islam, 30, 204, 207–8,
   209, 210, 223
Stuxnet operation (2009–10),
   21–2, 31–2, 75–7, 93, 94,
   105, 135–6, 202, 215, 253
swarming tactics, 241
Syrian War (2011–), 160
terrorism, use of, 207, 214,
   217
Twitter hack (2009), 216
United States, relations with,
   203, 207, 210

Witt defection (2013), 205–6
Iran Cyber Army, 215–16
Iraq
   Gulf War I (1990–91), 38,
      56–7, 60, 117, 149
   Gulf War II (2003–11), 30, 34,
      209–10
   Iran War (1980–88), 208
   Tuwaitha strike (1981), 202
Islamic State, 1–2, 17, 128, 138,
   161, 168
Israel, 40–41, 43, 66, 73, 243,
   250
   Arab–Israeli War (1973), 53–4
   Eilat sinking (1967), 53–4
   Hamas, campaigns against, 252
   information warfare, 59
   Iran, relations with, 203, 207,
      208, 209, 210, 216, 217,
      225
   Kaibar strike (2007), 32, 35,
      40–41, 202
   Lebanon conflict, 34, 216
   Operation Cast Lead (2009),
      216
   Stuxnet operation (2009–10),
      21–2, 31–2, 75–7, 93, 94,
      105, 135–6, 202, 215, 253
   Tuwaitha strike (1981), 202
   water source sabotage (2020),
      225
ITSecTeam, 211, 212
itsoknoproblembro, 218
Izz ad-Din al-Qassam Cyber
   Fighters, 210

J12.0 messages, 194
jamming, 8, 40–41, 46, 49–53,
   55, 65, 132, 193, 241, 248
Japan, 176, 184, 192, 200
Joint Chiefs of Staff, 55

INDEX

INDEX

INDEX